SpringerWienNewYork

Sandra Häuplik-Meusburger

Architecture for Astronauts

An Activity-based Approach

SpringerWienNewYork

Dr.-Ing. DI Arch.SBA
Sandra Häuplik-Meusburger
Institute for Architecture and Design,
Vienna University of Technology, Vienna, Austria

SpringerWienNewYork is part of
Springer Science+Business Media
springer.at

Concept by the author
Layout: studio d*luxe, 1150 Vienna, Austria
Printing: Strauss GmbH,
69509 Mörlenbach, Germany

Cover illustration: CUPOLA VIEW,
L. Cahora Bassa, Zimbabwe (ISS023-E-32759).
Image taken by the ISS Expedition 23 Crew.
Image courtesy of the Image Science & Analysis Laboratory, NASA Johnson Space Center *The Gateway to Astronaut Photography of Earth*, http://eol.jsc.nasa.gov. Cover line drawings by the author

Printed on acid-free and
chlorine-free bleached paper
SPIN: 80022673

With 250 (partly coloured) Figures

Library of Congress Control Number:
2011924551

ISBN 978-3-7091-0666-2 SpringerWienNewYork

ABSTRACT

This book is the result of researching the interface between people, space and objects in an extra-terrestrial environment.

The first part of this book introduces all relevant American and Russian realized extra-terrestrial habitats: **The Apollo Spacecraft and Lunar Module, the Space Shuttle Orbiter, and the Space Stations; Salyut, Skylab, Mir, as well as the International Space Station.** It provides an overview of the architecture and configuration, highlights specific issues concerning the interior layout and compares the spatial and time allocation of human activities. Drawings and diagrams facilitate comparison.

The main part of the book concentrates on the investigation of the relationship between the environment and its users. This method evaluates and summarizes all the selected habitats by means of the human activities in relation to the characteristics of the built environment: **Sleep, Hygiene, Food, Work and Leisure**. In addition to analyzing the available data, it integrates the astronauts' personal experiences into the evaluation. The author conducted structured interviews with nine astronauts with a special focus on human activity.

The book confronts the findings from relevant literature and analysis - based on crew transcripts, spacecraft drawings and mission images - with the personal experiences of the users: the astronauts and cosmonauts.

To facilitate orientation and to ease comparison with architectural drawings and diagrams each category was assigned a specific colour. Design directions for each category conclude each chapter.

CONTENTS

FOREWORD

A good test for a book is always how much you enjoy revisiting it after some time has passed. I found Sandra Häuplik-Meusburger's doctorate paper, now her book, delightfully fresh every time I pick it up.

This is in a large part due to the very fine and clear in-book 'navigation system'. The five human functions are given a strong primary colour tab on the left and right side of each page; appropriately blue for sleep, yellow for hygiene, green for food preparation, red for work and indigo for leisure.

This excellent visual aid is then used in each plan or cross section for all of the six mission vehicles analysed, Apollo, Salyut, Skylab, Shuttle, Mir and The International Space Station.

The result is that what could be a highly complex comparative task, the analysis of habitation in different spacecraft, becomes surprisingly simple, graphically clear and enjoyable for such a wide scan of information. The ability to access and compare specific data, for example astronaut comments for 'leisure' or 'hygiene' etc in space, is fast and clear and does not require a whole book knowledge. Page six gives an overview of the book structure in a clear concise single diagram.

The line drawings, at the bottom of every section's summary page are elegantly minimal with a fine graphic system which follows the traditions of the great post war graphic designer Otl Aicher.

The book is a lesson to all of us, nothing even the intricacies of habitation in space need be made over complex. The best teachers are those who can un-complicate issues and Sandra Häuplik-Meusburger has pioneered an excellent communication method with this book.

There is huge potential for improving space craft habitability, but the briefing and skills required to make this a reality for students, graduates, architects and designers is formidable. Sandra Häuplik-Meusburger gives us some insight in her interviews with nine astronauts about their experiences of life in micro gravity. These are an invaluable personalized addition to the book, from the psychological importance of windows for views of earth, the need to provide volume for enjoying the dynamic possibilities of movement and sports to the frustrations of losing things in the gentle exhaust airflow in micro gravity vehicles.

Issues of usability and livability with special restraint for the human at work or sleep etc. and precise stowage compartments for tools and objects are covered within each of the five habitation categories and with their own concluding sections.

Sandra Häuplik-Meusburger makes it clear that the design dream for future interiors of space vehicles must be to create an essentially uncluttered environment where astronauts can nevertheless move with ease using surface integrated restraints, precisely concealed storage and lighting. She opens the way to this by the minimalist and clear presentation of the book and her concluding pages within each of the five activity groupings, summary observations, comparisons and directions.

Architecture for Astronauts will be enjoyed by aspiring young and old people, students, architects, designers and teachers of all disciplines and no doubt will be emulated by many who are fascinated by the prospects of developing innovation for future compact living spaces on earth.

Sandra Häuplik-Meusburger balances her wide knowledge of space technology facts with psychological issues, suggests elegant solutions for living in micro gravity, space sports and games.

She sub-titles the book 'an activity-based approach' and as so often happens when we move our focus of attention we are inspired to revisit our preconceptions and create new and innovative solutions for life in space and for future generations on earth.

Richard Horden
London, January 2011

AUTHOR'S PREFACE

Since long before the dawn of the Space Age, the idea of conquering outer space has inspired architects to design innovative habitats and living spaces. These avant-garde concepts influenced a generation of architects. Examples are the projects 'Plug-in City' and 'Walking City' by Archigram (Cook, 1991), 'Oasis 7' by Haus-Rucker-Co (Ortner, 1967-92), and 'Villa Rosa' by Coop Himmelb(l)au (Coop Himmelb(l)au, 1967 -).

Today, we have the unique chance to look back at a series of realized space habitats with the presently orbiting International Space Station. Vision has become reality, and now we have the opportunity to look forward to the next generations of space habitats.

Extra-terrestrial habitats can be considered as "new" building types. A number of extra-terrestrial habitats have been occupied over the last 40 years of space exploration by varied users over long periods of time. This experience offers a fascinating field to investigate the relationship between the built environment and its users. Living and working in such habitats means being potentially vulnerable to very harsh environmental, social, and psychological, conditions. With the stringent technical specifications for launch vehicles and transport into space, a very tight framework for the creation of habitable space is set. These constraints result in a very demanding "partnership" between the habitat and the inhabitant.

Habitability and human factors are important determinants for the design of any inhabited building type or human-used object, but in a confined and isolated environment they become especially critical. Beyond Earth, only a habitat secures the basic requirements of humans' existence. Isolated from the Earth, inhabitants live for a long time in a small and confined environment, completely dependent on mechanical and chemical life support systems. Therefore, this building type demands particularly careful planning, designing, and building. Especially when the relationship between the built environment and its users is pre-determined to such an extent, habitability becomes an important design issue.

The author uses the term 'habitability' as a general term to describe the suitability of a built habitat for its inhabitants in a specific environment and over a certain period of time. Set into the space context, habitability can be understood as the measure of how well the (built) environment supports human health, safety and well being to enable productive and reliable mission operation and success (cf. Cohen, 2011).

To accomplish this task, this thesis develops a new framework for a design-in-use study, which differs from the usual analysis in architecture (that places the building first) in that it assigns human activity to a superior position. The human comes to the fore, because first, it is extremely complex and expensive to take a human being off the planet, and second, being there they have to optimally use the short time to fulfil the assigned tasks 100%. Therefore, this 'up-valuation' is not a question of comfort, but rather one of high mission priority.

Humans are explorers. The 'if' is not the question, the 'if' is a given. But until then many research questions still need to be solved.

This book takes the approach of integrating the vast experiences of the users with technical requirements to evaluate the totality of the space habitats. This work is a first step towards a broad evaluation of lessons learned towards an even more human-orientated design approach for space exploration.

In addition, this book offers insights to one 'extreme environment' inhabited by humans. The potential of using this knowledge for planning and building in other environments where humans largely depend on their habitat (such as polar, desert or underwater areas) but also for compact living and working environments on Earth seems obvious.

Sandra Häuplik-Meusburger
Wien, 2011

"Life aims at the harmony between us and the outside world. In space, the human body and spirit are uniquely free to float and reflect high above Earth, in weightlessness. We should learn to adapt to this new physical and psychological environment because there resides our far future."

François Clervoy,
Paris, 2010

ACKNOWLEDGEMENTS

Lots of people have contributed to my work. It is impossible to name them all to thank them. Herewith I want to thank those people that have directly influenced the generation of this book. First of all I like to thank the following astronauts for sharing their personal experiences in such a unique environment with me and for providing personal images (in the order of the interviews):

Jean-François Clervoy, France (CNES/ESA); STS-66, November 3 - 14, 1994; STS-84, with docking to Mir, May 15 - 24, 1997; STS-103, December 19 - 27, 1999
Jean-Pierre Haigneré, France (CNES); Mir-Altair, July 1 - 22, 1993; Mir-Perseus, February 20 - August 28, 1999
Reinhold Ewald, Germany (DLR/ESA); Mir 97, February 10 - March 2, 1997
Michel-Ange Charles Tognini, France (ESA); Mir-Antares, July 27 - August 10, 1992; STS-93, July 23 - 27, 1999
Franz Viehböck, Austria; Mir-Austromir 91, October 2 - 10, 1991
Hans Wilhelm Schlegel, Germany (DLR/ESA); STS-55, April 26 - May 6, 1993; STS-122, February 7 - 20, 2008
Harrison "Jack" Schmitt, USA (NASA); Apollo 17, last lunar landing mission, Dec. 7-19, 1972
Gerhard Thiele, Germany (DLR/ESA); STS-99, February 11 - February 22, 2000
Dumitru-Dorin Prunariu, Rumania (Intercosmos); Soyuz 40, May 14 - 22, 1981, to Salyut 6

This work would not have been possible without their input. In addition I would like to thank:

Richard Horden, Architect (HCLA Architects, London) and Professor at the Technical University in Munich, this book is based upon my doctoral thesis and I would like to express my gratitude to him as my supervisor. He has been teaching and realizing architectural projects within the fields of 'Micro Architecture' and 'Space Architecture' over many years. His projects as well as his enthusiastic personality provided a high degree of motivation throughout the work; **Marc M. Cohen**, Architect in the US, previously working on Aerospace Architecture for the NASA-Ames Research Centre for the lectorate on the final book; **Manuela Aguzzi** and **Olivier Lamborelle**, Astronaut instructors at the European Astronaut Centre, Cologne for valuable discussions and reading parts of the draft book; **Verena Holzgethan**, Landscape designer and project partner for valuable discussions throughout the process of packing the book together; **Cuno Brullman**, Architect and Professor at the TU Vienna for being my second thesis adviser and his continuous interest in this topic; **Jens Dangschat**, Professor for Sociology at the TU Vienna; **Ulrike Schmitzer**, Science journalist; **Andreas Vogler**, Space architect from Munich, **Michael Seidel**, colleague at the Department HB 2 and **Iva Kovacic** from the Institute for Industrial Building and Interdisciplinary Planning at the TU Vienna for discussions in different phases of the book; **Maria Antonietta Perino**, Head of Advanced Exploration Programs, **Enrico Gaia**, R&D Coordinator (Business Segment Space Infrastructures & Transportation), **Giorgio Musso**, Head of Physical Architecture & Ergonomics Unit, **Domenico Tedone** and **Marinella Ferrino**, Physical Architecture & Ergonomics Unit and **Alessandro Bergamasco**, System Engineer of the PLR Study at Thales Alenia Space Italia, for discussion and providing information on different studies; **Marietta Steinhart** for the transcription of the interviews; **Kevin Dooley** for English language corrections; **Amine Khouni** for his support in finalizing the book; **Birgit Dejaco** for her support in finalizing the thesis draft; **Sylvia Kostenzer**, for improving the graphical concept to make it a beautiful book;

Further I would like to thank my colleagues and project partners that I have been working with over the last years.

Last but not least, the greatest thanks go to the ones that have been supporting my work and life: my family, friends and beloved husband.

A: Jean-François Clervoy
B: Jean-Pierre Haignére
C: Reinhold Ewald
D: Michel-Ange Charles Tognini
E: Franz Viehböck
F: Hans Wilhelm Schlegel
G: Harrison Schmitt
H: Gerhard Thiele
I: Dumitru-Dorin Prunariu

(credit A: NASA, Jean-François Clervoy; B: ESA/CNES; C, D, F: ESA - S. Corvaja; E: Franz Viehböck; G: NASA; H: NASA/ESA; I: Dumitru-Dorin Prunariu)

A:

B:

C:

D:

E:

F:

G:

H:

I:

PART 1

INTRODUCTION

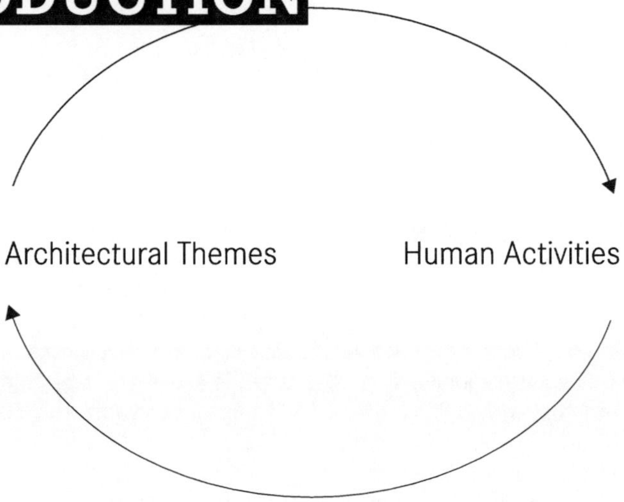

This chapter gives a short summary on the research methodology used and the structure of the book.

1.1 THE QUEST FOR HABITABILITY

"Early spacecraft had been designed to be operated and not lived in."
(Compton, et al., 1983 p. 130)

The term 'Habitability' or any similar concept that can describe the suitability of the environment for daily human life was not in the vocabulary of early spacecraft designers or engineers. Their often-used term 'man in a can' underlines that attitude. After the first space missions, when NASA and the Soviets were advancing their goals for long duration missions to prove that humans could live and work in space for extended periods, the habitable design of the interior became increasingly important.

In 1968, Raymond Loewy, a world-renowned industrial designer was hired as a habitability consultant for the Saturn-Apollo and Skylab projects. George E. Mueller, NASA Associate Administrator for Manned Space Flight at NASA Headquarters, seems to have been the driving force behind the inclusion of human factors in the design of the Skylab Space Station. Loewy suggested a number of improvements to the layout, such as the implementation of a wardroom, where the crew could eat and work together, the wardroom window, the dining table and the colour design, amongst others. The design process was difficult. The engineers regarded crew quarter design as part of their responsibility and the test pilot-astronauts who were reviewing the crew quarter concepts adopted the attitute that they cared even less about the design. It took some time until the improvements Loewy suggested were implemented (Compton, et al., 1983).

One controversial subject was the provision of a window in the Skylab Saturn Workshop's Wardroom area. The engineers argued it would weaken the structure, be too expensive, and take too long to develop. At first the window seemed not to be essential to mission success. Mueller asked Loewy for his opinion, on which he replied, that: "Not to have a window is unthinkable!" (Compton, et al., 1983 p. 137). Mueller answered: "Put in the window" (Compton, et al., 1983 p. 137) and the window, the wardroom and other changes were authorized.

George Mueller later said of the early mock-up: "Nobody could have lived in that thing for more than two months." (Compton, et al., 1983 p. 133).

According to B.J. Bluth and Martha Helppie, the Soviets, working on the Salyut space stations at the same time, acknowledged the emotional value in addition to the functional value of the interior design for mission success. The interior of the space stations had to be "comfortable, functional, harmonious and attractive" (B.J.Bluth, 1987 p. III-123).

Still habitability and the quality of the living and working environment of extra-terrestrial habitats was not considered of high priority in the past. In the meantime a variety of definitions for the term 'habitability' have been formulated.

In his book Bold Endeavours, anthropologist Jack Stuster refers to Kubis' (1965) definition of habitability: "habitability as the sum of interactions between operators and environment which include physical, physiological, psychological and social interactions" (Stuster, 1996 p. 40).

Psychologist Dr. James Wise provides a definition for spatial habitability: "Spatial habitability refers to the ways in which the volume and geometry of liveable space affect human performance, well-being and behaviour" (Wise, 1988 p. 6).

Mary Connors from NASA Ames simply states in her book Living Aloft – Human Requirements for

A: Astronaut Gregory C. Johnson, STS-125 pilot, uses a still camera at an overhead window on the aft flight deck of Space Shuttle Atlantis during flight day six activities

(credit A: NASA)

Extended Spaceflight, that "Habitability is a general term that connotes a level of environmental acceptability" (Connors, Harrison, Akins, 1999 p. 59).

The author uses the term 'habitability' as a general term to describe the suitability and value of a built habitat (house or spacecraft) for its inhabitants in a specific environment (Earth or Space) and over a certain period of time.

Effects of impaired habitability

The history of space exploration is full of reports about mishaps (cf. Shayler, 2000). Inappropriate or faulty design presents a threat to the crew's health and the overall mission. In his book Off the Planet Jerry Linenger reports an interesting strategy to overcome faulty design. In 1997, a backup life support device consisting of a solid-fuelled oxygen canister called an "oxygen candle" caused a fire to break out on the space station Mir. After a desperate struggle, the crew finally extinguished the fire. Hardware was damaged, but the crew was not injured. Linenger reports, that later they were told, that the canisters were 'now' safe to use. "They were deemed safe not because the cause of fire had been determined, but rather because mission control in Moscow now introduced the requirement that whenever we activated one of the canisters, we stand by with a fire extinguisher" (Linenger, 2000 p. 116).

In 2004 a poorly designed flexible air hose caused a leak at the International Space Station (Banks, 2004). The hose was located in the Destiny science module, close to an optical window for earth-observation. Due to a lack of appropriate handholds, the astronauts repeatedly held onto the air hose when looking out of the window. This unplanned practice finally resulted in a leaky hose, through which internal air left the station.

It has by now been widely acknowledged, that 'window gazing' is the number one leisure activity, and that astronauts and cosmonauts spend a lot of time in front of windows looking at the Earth. Previously Skylab astronaut Gerald Carr stated in 1974, that "if something is going to stick out and make a nice handhold, it's going to be used for a handhold" (NASA [Bull.1], 1974 p.76).

Although the consideration of habitability and human factors has been integrated in the design process of human spacecraft, there is still a requirement for improving habitability *[Fig. A]*. This precept is valid also for the design of commercial spacecraft.

If designed from a more human-orientated view rather than a solely engineering one, the window would have been provided with means to hold on that was structurally suited for this function.

A:

1.2 RESEARCH METHODOLOGY

RESEARCH OPPORTUNITY

Data that has particular relevance for the design of extra-terrestrial habitats is available from so called 'space analogues', such as experiences of underwater habitats (NASA [NEEMO], 2007), habitats in polar areas (Stuster, 1996; Harrison, 1991), space mission simulators (SSC RF) and other extreme terrestrial environments. This literature primarily provides information on behavioural, psychological and sociological factors based on experiences in an extreme Earth-based environment.

Additionally there is a lot of data available on extra-terrestrial habitats. Online information is available on the NASA technical report server and from other space agencies. Books and reports about lessons learned from past space stations and presently the International space station have been published. This kind of literature provides important insights on specific technical and mission-related issues.

Further information is available from personal experiences and anecdotes of astronauts in the form of books, interviews or online diaries. These kinds of resources provide information from an individual's point of view that is more qualitative than systematic.

Although a lot of specific data is available, it is 'spread' widely among diverse sources. If, for example, someone is looking for information on the design of specific equipment, one has to gather information from many different sources. For this research, a lot of time is needed and already some knowledge of where to find what. Furthermore, basic knowledge is required to read the available plans and images to recognize shortcomings and potentials because information from different sources may not be comparable.

SELECTION OF HABITATS

The case studies were selected according to the following criteria:

(1) Selected buildings had to be extra-terrestrial, implying that it is the most hostile environment in terms of physical, social and psychological means.

(2) Selected buildings had to be realized: completed, occupied, and utilized – in order to allow post-evaluation with personal feed-back from the users - the astronauts.

(3) Further habitats were selected that hosted a minimum crew of two;

(4) With mission lengths exceeding 30 days, to provide minimal interaction between crewmembers over a certain time within the built environment.

Selected case studies came from: the **Apollo Spacecraft** and **Lunar Module**, **the Space Shuttle Orbiter**, and **the Space Stations: Salyut, Skylab, Mir**, as well as **the International Space Station.**

The Apollo Spacecraft and Lunar Module did not fulfil the selection criteria (12 days), but were chosen because of their importance for current lunar mission architecture studies and because they were the only realized manned mission series to the lunar surface. The Space Shuttle Orbiter (7-17 days) did not fulfil the selection criteria but was chosen, because it has launched more crewmembers to space than any other spacecraft and was vital to constructing the ISS.

Selection of Extra-terrestrial Habitats *(credit: Author)*

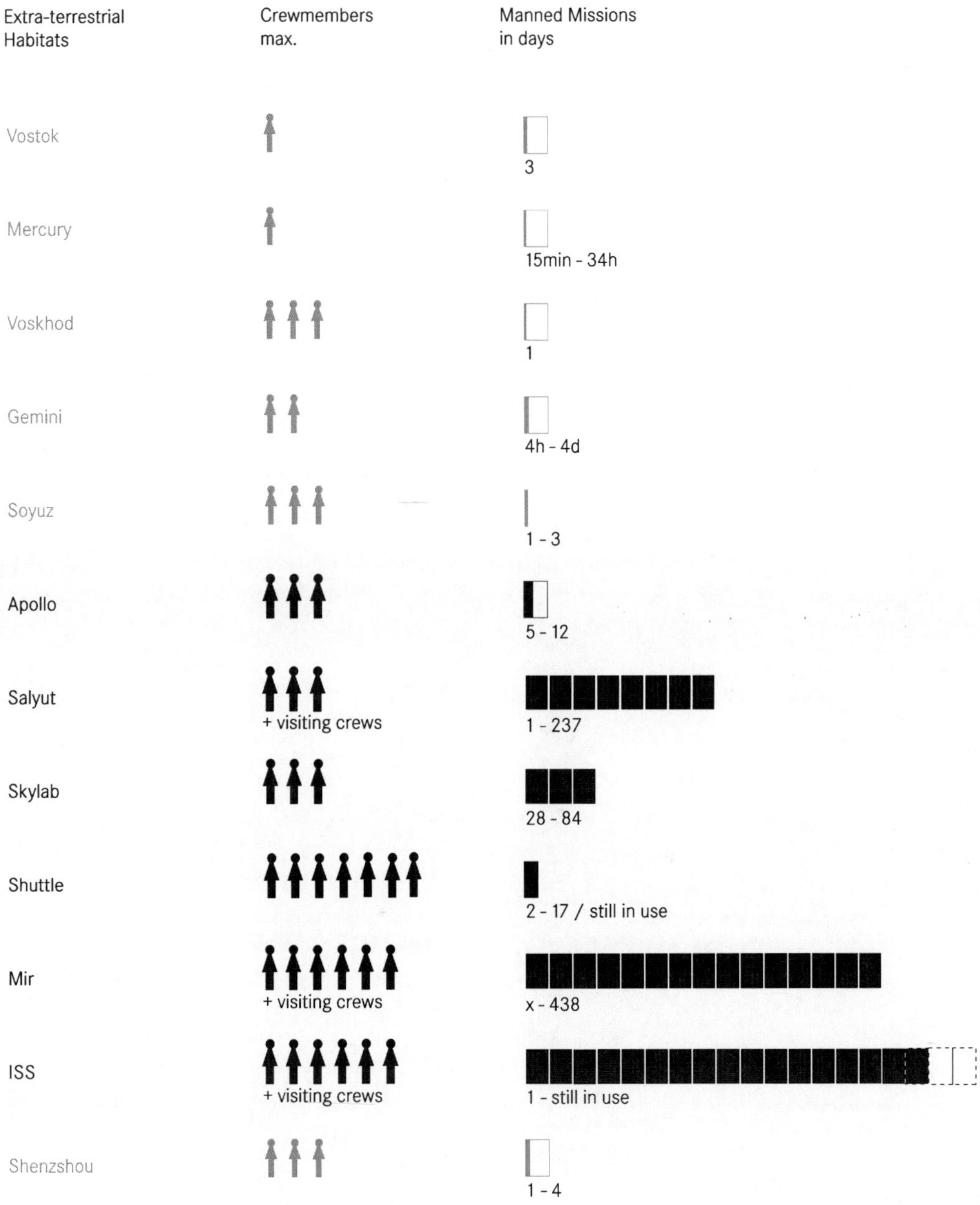

1.2 RESEARCH METHODOLOGY

METHODOLOGY AND WORK FLOW

The principal function of any habitat is to provide living and working environment for humans - as best as circumstances allow. Basic human requirements don't change in different environments. A human must sleep, go to the toilet, eat and be active in some way (cf. Maslow's hierarchy of needs). Therefore, a comparative analysis focusing on human activities within a built environment was chosen as the method for this research. *Figure A* shows the diagram of the workflow. The research and compilation of data according to human activities represents a new approach.

The Apollo Spacecraft and Lunar Module, the Space Shuttle Orbiter, and the Space Stations; Salyut, Skylab, Mir, as well as the International Space Station have been analysed according to the human activities: sleep, hygiene, food, work and leisure.

To make a consistent comparison, the same architectural themes were applied for every human activity. Information on astronaut's personal experiences from anecdotal references and from personal interviews offers an enlightening perspective. The methodology involved selecting the astronaut's statements and categorizing them according to the architectural themes.

A:

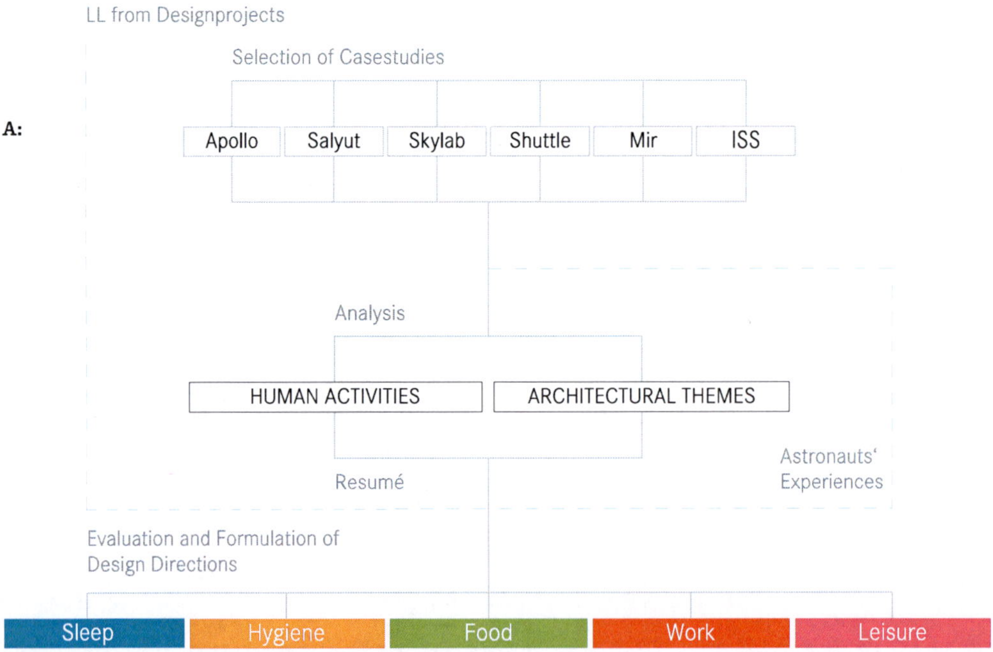

A: Methodology and Structure *(credit A: Author)*

INTRODUCTION HUMAN ACTIVITIES

The diverse human activities were grouped with the main human activities: **sleep, hygiene, food, work and leisure**. Sub-categories have been added where needed and could further be expanded when new facts or research directions emerge. To facilitate orientation and to ease comparison with architectural drawings and diagrams each category was assigned a specific colour:

The human activity category **sleep** (colour blue) includes the sub-activities rest; preparation for sleep, relaxation, and sleep.

The category **hygiene** (colour code yellow) is divided into the following sub-categories:
- Personal Hygiene includes the sub-activities: full and part body cleansing, change clothes
- Shower was a special activity on Salyut, Mir and Skylab
- Toilet includes the sub-activities: collect, store, process, and dispose of waste
- Housekeeping

The category **food** (colour code green) includes the sub-activities store, prepare, grow and consume food and drinks.

The category **work** (colour code red) includes the sub-activities: operation, work tasks, conducting experiments, communication, education and training.

The category **leisure** (colour code pink) includes the sub-activities: free-time, exercise activities and intimate behaviour. Although exercise is a scheduled activity on space stations it falls into the category leisure, because it is mostly considered a pleasant activity.

All sub-categories include associated translation paths and stowage areas.

SLEEP

rest, relaxation, sleep and storage

HYGIENE

PERSONAL HYGIENE

full and part body cleansing, clean and change clothes and storage

SHOWER, TOILET

HOUSEKEEPING

FOOD

store, prepare, grow, consume and storage

WORK

operations, worktasks, experiments, communication, education, training and storage

LEISURE

free-time activities, exercise, intimate behaviour and storage

1.2 RESEARCH METHODOLOGY

ARCHITECTURAL THEMES

An essential part of this research entailed finding and defining the architectural themes that describe and enable habitability in space. To make a consistent comparison, the methodology applies the same themes for every human activity in the analysis. The selection derives from relevant evaluation themes on architecture (cf. Van der Voordt, et al., 2002) and themes raised by astronauts. The comparison applies these themes: usability, livability, flexibility.

These themes may appear superficially similar to Vitruvius 'firmitas' (Firmness), 'utilitas' (Commodity), and 'venusitas' (Delight). However, they are quite different. Usability comes the closest to Palladio's utilitas. Livability might seem to relate to venusitas, but it has much more in common with the bottom levels of Maslov's pyramid (cf. Maslov, 1943; Cohen, Houk 2010). Finally, flexibility has nothing in common with firmitas, which addresses structural integrity and the ability to withstand environmental forces.

Usability

According to the ISO 9241-11 (1998): Guidance on Usability Standard, usability is defined as "the extent to which a product can be used by specified users to achieve specified goals with effectiveness, efficiency, and satisfaction in a specified context of use". According to the Webster-Merriam Dictionary Online 'usable' means (1) capable of being used and (2) convenient and practicable for use. The adapted noun-form is 'usability'.

The term 'Usability' is used in this book in the following sense: *The layout, configuration and design of extra-terrestrial habitats assure efficient, user-friendly and trouble-free habitation over a specified or planned period of time. This metric is also valid for associated equipment and subsystems.* Out of a number of factors that influence 'usability' of a habitat by its users, the following themes were selected:

- Availability and Equipment (available infrastructure)
- Spatial arrangement (spatial orientation, relations between activity zones)
- Object management (storage concept and management)
- Ergonomic safety (issues related to the ergonomic and safe use, including restraints)

Livability

According to the Webster-Merriam Dictionary Online 'Livability' means (1) survival expectancy or (2) suitability for human living. The term 'Livability' is often equated to 'Quality of Life'. In a report about the quality of life in Europe it is described that the "quality of life concept goes beyond the living conditions approach". Accordingly, three major characteristics are described in this concept (Albers, et al., 2004 p. 1): It refers to an (1) individuals' life situation, (2) it is a multi-dimensional concept and (3) it is measured by objective as well as subjective indicators.

The term 'Livability' is used in this book in the following sense: *The habitat provides maximum living space even within a minimal limited and socially isolated volume for the individual and the crew. Limitations include a low level of environmental comfort, restricted volume, and little opportunity to "go outside".* The following themes were selected:

- Territoriality and Privacy (individual versus social zones and use of infrastructure)
- Sensory perception (such as light, sound, temperature, humidity, astronaut's perception)
- External relations (physical and virtual relations to the 'outside; Windows, contact to ground)
- Internal relations (spatial relations to other activity zones)

A: Relationship between Architectural Themes and Human Activities

(credit A : Author)

Flexibility

The term 'Flexibility' is widely used, often defined in various fields, but hard to define in a general manner. In a presentation on the 'Architectural Design for Flexibility and Buildability to Facilitate Evolution' by the Fraunhofer Institute, the term 'Flexibility' is defined as "the degree, to which a system supports possible or future changes to its requirements" (Carbon, et al., 2010). According to the Webster-Merriam Dictionary Online 'flexible' means (1) capable of being flexed, (2) yielding to influence or (3) characterized by a ready capability to adapt to new, different, or changing requirements. The noun is 'flexibility'.

The term 'Flexibility' is used in this book in the following sense: *The habitat allows adjustments according to the requirements of the users, to changing mission tasks as well as unforeseen social and mission related changes*. The following themes were selected

- Spatial flexibility (variations in size and locations)
- Object flexibility (variations in usage)
- Individual flexibility (ergonomic and user orientated variations)

ASTRONAUTS' EXPERIENCES

In addition to available data, personal interviews with the real users were included in the study. Nine astronauts agreed to share their personal experiences as the users of their respective space habitats. The structured interviews were conducted with a special focus on human activity, from which new facts emerged. The findings from relevant literature and analysis based on drawings and images were compared to the personal experiences of users.

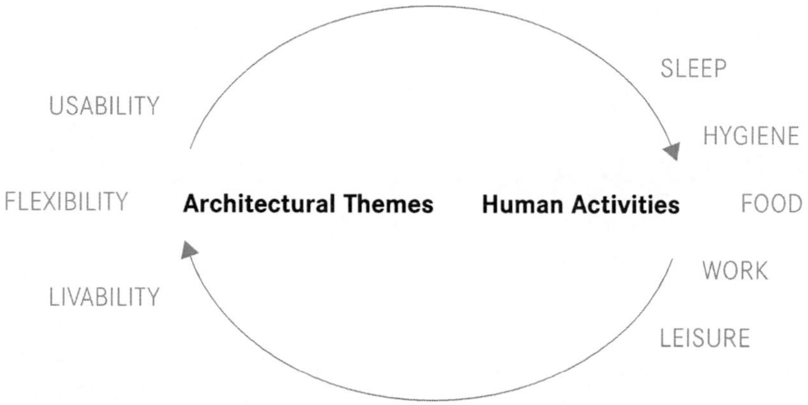

1.3 STRUCTURE OF THE BOOK

SLEEP

rest, relaxation, sleep and storage

HYGIENE

PERSONAL HYGIENE

full and part body cleansing, clean and change clothes and storage

SHOWER, TOILET

HOUSEKEEPING

FOOD

store, prepare, grow, consume and storage

WORK

operations, worktasks, experiments, communication, education, training and storage

LEISURE

free-time activities, exercise, intimate behaviour and storage

Part 1 gives a short summary on the research methodology used and the structure of the book.

Part 2 of this book provides the background to this work as well as an introduction into the commonalities and differences of the physical and social environment. It investigates the consequences for the design and building of an extra-terrestrial habitat and the living and working environment of the user.

Part 3 introduces the characteristics of the selected extra-terrestrial habitats (Apollo Spacecraft and Lunar Module, Salyut Space Station, Skylab Space Station, Space Shuttle Orbiter, Mir Space Station and ISS International Space Station).

This introduction includes the basic mission-related objectives, the general configuration and layout of the extra-terrestrial habitat as well as the time and spatial allocation of human activities. It presents the comparison of collected data from the inhabited American and Russian habitats with drawings and diagrams.

Part 4 to 8 presents the comparative post analysis of the selected case studies according to the human activities.

Each chapter summarizes the research results and important issues that significantly influence the habitability system. They highlight future potentials for each category of human activity. The results of the study were further formulated as design directions. They are expressed as universal headlines that are valid for the planning of compact habitats, wherein human activities are in the foreground.

1.4 REFERENCES

Albers, Jens, et al. 2004. Quality of Life in Europe - First European Quality of Life Survey 2003. Luxembourg : European Foundation for the Improvement of Living and Working Conditions, 2004. ISBN 92-897-0260-5.

B.J.Bluth, M. Helppie. 1987. Soviet Space Stations as Analogs. NASA Headquarters Washington, D.C., USA : NASA Grant, 1987. NAGW-659.

Banks, Jim. 2004. Space.com. [Online] 23 01 2004. [Cited: 10 07 2006.] http://www.space.com/missionlaunches/exp8_update_040123.html.

Carbon, Ralf and Naab, Matthias. 2010. Architectural Design for Flexibility and Buildability to Facilitate Evolution - Design for Future 2010. Presentation on 03.05.2010. Bad Honnef, Germany : Fraunhofer Institute IESE, 2010.

Cohen, Marc M. and Houk Paul C. 2010. Framework for a Crew Productivity Figure of Merit for Human Exploration. Anaheim, California, USA : AIAA Space 2010 Conference & Exposition, 2010. AIAA 2010-8846

Cohen, Marc M. 2011. not published. Email, 2011.

Clervoy, Jean-François. 2009. [interv.] Sandra Häuplik-Meusburger. Transcript not published. Paris, France, June 2009.

Compton, W. David and Charles D., Benson. 1983. Living and Working in Space, A History of Skylab. United States : National Aeronautics and Space Administration, The NASA History Series, 1983. ISBN-10: 0160041465.

Connors, Mary M., Harrison, Albert A. and Akins, Faren R. 1999. Living Aloft, Human Requirements for Extended Spaceflight. Washington, D.C., USA : NASA, Ames Research Center, 1999. SP-483 / ISBN: 1-4102-1983-6.

Cook, Peter, [ed.]. 1991. Archigram. Basel, Switzerland : Birkhäuser Verlag , 1991.

Coop Himmelb(l)au. 1967 -. Coop Himmelb[lau). [Online] Wolf D. Prix, W. Dreibholz & Partner ZT GmbH, 1967 -. http://www.coop-himmelblau.at/site/.

Harrison, Albert A. 1991. From Antarctica to Outer Space - Life in Isolation and Confinement. New York, USA : Springer, 1991. ISBN-13: 978-0387973104.

ISO 9241-11. 1998. Definition Usability. [Online] ISO 9241-11: Guidance on Usability (1998), 1998. http://www.usabilitynet.org/tools/r_international.htm#9241-1x.

Linenger, Jerry M. 2000. Off The Planet - Surviving five periolous months aboard the Space Station Mir. USA : McGraw-Hill, 2000. ISBN 0-07-136112-X.

Maslow, Abraham Harold. 1943. A Theory of Human Motivation. Psychological Review. 50, 1943, Vols. ‚Classics in the History of Psychology‘, an internet resource developed by Christopher D. Green, York University, Toronto, Ontario, http://psychclassics.yorku.ca/Maslow/motivation.htm, pp. 380-396.

Merriam-Webster, Incorporated. 2011. Merriam Webster Dictionary. [Online] 2011. http://www.merriam-webster.com/dictionary.

Morgan, Morris Hicky. 1914. Vitruvius: The Ten Books on Architecture. London, UK : Oxford University Press. [Online] 2006. http://www.gutenberg.org

NASA [Bull.1]. 1974. Skylab Experience Bulletin No.1: Translation Modes and Bump Protection. Houston, Texas, USA : NASA, Lyndon B. Johnson Space Center, 1974. JSC-09535.

NASA [NEEMO]. 2007. NEEMO Nasa Extreme Environment Mission Operations. [Online] NASA, 5 September 2007. [Cited: 12 December 2008.] http://www.nasa.gov/mission_pages/NEEMO/.

Ortner. 1967-92. Haus-Rucker-Co Splendid Blend. [Online] 1967-92. [Cited: 06 January 2011.] http://www.ortner.at/haus-rucker-co/de/haus-rucker.html.

SSC RF. MARS-500. [Online] SSC RF - Institute of biomedical problems of RAS, Moscow, Russia. [Cited: 5 July 2009.] http://www.imbp.ru/Mars500/Mars500-e.html.

Shayler, David J. 2000. Disasters and Accidents in Manned Spaceflight. UK : Springer Praxis Books with Praxis Publishing, 2000. ISBN 978-1-85233-225-9

Stuster, Jack. 1996. Bold Endeavors: Lessons from Polar and Space Exploration. USA : US Naval Institute Press, 1996. ISBN-10: 155750749X.

Van der Voordt, Theo and Van Wegen, Herman. 2002. Ex Post Evaluation of Buildings. [book auth.] T.M. De Jong and D.J.M Van der Voordt. Ways to Study and Research Urban - Architectural and Technical Design. Delft, NL : DUP Science, Delft University Press, 2002, pp. 151-158.

Wise, James. 1988. The Quantitative Modelling of Human spatial Habitability. Ames Reseach Center, USA : NASA Contractor Report, 1988. 177501.

PART 2:

BACKGROUND

This chapter gives a short introduction to selected environmental, social and psychological issues and their consequences for the habitat and its users.

BACKGROUND

2.1 THE ENVIRONMENT

	EARTH	OUTER SPACE
Diameter	12.756 km	-
Period of Revolution (Year)	365 Earth days	Not applicable
Period of Rotation (Day)	24 hours	Not applicable
Gravity	9,8 m/s^2	Microgravity
Mean Temperature	15 °C max. +60° C min. -89° C	- max. +200 °C (radiant energy) min. -270 °C (cosmic background radiation)
Lenght of day	24.0 hours	-
Atmospheric pressure	1 bar (N2, O2, C02)	0 bar (Vacuum)
Presence of Water	70.8 % of surface is covered with water	Known to exist in comets
Dust	Exists in the atmosphere, generally not harmful, expect for allergies	Exists but minimal
Radiation	Natural protection by the Earth's atmosphere equivalent to about 1000g/cm2 and the Van Allen Belt	Exposure to Solar Particle Events and Galactic Cosmic Rays
Other Specific Characteristics	-	Extreme bright light and glare; Exposure to micrometeoroids

Comparison of Evironmental Parameters

sources: (NASA [LCROSS], 2010; NASA [Moon], 2010; Nasa [Mars], 2010; Cohen, 2009; Williams, David Dr. R. , 2007; Hamilton, 1995-2009; Eckart, 1999)

MOON	MARS
3475 km	6792 km
28 Earth days	792 Earth days 26 months
28 Earth days 672 hours	~24,66 hours
1.62 m/s^2; 1/6 of Earth	3.69 m/s^2; 1/3 of Earth
-20 °C (much colder at the poles in deep craters) max. +123 °C (possibly up to 140 °C in some locations on the equator) min. -233 °C	-65 °C max. 20 °C min. -140 °C
708.7 hours	24.7 hours
0 bar (almost a perfect vacuum)	0.01 bar
Water in the deep, permanently shadowed craters at the poles	Found in a variety of „special regions“
Pervasive on the surface: abrasive, sharp, potentially toxic, electromagnetic cling, lofts above the surface	Pervasive on the surface, very fine grain, dust storms in Mars atmospheric winds, potentially toxic and abrasive
Exposure to Solar Particle Events and Galactic Cosmic Rays; Mass of the surface gives “half protection” but the surface also generates secondary thermal neutrons from bombardment	Atmosphere gives about 30g/cm equivalent protection; Mass of the planet gives about “half shielding”
Extreme bright light and glare; Exposure to micrometeoroids	-

2.1 THE ENVIRONMENT

SELECTED ENVIRONMENTAL ISSUES

Extra-terrestrial habitats are the "houses and vehicles" where people live and work beyond Earth: non-planetary habitats such as a spacecraft or space station; and planetary habitats such as a base or vehicle on the Moon or Mars. These building types are set up in environments different from the one on Earth and can be characterized as 'extreme environments.' Multiple requirements arise for the architecture and design of such a habitat.

The physical environment in space and on planetary bodies is different compared to Earth. It is inhospitable. In the conference proceedings 'Human Performance, Situation Awareness and Automation', Kring, et al. (2000 p. 119) characterized extreme environments as "settings that possess extraordinary technological, social and physical components that require significant human adaptation for successful interaction and performance".

The following introduction describes some of the main environmental issues relevant for extra-terrestrial habitats. Only a few will be named but not explained in detail; They are listed for the principal understanding of the environmental setting of extra-terrestrial habitats.

Microgravity

Gravity is an invisible force of nature that designates our life on Earth. The mass of an object is everywhere in the universe the same, because it is a property of an object. But the weight is the result of gravity. It is the "gravitational force of attraction between the object and the Earth" (ESA [Gravity], 2008). On Earth this acceleration is about $9.8 m/s^2$ and is called 1G.

'Zero-G' or 'Weightlessness' is a condition experienced while in 'free-fall', in which the effect of gravity is cancelled by the inertial force resulting from orbital flight (Britannica, 2011), thus no acceleration is experienced. An orbital space craft in a circular orbit travels with a certain velocity parallel to the (Earth's) surface, while 'falling down' at the same time, thus creating very small gravity forces. Such a condition is called 'Microgravity'.

The gravity on the Moon is about one-sixth and the gravity on Mars about one-third of that on Earth. The reduction in gravity can cause severe effects on the human body. Physiological effects of microgravity include calcium loss, fluid shifts, skeletal changes, muscle mass loss and vestibular alterations (NASA [MSIS], 1995 p. 178). We have evolved within earth gravity. A change in gravity affects a wide range of human activities, like body movement, posture and locomotion and provides a dramatic challenge to human physiology.

During the Skylab missions, it became evident that workstation designers have to take microgravity conditions into account. Following this experience, in 1975 an analysis of positions in microgravity was conducted. At that time, it was believed, that "there is a definable relaxed body posture in zero-g and that the eligible flight crew population can be fitted to that posture" (NASA [Bull.17], 1975 p. 2).

A graphic with precise angels on 'the microgravity body posture' (Neutral Body Posture, NMP) is documented in the Man-Systems Integration Standard (1985) and also in the newly published Human Integration Design Handbook (NASA [HIDH], 2010 p. 50). This graphic is widely used as a basis for designing living and working environments in space. However the data leading to that graphic is based only on the measurements of 3 (!) male American subjects (Skylab 4 crew) and 12 photos (NASA [Bull.17], 1975 p. 13). An additional survey between STS-57 crew members

A: Neutral body orientation on Earth and in microgravity *(credit A: Author, based on NASA-STD-3000)*

showed, that "no single crewmember exhibited the typical NBP called out in the MSIS" (Mount, et al., 2003 p. 6).

Figure A shows a comparison between the neutral body orientation on Earth and the one in microgravity. The head is tilted down; ankle, knee and hip height increase; elbow, wrist and shoulder are raised and elbows are abducted (NASA [HIDH], 2010 p. 49). However angles vary according to the individual.

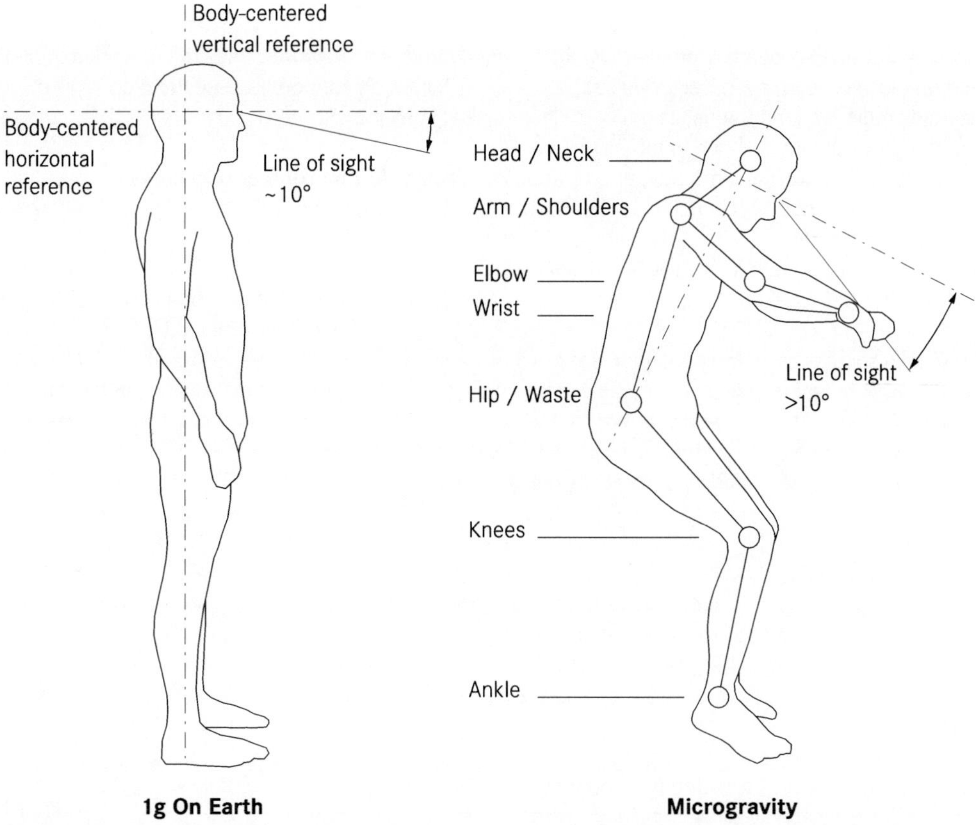

2.1 THE ENVIRONMENT

Atmosphere
"Humans are accustomed to breathing an atmosphere that contains 21% oxygen by volume at sea-level" (NASA [MSIS], 1995 p. 146). In outer space and on the planetary bodies Moon and Mars the available atmospheric conditions cannot naturally support human survival. Additional technical infrastructure is needed.

Radiation
The Earth is protected from hazardous radiation by its magnetic field and the atmosphere. Beyond the Earth's magnetic shield humans are exposed to ionizing and non-ionizing radiation. "Ionizing radiation, which breaks chemical bonds in biological systems, can have immediate (acute) as well as latent effects, depending on the magnitude of the radiation dose absorbed, the species of ionizing radiation, and the tissue affected. Non-ionizing radiation (consisting of different types of electromagnetic radiation) are generally not energetic enough to break molecular bonds in biological systems but, with sufficient intensity, can produce adverse biological effects" (NASA [MSIS], 1995 p. 248).

The main sources of ionizing radiation are galactic cosmic radiation and solar particle events, amongst others. They present a hazard for the biological system as well as for materials and equipment. For habitats beyond the Earth's magnetic shield additional protection is needed.

Micrometeoroids
Micrometeoroids are very small meteoroids - tiny pieces of rocks or debris - that can reach a high velocity in deep space. Thus they present a high threat of damaging spacecraft systems and the hull of the space habitat.

The Earth is naturally protected from meteoroids by its atmosphere. Only remainings from meteoroids that are not burned up completely reach as meteorites the surface of the Earth. In deep space, during transit or on the surface of a planetary body without an atmosphere, such as the Moon there is no natural protection from micrometeoroids.

Thermal Environment
Compared to Earth, outer space and planetary body environments are subject to very high temperature differences, as well as high and low temperatures. According to NASA, "the cabin heating, circulation, and cooling systems need to be designed to maintain human comfort by controlling the atmospheric parameters of gas, temperature, velocity, pressure, and humidity. Radiant heat sources must be identified and their impact on comfort assessed and controlled" (NASA [MSIS], 1995 p. 301).

Specific Environmental Issues
Each designated environment has specific environmental issues.

On the Moon for example one of the delicate issues that need to be tackled is lunar dust. It is very fine and adhesive. Some of the astronauts complained about it, particularly on the longest mission on the lunar surface, Apollo 17 during the Apollo missions.

A: First picture of the Earth and the Moon in a single frame, taken by the spacecraft Voyager 1 in 1977

(credit A: NASA)

Further Reading

Human Spaceflight: Mission Analysis and Design by Wiley J. Larson and Linda K. Pranke (Larson, et al., 2000)

The Lunar Base Handbook by Peter Eckart (Eckart, 1999)

Human Integration Design Handbook by NASA (NASA [HIDH], 2010)

Man-Systems Integration Standards by NASA (NASA [MSIS], 1995)

A:

2.2 THE USERS

On 12 April 1961 Yuri Alexandrovitch Gagarin became the first human to travel into space. The mission lasted 1 hour and 48 minutes. People went to space - at the beginning only for some minutes, then hours, days, weeks, months, up to a year with a current record of 438 days by Mir cosmonaut Valeri Polyakov in 1994. One day people will overcome microgravity and hypogravity to stay 'there' much longer.

By April 2010, 517 humans from 38 countries had travelled to space. 23 astronauts left Earth orbit and went to deep space, 12 people walked on the lunar surface. The list on the right side provides an alphabetical list of all space travellers (Wikipedia [space travellers], 2010).

The astronauts interviewed for this work are highlighted.

1. Joseph M. Acaba
2. Loren Acton
3. James Adamson
4. Viktor M. Afanasyev
5. Thomas Akers
6. Toyohiro Akiyama,
7. Vladimir Aksyonov
8. Sultan Salman Al Saud
9. Edwin Buzz Aldrin
10. Aleksandr Panayotov Aleksandrov
11. Aleksandr Pavlovich Aleksandrov
12. Andrew M. Allen
13. Joseph P. Allen
14. Scott Altman
15. William Anders
16. Clayton Anderson
17. Michael P. Anderson
18. Anousheh Ansari
19. Dominic A. Antonelli
20. Jerome Apt
21. Lee Archambault
22. Neil Armstrong
23. Richard R. Arnold
24. Anatoly Artsebarsky
25. Yuri Artyukhin
26. Jeffrey Ashby
27. Oleg Atkov
28. Toktar Aubakirov
29. Sergei Avdeyev
30. James Bagian
31. Ellen Baker
32. Michael Baker
33. Aleksandr Balandin
34. Michael Barratt
35. Daniel Barry
36. John-David F. Bartoe
37. Yuri Baturin.
38. Patrick Baudry
39. Alan Bean.
40. Robert L. Behnken
41. Ivan Bella, first Slovak in space.
42. Pavel Belyayev
43. Georgi Beregovoi
44. Anatoli Berezovoy
45. Brian Binnie.
46. John Blaha
47. Michael J. Bloomfield
48. Guion Bluford.
49. Karol Bobko
50. Eric A. Boe
51. Charles Bolden
52. Roberta Bondar
53. Frank Borman,
54. Stephen G. Bowen
55. Kenneth Bowersox
56. Charles Brady
57. Vance Brand
58. Daniel Brandenstein
59. Randolph Bresnik
60. Roy Bridges.
61. Curtis Brown
62. David M. Brown
63. Mark Brown
64. James Buchli
65. Jay Buckey
66. Nikolai Budarin
67. Daniel Burbank
68. Daniel Bursch
69. Valery Bykovsky
70. Robert Cabana
71. Tracy Caldwell
72. Charles Camarda
73. Kenneth Cameron
74. Duane Carey
75. Scott Carpenter
76. Gerald Carr
77. Sonny Carter
78. John Casper
79. Christopher J. Cassidy
80. Robert Cenker
81. Gene Cernan
82. Gregory Chamitoff
83. Franklin Chang-Diaz
84. Kalpana Chawla
85. Maurizio Cheli
86. Leroy Chiao
87. Kevin Chilton
88. Jean-Loup Chrétien
89. Laurel B. Clark
90. Mary Cleave
91. **Jean-François Clervoy**
92. Michael Clifford
93. Michael Coats
94. Kenneth Cockrell
95. Catherine Coleman
96. Eileen Collins
97. Michael Collins
98. Pete Conrad
99. Gordon Cooper
100. Richard Covey
101. Timothy Creamer
102. John Creighton
103. Robert Crippen
104. Roger Crouch
105. Frank Culbertson
106. Walter Cunningham
107. Robert Curbeam
108. Nancy Currie
109. Nancy Jan Davis
110. Lawrence J. DeLucas
111. Frank De Winne
112. Vladimir N. Dezhurov
113. Georgi Dobrovolski
114. Takao Doi
115. B. Alvin Drew
116. Brian Duffy
117. Charles Duke
118. Bonnie J. Dunbar
119. Pedro Duque
120. Samuel T. Durrance
121. James Dutton
122. Lev Dyomin
123. Vladimir Dzhanibekov
124. Joe Edwards
125. Donn F. Eisele
126. Anthony W. England
127. Joseph H. Engle
128. Ronald Evans
129. **Reinhold Ewald**
130. Léopold Eyharts
131. John Fabian
132. Muhammed Faris
133. Bertalan Farkas
134. Jean-Jacques Favier
135. Fèi Jùnlóng
136. Konstantin Feoktistov
137. Christopher Ferguson
138. Martin J. Fettman
139. Andrew J. Feustel
140. Anatoli Filipchenko
141. Michael Fincke
142. Anna Fisher
143. William Frederick Fisher
144. Klaus-Dietrich Flade
145. Michael Foale
146. Kevin A. Ford
147. Michael Foreman
148. Patrick Forrester
149. Michael Fossum
150. Stephen Frick
151. Dirk Frimout
152. Christer Fuglesang
153. Charles Fullerton
154. Reinhard Furrer
155. Francis Gaffney
156. Yuri Gagarin
157. Ronald Garan
158. Dale Gardner
159. Guy Gardner
160. Jake Garn
161. Marc Garneau
162. Owen Garriott
163. Richard Garriott
164. Owen Garriott
165. Charles Gemar
166. Michael Gernhardt
167. Edward Gibson
168. Robert L. Gibson
169. Yuri Gidzenko
170. Yuri Glazkov
171. John Glenn
172. Linda Godwin
173. Michael T. Good
174. Viktor Gorbatko
175. Richard Gordon
176. Dominic Gorie
177. Ronald Grabe
178. Georgi Grechko
179. Frederick Gregory
180. William Gregory
181. Stanley Griggs
182. Virgil Grissom
183. John Grunsfeld
184. Aleksei Gubarev
185. Umberto Guidoni
186. Jügderdemidiin Gürragchaa
187. Sidney Gutierrez
188. Chris Hadfield
189. Claudie Haigneré
190. **Jean-Pierre Haigneré**
191. Fred Haise
192. James Halsell
193. Kenneth Ham
194. Lloyd Hammond
195. Gregory Harbaugh
196. Bernard Harris
197. Terry Hart
198. Henry Hartsfield
199. Frederick Hauck
200. Steven Hawley
201. Susan Helms
202. Karl Henize
203. Thomas Hennen
204. Terence Henricks
205. Miroslaw Hermaszewski
206. Jose Hernández
207. John Herrington
208. Richard Hieb
209. Joan Higginbotham
210. David Hilmers
211. Kathryn Hire
212. Charles Hobaugh
213. Jeffrey Hoffman
214. Scott Horowitz
215. Akihiko Hoshide
216. Millie Hughes-Fulford
217. Douglas G. Hurley
218. Rick D. Husband
219. James Irwin

List of space travellers, interviewed persons highlighted source: (Wikipedia [space travellers], 2010)

220. Aleksandr Ivanchenkov
221. Georgi Ivanov
222. Marsha Ivins
223. Sigmund Jähn
224. Mae Jemison
225. Tamara E. Jernigan
226. Brent W. Jett, Jr.
227. Jing Haipeng
228. Gregory C. Johnson
229. Gregory H. Johnson
230. Thomas D. Jones
231. Leonid Kadenyuk
232. Alexander Kaleri
233. Janet Kavandi
234. James Kelly
235. Mark E. Kelly
236. Scott Kelly
237. Joseph Kerwin
238. Yevgeny Khrunov
239. Robert S. Kimbrough
240. Leonid Kizim
241. Pyotr Klimuk
242. Vladimir Komarov
243. Yelena V. Kondakova
244. Oleg Kononenko
245. Timothy L. Kopra
246. Mikhail Korniyenko
247. Valery Korzun
248. Oleg Kotov
249. Vladimir Kovalyonok
250. Konstantin Kozeyev
251. Kevin Kregel
252. Sergei Krikalev
253. Valeri Kubasov
254. André Kuipers
255. Aleksandr Laveykin
256. Guy Laliberté
257. Wendy Lawrence
258. Vasili Lazarev
259. Aleksandr Lazutkin
260. Valentin Lebedev
261. Mark C. Lee
262. David Leestma
263. William B. Lenoir
264. Aleksei Leonov
265. Frederick W. Leslie
266. Anatoli Levchenko
267. Byron Lichtenberg
268. Don Lind
269. Steven Lindsey
270. Jerry Linenger
271. Richard Linnehan
272. Gregory Linteris
273. Liu Boming
274. Paul Lockhart
275. Yuri Lonchakov
276. Michael Lopez-Alegria
277. John Lounge
278. Jack Lousma
279. Stanley G. Love
280. Jim Lovell
281. G. David Low
282. Edward Lu
283. Shannon Lucid
284. Vladimir Lyakhov
285. Steven MacLean
286. Sandra Magnus
287. Oleg Makarov
288. Yuri Malenchenko
289. Franco Malerba
290. Yuri Malyshev
291. Gennadi Manakov
292. Musa Manarov
293. Thomas H. Marshburn
294. Michael Massimino
295. Richard Mastracchio
296. Thomas Kenneth Mattingly II
297. K. Megan McArthur
298. William S. McArthur
299. Jon McBride
300. Bruce McCandless II
301. William C. McCool
302. Michael McCulley
303. James McDivitt
304. Donald McMonagle
305. Ronald McNair
306. Carl Meade
307. Bruce Melnick
308. Pamela Melroy
309. Leland D. Melvin
310. Mike Melvill
311. Ulf Merbold
312. Ernst Messerschmid
313. Dorothy Metcalf-Lindenburger
314. Edgar Mitchell
315. Abdul Ahad Mohmand
316. Mamoru Mohri
317. Barbara Morgan
318. Lee Morin
319. Boris Morukov
320. Chiaki Mukai
321. Richard Mullane
322. Talgat Musabayev
323. Story Musgrave
324. Steven R. Nagel
325. Bill Nelson
326. George Nelson
327. Rodolfo Neri Vela
328. Paolo A. Nespoli
329. James H. Newman
330. Claude Nicollier
331. Niè H Ishèng
332. Andriyan Nikolayev
333. Soichi Noguchi
334. Carlos I. Noriega
335. Lisa Nowak
336. Karen Nyberg
337. Bryan O'Connor
338.Ellen Ochoa
339. Wubbo Ockels
340. William Oefelein
341. John D. Olivas
342. Gregory Olsen
343. Ellison Onizuka
344. Yuri Onufrienko
345. Stephen Oswald
346. Robert Overmyer
347. Gennady Padalka
348. William Pailes
349. Scott Parazynski
350. Ronald A. Parise
351. Robert Parker
352. Nicholas Patrick
353. Viktor Patsayev
354. James Pawelczyk
355. Julie Payette
356. Gary Payton
357. Philippe Perrin
358. Donald Peterson
359. Donald Pettit
360. Pham Tuân
361. John Phillips
362. William Pogue
363. Alan G. Poindexter
364. Mark Polansky
365. Alexander Poleshchuk
366. Valeri Polyakov
367. Marcos Pontes
368. Leonid Popov
369. Pavel Popovich
370. Charles Precourt
371. **Dumitru-Dorin Prunariu**
372. Ilan Ramon
373. William Readdy
374. Kenneth Reightler
375. James F. Reilly
376. Garrett Reisman
377. Thomas Reiter
378. Vladimir Remek
379. Judith Resnik
380. Paul W. Richards
381. Richard N. Richards
382. Sally Ride
383. Stephen Robinson
384. Roman Romanenko
385. Yuri Romanenko
386. Kent Rominger
387. Stuart Roosa
388. Jerry L. Ross
389. Valeri Rozhdestvensky
390. Nikolay Rukavishnikov
391. Mario Runco, Jr.
392. Valery Ryumin
393. Albert Sacco
394. Gennadi Sarafanov
395. Robert Satcher
396. Viktor Savinykh
397. Svetlana Savitskaya
398. Wally Schirra
399. **Hans Schlegel**
400. **Harrison Schmitt**
401. Rusty Schweickart
402. Dick Scobee
403. David Scott
404. Winston Scott
405. Paul Scully-Power
406. Richard Searfoss
407. Rhea Seddon
408. Ronald Sega
409. Piers Sellers
410. Aleksandr Serebrov
411. Vitali Sevastyanov
412. Yuri Shargin
413. Salizhan Sharipov
414. Rakesh Sharma
415. Helen Sharman
416. Vladimir Shatalov
417. Brewster Shaw
418. Alan Shepard
419. William Shepherd
420. Sheikh Muszaphar Shukor
421. Georgi Shonin
422. Loren Shriver
423. Mark Shuttleworth
424. Charles Simonyi
425. Aleksandr Skvortsov
426. Donald Slayton
427. Steven Smith
428. Anatoly Solovyev
429. Vladimir Solovyov
430. Sherwood Spring
431. Robert Springer
432. Thomas Patten Stafford
433. Heidemarie Stefanyshyn-Piper
434. Robert Stewart
435. Susan Still
436. Nicole Stott
437. Gennady Strekalov
438. Frederick Sturckow
439. Kathryn Sullivan
440. Maksim Surayev
441. Steven Swanson
442. John Swigert
443. Arnaldo Tamayo Méndez
444. Daniel Tani
445. Joseph Tanner
446. Valentina Tereshkova
447. Norman Thagard
448. **Gerhard Thiele**
449. Robert Thirsk
450. Andrew Thomas
451. Donald Thomas
452. Kathryn Thornton
453. William Thornton
454. Pierre Thuot
455. Dennis Tito
456. Gherman Titov
457. Vladimir Titov
458. **Michel Tognini**
459. Valery Tokarev
460. Sergei Treschev
461. Eugene Trinh
462. Richard Truly
463. Bjarni Tryggvason
464. Vasili Tsibliyev
465. Mikhail Tyurin
466. Yury Usachev
467. Lodewijk van den Berg
468. James van Hoften
469. Vladimir Vasyutin
470. Charles Veach
471. **Franz Viehböck**
472. Alexander Viktorenko
473. Terry Virts
474. Pavel Vinogradov
475. Roberto Vittori
476. Igor Volk
477. Sergey Volkov
478. Vladislav Volkov
479. Alexander Volkov
480. Boris Volynov
481. James Voss
482. Janice Voss
483. Koichi Wakata
484. Rex Walheim
485. Charles Walker
486. David M. Walker
487. Joseph A. Walker
488. Shannon Walker
489. Ulrich Walter
490. Carl Walz
491. Taylor Wang
492. Mary Weber
493. Paul Weitz
494. James Wetherbee
495. Edward White
496. Douglas H. Wheelock
497. Peggy Whitson
498. Terrence Wilcutt
499. Dafydd Williams
500. Donald Williams
501. Jeffrey Williams
502. Sunita Williams
503. Barry Wilmore
504. Stephanie Wilson
505. Peter Wisoff
506. David Wolf
507. Alfred Worden
508. Naoko Yamazaki
509. Yáng Lìwei
510. Boris Yegorov
511. Aleksei Yeliseyev
512. Yi So-yeon
513. John Young
514. Fyodor Yurchikhin
515. Sergei Zalyotin
516. George D. Zamka
517. Zhai Zhigang
518. Vitaliy Zholobov
519. Vyacheslav Zudov

2.2 THE USERS

SELECTED PHYSIOLOGICAL, SOCIAL AND PSYCHOLOGICAL ISSUES

The social and psychological environment in Space Habitats differs from the norm on Earth.

Today, on-board the International Space Station, people from different countries live and work together within a confined environment for several months. Aside from physical survival and functionality of the technical infrastructure, the physiological, social and psychological issues emerge as equally critical for mission success.

In anticipation of these human issues, crew selection, preparation, training and support play an important role for the success of a mission. The following introduction discusses some of the main human-related issues relevant for extra-terrestrial habitats. Only a few will be named and not explained in detail, but are listed for the understanding of the relationship between the users and their habitat.

Physiological Challenges
The human being has evolved in a 1G environment and is especially 'designed for' it. A change of gravity conditions has an impact on the physiological system. Weightlessness rapidly alters cardiovascular, bone, and hormonal physiology. The astronauts' heart rate and blood pressure decrease in space as does the variability in heart rate and blood pressure. In space astronauts' bones lose calcium and strength, their muscles lose mass. Changes in spatial orientation, movement, and sensory perception are among the most important aspects to consider for the planning of the habitat. (NASA [HIDH], 2010; Connors, et al., 1985)

Isolation and Confinement
In an extra-terrestrial habitat a small group of people live within a confined and often cramped space. People are separated from their usual surroundings. Mary M. Connors (Connors, et al., 1985 p. 10) explained the space environment as follows:

"Space environments are characterized by (1) Isolation (a separation from the normal or daily physical and social environment), (2) Confinement (restriction within a highly limited and sharply demarcated physical and social environment), (3) Deprivation, and (4) Risk. Such environments may be expected to place heavy demands on astronauts' psychological and social resources."

All these constraints affect directly the habitability and performance of the crew.

Human Interactions

Living and working in space is more complex than on Earth. Mission tasks take priority and next to it, people have to get along with each other for a long time. The amount of available contacts is limited to the crew size, except for communication to the ground. Astronauts come from different countries with a diverse cultural background. High levels of personal skills are required and people cannot simply 'go outside' to get away for a break in routine.

Specific Stressors

Psychological health and well-being are important for humans on Earth, as well as anywhere else. But in an isolated, confined and restricted environment, such as in extra-terrestrial habitats, some stressors are amplified. Various stressors have severe implications on human well-being and performance. In addition to the stressor of living in a harsh and dangerous environment, other stressors include lighting, temperature, humidity, odour, sound, vibrations, and workload among others (Connors, et al., 1985). The cumulative effects can lead to degraded performance, decreased attention, and anxiety.

With all of the very precise physiological, psychological, social and technological specifications connected to limitations such as transport, a very tight framework for the creation of living and working spaces is set.

Further Reading

Living Aloft - Human Requirements for Extended Spaceflight by Mary M. Connors, Albert A. Harrison and Faren R. Akins
(Connors, et al., 1985)
Bold Endeavours: Lessons from Polar and Space Exploration by Jack Stuster
(Stuster, 1996)
Space Psychology and Psychiatry by Nick Kansas and D. Manzey
(Kanas N., 2003)
Neuroscience in Space by Gilles Clément and Millard F. Resche
(Clément, et al., 2008)
Human Integration Design Handbook by NASA
(NASA [HIDH], 2010)

2.3 CONSEQUENCES FOR DESIGN

Form, Volume and Mass

In order to operate a space station in orbit or a habitat on the Moon and Mars, an adequate launch system and a spacecraft are required. The habitat size and mass depend upon the launcher system used. In terms of engineering and economic issues, there is a severe limitation on mass and volume. *Figure A* shows the relation of the size of the launch vehicle to the Apollo spacecraft. The flight system was made for only one flight, to land on the lunar surface; only the Command and Service Module CSM) came back to Earth. In addition, a big portion of the available space within the Apollo spacecraft was occupied with technology (structure, life support systems, propulsion systems, power systems, etc.).

So far no space station module has had a larger diameter than 6.6m - 6.25m interior diameter (Skylab Space Station). To create big space station complexes, such as the Mir space station and currently the International Space Station, a modular architecture approach was applied. Also on the table is the use of inflatable habitats to minimize volume and weight during launch, and maximize living space at the desired destination (cf. Transhab, Bigelow).

A protective pressure vessel

Any extra-terrestrial habitat must provide a livable atmosphere with a certain internal pressure. The outer space and planetary environments on the Moon and Mars are zero - to low-pressure environments; therefore these habitats need to be constructed as pressure vessels. To protect humans and equipment from radiation and micro-meteorites additional protection layers and technologies have to be applied. Further habitats must be thermo-regulated in order to maintain an even and comfortable inside temperature.

A:

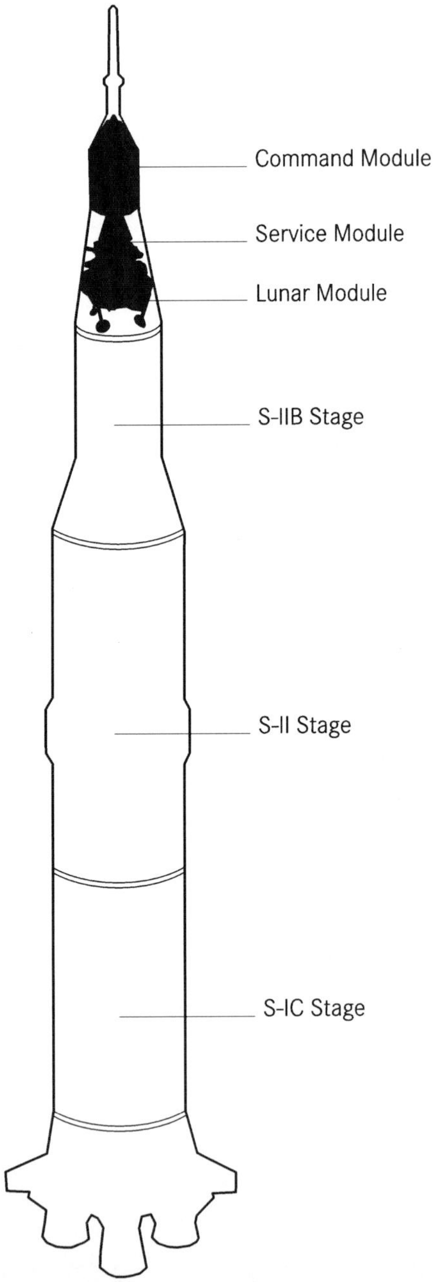

A: Saturn V Launch Vehicle with Command and Service Module and Lunar Module

(credit A: Author)

The Apollo CSM used the 'Barbecue-roll' technique. During trans-earth and trans-lunar coast the spacecraft would roll slowly about its axis to prevent localized overheating (Butler, 2002). The passive thermal control system of the International Space Station is provided by multi layer insulation (MLI) of the skin; a special surface coating and shell heaters to keep the temperature of the structure above dew point. The active thermal control is provided to actively control temperature via heat exchangers and water pumps. The ISS radiators reject the waste heat to the blackness of space.

Self-Sufficiency

According to the 'Organisation for Economic Co-operation and Development' (OEDC) Glossary of Statistical Terms, self-sufficiency "reflects the extent of participation in the economy and society and how well individuals are able to get through daily life on their own" (OECD, 2007). The Longman Dictionary of Contemporary English defines self-sufficient as "being able to provide all the things you need without help from other people" (LDoCE) Synonyms are self-direction, self-reliance, autonomy (Princeton University, Farlex Inc., 2003-2008).

In the case of extra-terrestrial habitats self-sufficiency refers to the basic requirements of the habitat and its users, when outside support is difficult, not favoured or not possible and people have to rely on themselves to survive. A high degree of self-sufficiency is a requirement for the operation of extra-terrestrial habitats.

Habitable Volume

One basic parameter when designing a habitat is the available space and the allocation of the required habitable volume. The same is valid for the design of an extra-terrestrial habitat.
The history of space station design is full of comments regarding low habitability, as Jerry Linenger puts it, while relocating the Soyuz to the Mir space station: "We were stuffed in the capsule [Salyut] like sardines in a can" (Linenger, 2000 p. 94).

The largest habitable volume within one module was provided by the Skylab space station, as its architecture was formed by a converted third stage of a Saturn V moon rocket. Although most of the Skylab astronauts recommended that any future space station have ample open interior volume" (Kitmacher, 2002), the budget, transportation, technological and political constraints form a tight concept for the planning of habitable space. So, the later Salyuts, Mir and the ISS were not as volumetrically commodious as Skylab.

The first Soviet and Russian space stations were based on the maximum fairing diameter of the Proton rocket, thus leading to a maximum diameter of 4.35 meters of the Salyut and Mir modules. Also for the first concepts of the International Space Station second and third stages of the Saturn V rockets with a diameter of 7 and 10 meters were considered. For political reasons this never happened, as according to Kitmacher "the space shuttle became NASA's top priority", and thus "modules of the space stations were based on the 4.5m interior clear fairing diameter of the space shuttle" (Kitmacher, 2002).

2.3 CONSEQUENCES FOR DESIGN

Constance Adams (1999) clarified the term "habitable volume". To her, habitable volume equals the total free volume available for human habitation and use. Thus it is a part of the total pressurized volume, which contains life support systems, workstations, stowage and other. A comparison between the numbers describing habitable volume is sometimes confusing. According David Portree (1995), the Mir core module had a habitable volume of $90m^3$. But when you look at the pictures showing the interior of the space station, one can easily see that the station was cramped and a lot of this habitable volume could not be used because of storage problems.

Over time, many studies have tried to ascertain an exact number of how much spatial volume a person needs. Marc M. Cohen analyzed that pressurized volume increases as a function of mission duration, both as a power curve and a logarithmic curve. He also showed that this volume trend does not level off but continues to rise (Cohen, 2008). However, so far no mutual consent on a specific number has occurred. Another challenge for the calculation of appropriate habitable space is that different mission scenarios, people and environments impose different spatial requirements. Due to economic and technological restraints, available space is limited, but especially because of the limitations, the available space should be designed very carefully, every cm^3 supporting the human being. It seems that finding an exact number is not possible or even practical; instead we might have to consider the space-human relationship and integrate it into the design process.

Anthropometric Design

In 1987 the NASA-STD-3000 (revised in 1995 as NASA Std-3000B) set a guideline for the design and sizing of space modules. "Design and sizing of space modules should ensure accommodation, compatibility, operability, and maintainability by the user population. Generally, design limits are based on a range of the user population from the

A:

A: STS-116 astronaut William A. Oefelein restrains himself in order to assist Robert L. Curbeam and Christer Fuglesang in preparing for an extravehicular activity in the Quest Airlock of the International Space Station

(credit A: NASA)

5th percentile values for critical body dimensions, as appropriate. The use of this range will theoretically provide coverage for 90% of the user population for that dimension." (NASA [MSIS], 1995 p. 19)

Adequate gravity regime and ergonomic design is required for living and working spaces. In addition the requirement of training the astronauts in special training centres on Earth for their mission in microgravity restricts the design.

Restraints

"Man uses his arms and legs to complete a force coupe in zero-g much the same as he does on Earth. The major differences occur in how the man is restrained. On Earth, he has gravity holding his feet to the floor. In zero-g he must have restraints for that purpose." (NASA [Bull. 7], 1974 p. 2)

In order to stabilize the body and hold the human body in a stable position in microgravity to perform basic functions, such as sleep, dining, toilet, computer work, experiments and photography, etc restraints are needed *[Fig. A]*.

There are multiple kinds of restraints, such as handhold restraints, waist restraints, torso restraints and foot restraints. "Without proper restraints, a crewmember's work capabilities will generally be reduced and the time to complete tasks increased." (NASA [HIDH], 2010)

Further Reading

Publications by the AIAA/Space Architecture Technical Committee for education, research and commercial aerospace architecture publications, online: spacearchitect.org (SATC, 2010)

2.4 REFERENCES

Adams, Constance M. 1999. Habitability as a Tier One Criterion in Advanced Space Vehicle Design: Part One - Habitability. Denver, CO, USA : SAE, 1999. SAE-1999-01-2137.

Bigelow. 1999-2010. [Online] 1999-2010. http://www.bigelowaerospace.com/.

Butler, Aaron M. 2002. NASA Mission Control Anwers Your Questions. [Online] 01 2002, http://spaceflight.nasa.gov/feedback/expert/answer/mcc/sts-113/11_23_20_01_179.html.

Britannica. 2011. Encyclopaedia Britannica. [Online] 2011. [Cited: 9 January 2011.], http://www.britannica.com/EBchecked/topic/638979/weightlessness.

Clément, Gilles and Reschke, Millard F. 2008. Neuroscience in Space. NY, USA : Springer, 2008. ISBN 978-0-387-78949-1.

Cohen, Marc M. 2009. From Apollo LM to Altair: Design, Environments, Infrastructure, Missions, and Operations. Pasadena, California, USA : AIAA SPACE 2009 Conference & Exposition, 2009. AIAA 2009-6404.

Cohen, Marc M. 2008. Testing the Celentano Curve: An Empirical Survey of Predictions for Human Spacecraft Pressurized Volume. Warrendale, Pennsylvania, USA : SAE International Journal of Aerospace, 2008. 08-ICES-0046.

Connors, Mary M., Harrison, Albert A. and Akins, Faren R. 1985. Living Aloft, Human Requirements for Extended Spaceflight. Washington, D.C., USA : NASA, Ames Research Center, 1985. SP-483 / ISBN: 1-4102-1983-6.

Eckart, Peter. 1999. The Lunar Base Handbook - An Introduction to Lunar Base Design, Development and Operations. U.S.A. : The McGraw-Hill companies, Inc., 1999. Vol. Space Technology Series. ISBN 0-07-240171-0.

ESA [Gravity]. 2008. Materials Science in Space - What is Gravity. [Online] European Space Agency, 22 February 2008, http://www.spaceflight.esa.int/users/materials/introduction/gravity/gravity.html.

Hamilton, Calvin J. 1995-2009. Views of the Solar System. [Online] 1995-2009. [Cited: 12 December 2008.] http://www.solarviews.com/eng/moon.htm#stats.

Kanas, Nick M.D. und Manzey, Dietrich Ph.D. 2003. Space Psychology and Psychiatry. Dordrecht, The Netherlands : Springer/ Microcosm Press, 2003. ISBN 978-1-4020-6769-3.

Kitmacher, Gary H. 2002. Design of the Space Station Habitable Modules. Houston, Texas, USA : The World Space Congress, 2002. p. 18. IAC-02-IAA.8.2.04.

Kring, Jason, et al. 2000. Human Performance in Extreme Environment: From the Battlefield to the Final Frontier. [book auth.] Dennis A. Vincenzi, Mustapha Mouloua and Peter A. Hancock. [ed.] Psychology Press. Human Performance, Situation Awareness and Automation - Current Research and Trends. USA : HPSAA II, 2000, Vol. Volume II.

Larson, Wiley J. and Pranke, Linda K. 2000. Human Spaceflight - Mission Analysis and Design. New York, USA : Mc Graw Hill, 2000. ISBN 0-07-236811-X.

LDoCE. [Online] Longman Dictionary of Contemporary English. [Cited: 20 08 2009.], http://www.ldoceonline.com/dictionary/self-sufficient.

Linenger, Jerry M. 2000. Off The Planet - Surviving five periolous months aboard the Space Station MIR. USA : McGraw-Hill, 2000. ISBN 0-07-136112-X.

Mount, Frances E., Whitmore, Mihriban and Stealey, Sheryl L. 2003. Evaluation of Neutral Body Posture on Shuttle Mission STS-57 (SPACEHAB-1). Houston, Texas, USA : NASA, Lyndon B. Johnson Space Center, 2003. NASA TM-2003-104805.

NASA [Bull.7]. 1974. Skylab Experience Bulletin No. 7: An Overview of IVA Personal Restraint Systems. Houston, Texas, USA : NASA, Lyndon B. Johnson Space Center, 1974. JSC-09541.

NASA [Bull.17]. 1975. Skylab Experience Bulletin No. 17: Neutral Body Posture in Zero-G. Houston, Texas, USA : NASA, Lyndon B. Johnson Space Center, 1975. JSC-09551.

NASA [HIDH]. 2010. Human Integration Design Handbook (HIDH). Washington DC, USA : NASA JSC, Baseline Janurary 27, 2010, 2010. 20546-0001.

NASA [LCROSS]. 2010. Lunar Crater Observation and Sensing Satellite. [Online] October 2010. [Cited: 8 January 2011.], http://lcross.arc.nasa.gov/.

Nasa [Mars]. 2010. Solar System Exploration - Facts & Figures. [Online] 8 October 2010. [Cited: 8 January 2011.], http://solarsystem.nasa.gov/planets/profile.cfm?Object=Mars&Display=Facts&System=Metric.

NASA [Moon]. 2010. Solar System Exploration - Facts & Figures. [Online] 8 November 2010. [Cited: 10 January

2011.], http://solarsystem.nasa.gov/planets/profile.cfm?Object=Moon&Display=Facts&System=Metric.

NASA [MSIS]. 1995. Man-Systems Integration Standards, Volume 1, Revision B,. July 1995. NASA-STD-3000.

NASA [TransHab]. Technical Support Package: Design Concepts for the ISS TransHab Module. Houston, Texas, USA : NASA, Lyndon B. Johnson Space Center. MSC-23090.

OECD. 2007. OECD - Self-sufficiency. [Online] OECD, 2007, Society at a Glance: OECD Social Indicators, 2006 edition, OECD, Paris. , 23 July 2007. [Cited: 5 August 2008.], http://stats.oecd.org/glossary/detail.asp?ID=7324.

Portree, David S.F. 1995. MIR Hardware Heritage. Houston, Texas, USA : NASA, Lyndon B. Johnson Space Center, Information Services Division, 1995. NASA RP 1357.

Princeton University, Farlex Inc. 2003-2008. The Free Dictionary: self-sufficiency. [Online] 2003-2008. http://www.thefreedictionary.com/self-sufficiency.

SATC. 2010. AIAA Spacearchitect. [Online] Space Architecture Technical Committee (SATC) of the American Institute of Aeronautics (AIAA), 2010. [Cited: 12 December 2010.], http://www.spacearchitect.org/.

Stuster, Jack. 1996. Bold Endeavors: Lessons from Polar and Space Exploration. USA : US Naval Institute Press, 1996. ISBN-10: 155750749X.

Wikipedia [space travellers]. 2010. List of space travellers by name. [Online] 6 December 2010. [Cited: 10 December 2010.] http://en.wikipedia.org/wiki/List_of_space_travelers_by_name.

Williams, David Dr. R. . 2007. Planetary Fact Sheet - Metric. [Online] NASA Goddard Space Flight Center, 29 November 2007. [Cited: 17 January 2008.] http://nssdc.gsfc.nasa.gov/planetary/factsheet/index.html.

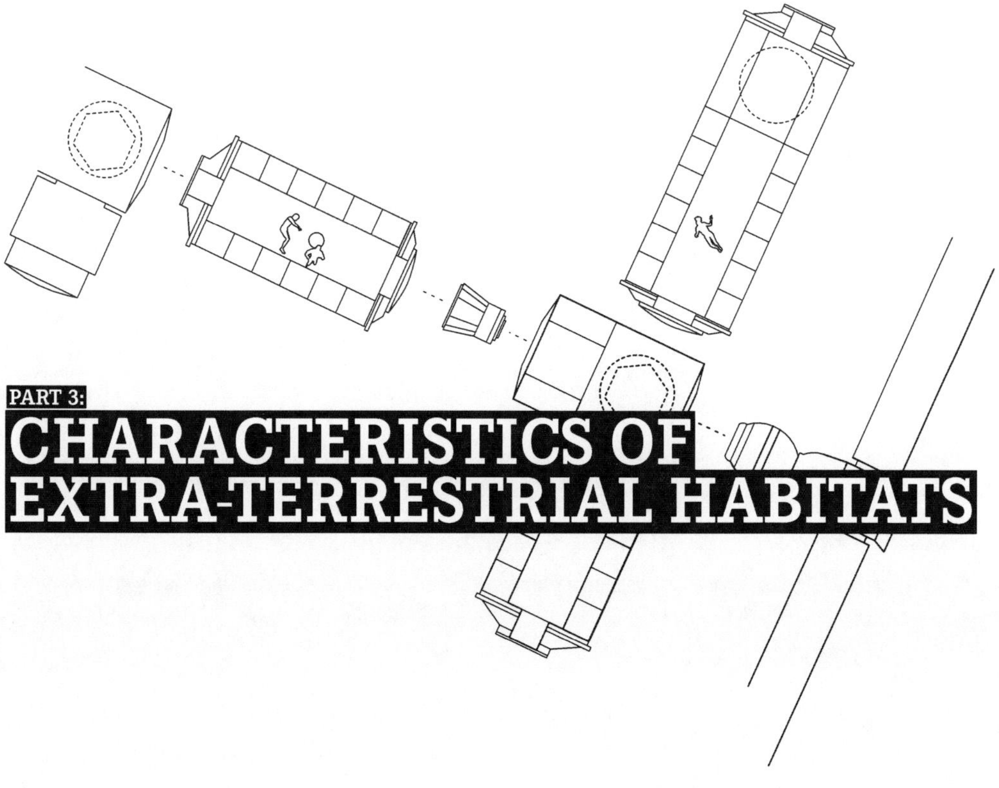

PART 3:

CHARACTERISTICS OF EXTRA-TERRESTRIAL HABITATS

This chapter provides an overview on key characteristics of the Apollo Spacecraft and Lunar Module, the Space Shuttle Orbiter, and the Space Stations; Salyut, Skylab, Mir, as well as the International Space Station.

3.1
OVERVIEW

APOLLO **SALYUT** **SKYLAB**

This diagram shows a chronological spaceflight overview of the Apollo Spacecraft and Lunar Module, the Space Shuttle Orbiter, and the Space Stations; Salyut, Skylab, Mir, as well as the International Space Station.

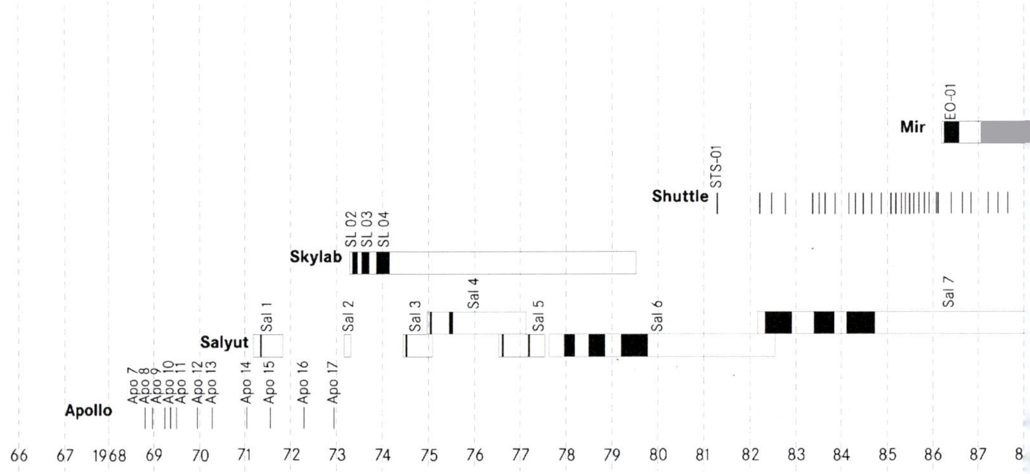

A:

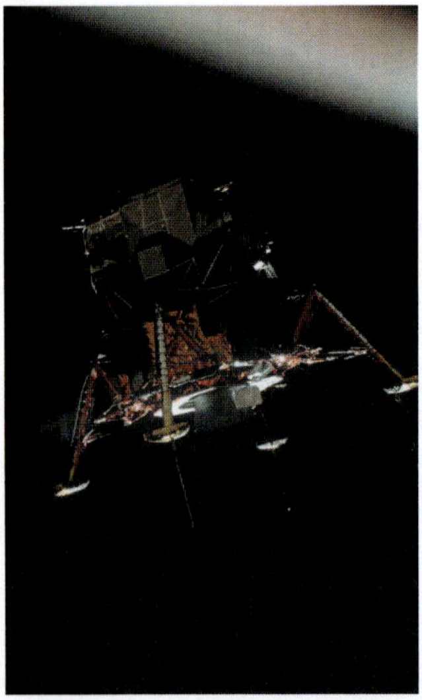

B:

C:

A: Apollo 11 Lunar Module Eagle, 1969
B: Salyut Space Station
C: Skylab Space Station, 1974
D: Space Shuttle Atlantis departing Mir Space Station, 1995
F: International Space Station, 2009

(credit A: NASA, Michael Collins; B: Spacefacts, J. Becker; C: NASA; D: NASA, photo taken by Mir 19 crew, E: NASA, photo taken by STS-129 crew)

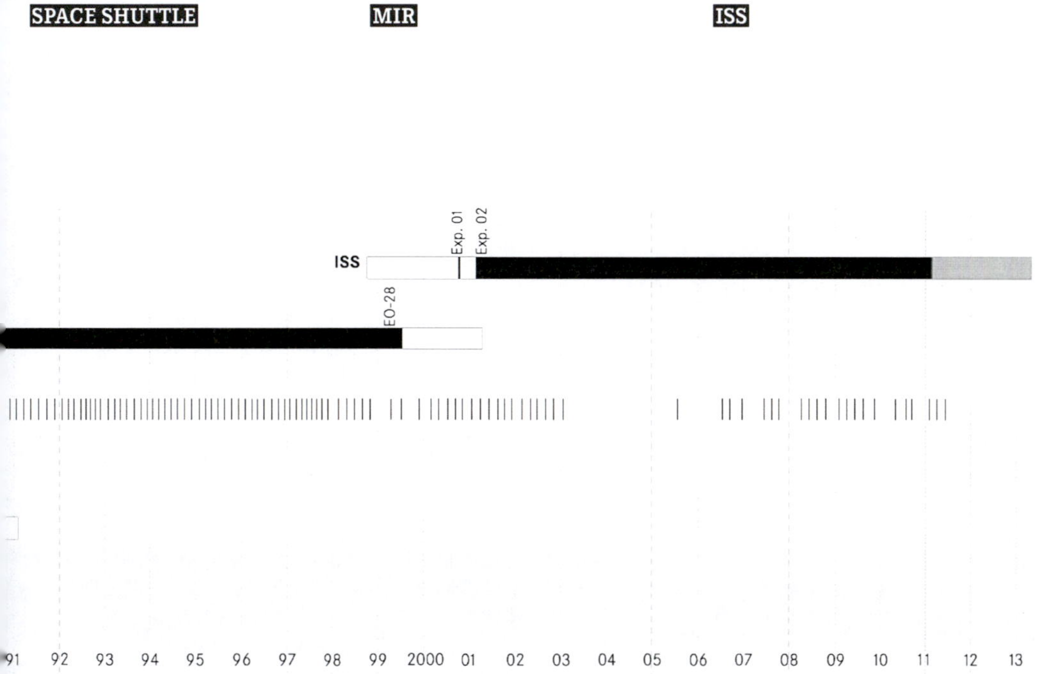

D:

E:

3.2 APOLLO SPACECRAFT

A: Apollo 16 Lunar Module Orion and Lunar Roving Vehicle *(credit: NASA)*

OBJECTIVE

The primary goal of the Apollo missions was "landing a man on the moon and returning him safely to Earth" in competition with the former USSR (Kennedy, May 25, 1961).

Nine years later, in 1969 Neil Armstrong set the first foot on the moon and said the famous words, "That's one small step for a man, a giant leap for mankind". The Apollo 11 Lunar Module, named "Eagle" became the first habitat on the Moon.

Each mission of the Apollo series had important scientific mission objectives that grew and developed with each successive mission. The last lunar landing mission was Apollo 17 in 1972. An overview of the Apollo missions is listed in the appendix.

Use
First manned mission to the Moon
American
Operational: 1967 - 1972
Inhabited: 1968 - 1972
11 missions, 6 lunar landings
Crew size: 3 in launch, Earth orbit, trans-lunar injection and cruise, lunar orbit insertion, low lunar orbit; 2 in the Lunar Module (LM) for descent, landing, ascent from the surface, rendezvous and docking with the CSM while the CSM pilot remained in the CM; 3 return to Earth.

Architecture
Launch Vehicle: Saturn IB, Saturn V
Crew Transportation: Saturn V, Apollo CSM
Elements: Command Module (CM), Service Module (SM), Lunar Module (LM, lander), Lunar Roving Vehicle (LRV)

Pressurized Volume: CM 10.34m^3, LM 6.65m^3
Diameter: CM 3.9m, LM max. 4.2m
Height: CM 3.65m, LM 7.04m
Mass: CM 5.6t, SM 23.3t, LM 15.1-16.4t

Selected Missions
Apollo 8:
1968, first manned flight in deep space
Apollo 11:
1969, first manned lunar landing
Apollo 15:
1971, first mission with Lunar Rover
Apollo 17:
1972, last manned mission to the Moon

* Sources are listed in the Appendix

UNITED
STATES

3.2 APOLLO SPACECRAFT

CONFIGURATION AND LAYOUT

The main architectural elements of the Apollo System consisted of the Command and Service Module *[Fig. A]* and the Lunar Module *[Fig. B]*. The Lunar Rover was integrated with the Apollo 15 mission.

The **Command and Service Module** (CSM), was used for the transportation of the crew to the lunar orbit and back to Earth. The CSM provided space for three astronauts and comprised of two compartments.
The **Command Module** (CM) provided the living space for the crew and was equipped with the operational equipment for the spacecraft. It was a conical pressure vessel with about $6m^3$ habitable volume with two hatches and five windows. The side hatch was used for EVA activities.
The **Service Module** (SM) housed the main propulsion systems and the consumables. It was an unpressurized cylindrical fairing that housed fuel and oxidizer tanks and the engines that performed the LOI and TEI burns, plus a great array of equipment.

The **Lunar Module** (LM), was used for crew transportation from the lunar orbit to the lunar surface. The LM provided space for two astronauts and had a habitable volume of about $6.6m^3$. The LM was a two-stage vehicle and consisted of an upper ascent stage and a lower descent stage. (Portree, 1995)The decent stage was an octagonal prism with four landing pads. It was mainly used for storage and carried the landing rocket, water, oxygen and other tanks as well as the scientific experiments and the Lunar Rover (Apollo 15, 16 and 17). It was left on the lunar surface. The ascent stage had an irregular geometry and provided the habitable volume for crew accommodation. It was used for lift-off to return to and dock to the CSM. The ingress/egress hatch was in the middle of the bottom floor. The ascent stage carried the environmental control system, the communication system and an optical telescope.

A:

B:

A: Apollo 15 CSM Endeavour in lunar orbit, photo taken from the LM
B: Edwin E. Aldrin infront of the Apollo 11 Lunar Module Eagle
C: Graphic, Command Module and Lunar Module interior layout

(credit A, B: NASA; C: Author, based on NASA documents)

C:

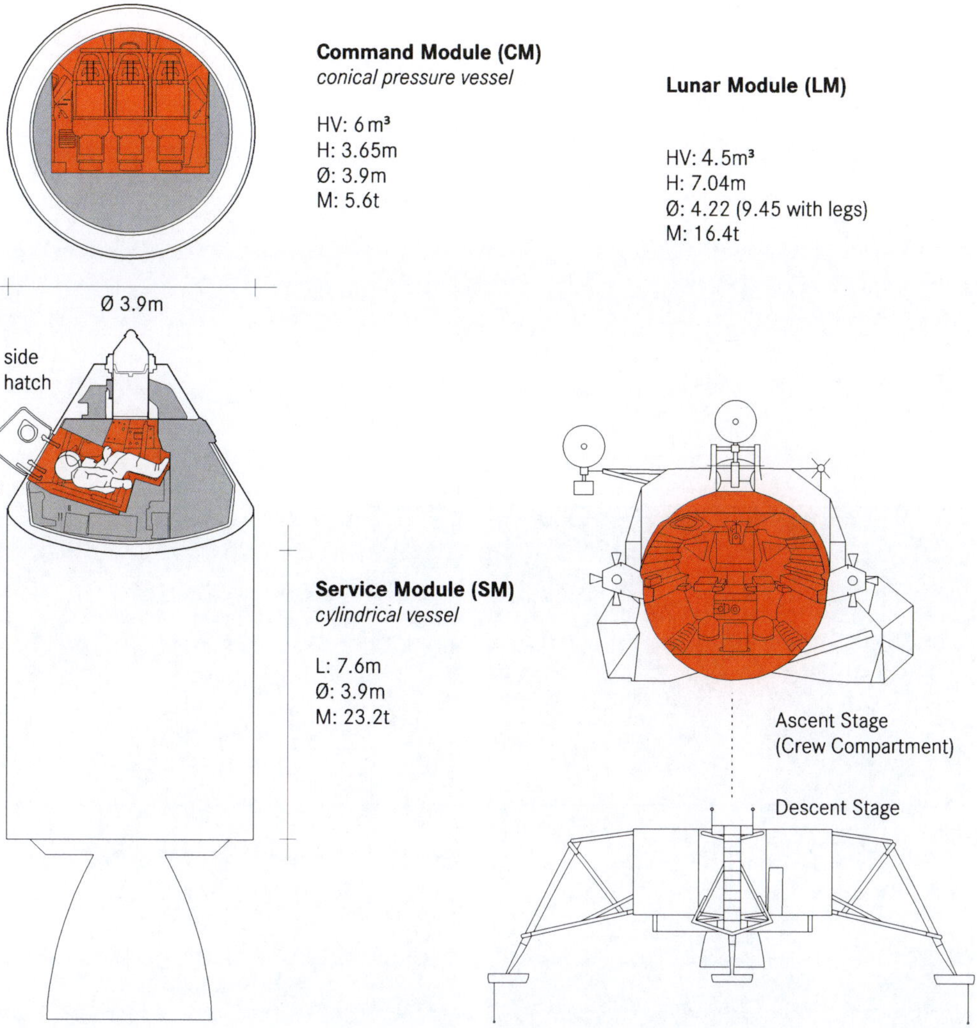

3.2 APOLLO SPACECRAFT

CONFIGURATION AND LAYOUT

The **Lunar Roving Vehicle** (LRV) was a deployable electric lunar roving vehicle for traversing the lunar surface. The folding and packaging concept is a fantastic example of genius engineering at that time. During transit it was stowed in one quadrant of the Lunar Module *[Fig. B].* On the lunar surface astronauts removed the outside insulation blankets, lowered the rover from the storage quadrant and unfolded it manually by pulling straps.

"We had straps, we walked away from the equipment bay, where it was stowed and we just walked away and used our own energy to unfold it and we turned around and it was sitting on the Moon." (Schmitt, 2009)

The **spacesuit** of the Apollo astronauts was designed "to protect the crewman in a low-pressure, micrometeoroid and thermal environment to provide comfort, mobility, dexterity, and a specified unobstructed range of vision during lunar surface of free-space operations outside of the spacecraft"
(NASA, 1971 p. 19).

A:

B:

A: Apollo 17 astronaut Eugene A. Cernan with the Lunar Roving Vehicle
B: Deployment Test of the Lunar Roving Vehicle (Apollo 16)
C: Graphic, Lunar Roving Vehicle

(credit A, B: NASA; C: Author, based on NASA documents)

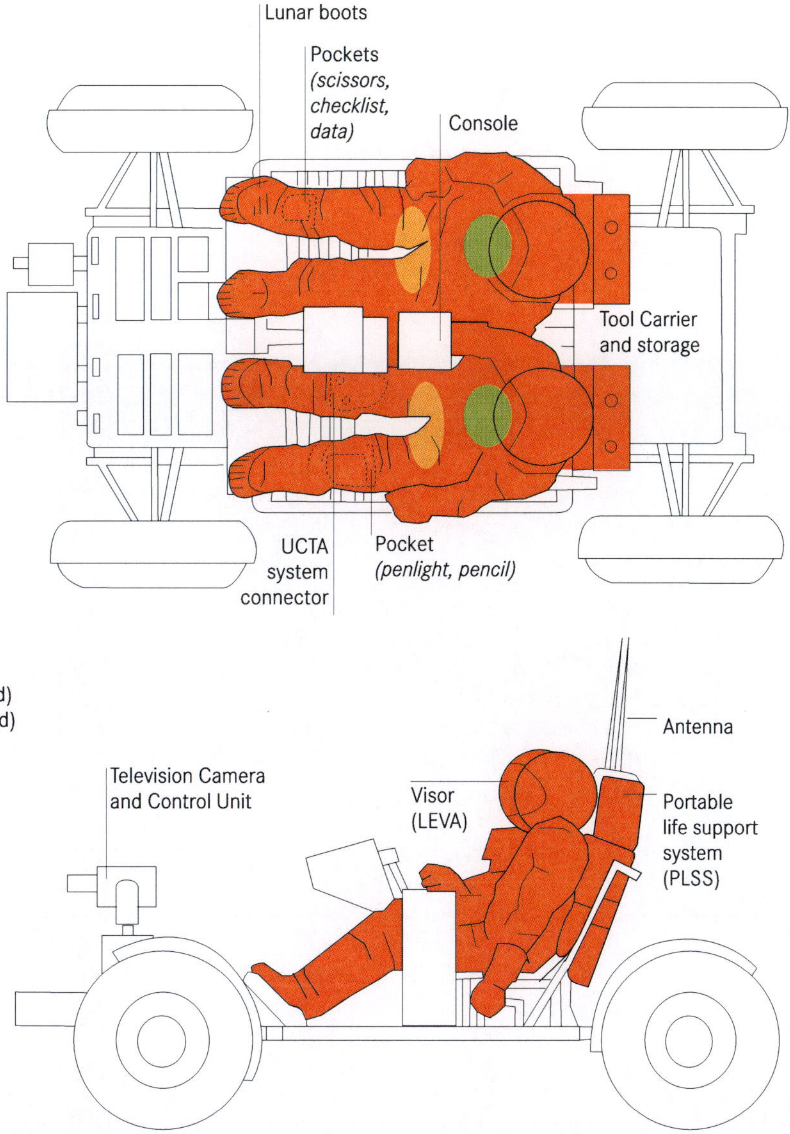

C:

Lunar Rover

L: 3.1 m (deployed)
1.5m (collapsed)
W: 1.83m
H: 1.14m (max)
M: 210kg

3.2 APOLLO SPACECRAFT

TIME ALLOCATION OF HUMAN ACTIVITIES

The Apollo missions were highly work-oriented. The times allocated for activities besides work were minimal or non-existent. References show that not until the Apollo 15 missions were the lunar stays long enough for the crew to get some sleep (Jones, 1995).

The Apollo astronauts spent about 8 hours in their spacesuit per planned EVA while on the lunar surface. It was designed for a 5-hour excursion without re-supply of consumables (NASA, 1971). The later Apollo missions included multiple EVA sorties, culminating in three EVAs on Apollo 17.

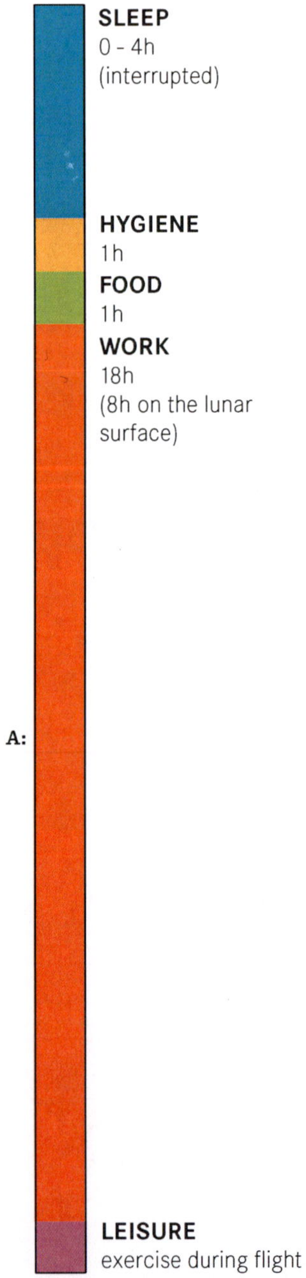

A: Graphic, Time Allocation of Human Acitivites: Apollo
B: Graphic, Spatial Allocation of Human Activities: Apollo *(credit A, B: Author)*

SPATIAL ALLOCATION OF HUMAN ACTIVITIES

The Command and Service Lunar Module and the Lunar Module had no spatial separation of functions. All human activities took place in the same volume. Within the small volume the controls and displays were laid out very carefully according to the crew seating in the CM and the windows of the LM.

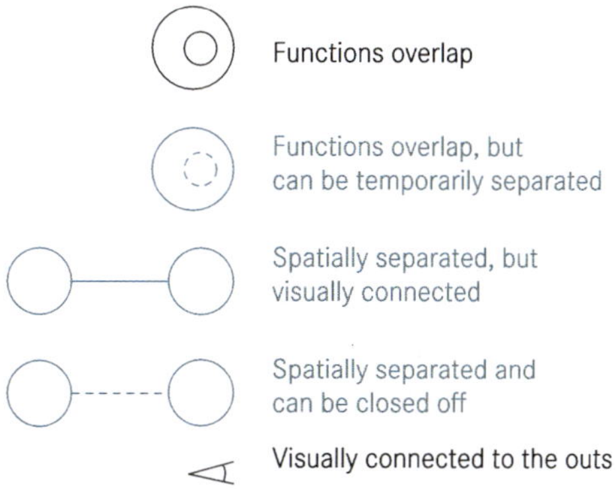

B:

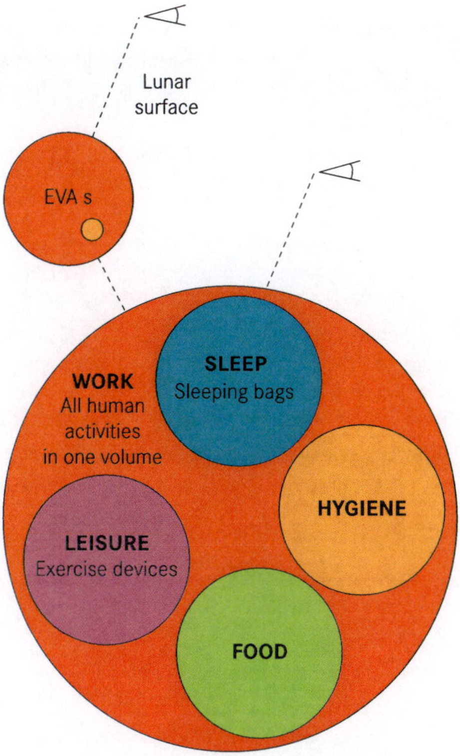

3.3 SALYUT SPACE STATION

A: Soviet orbital station Salyut 6
(credit: Dumitru-Dorin Prunariu)

OBJECTIVES

The Soviet Salyut program [russ. Salute, Firework] can be considered the first space station program. On 19 April 1971 the Soviet Union launched its first space station Salyut1 (DLR, 2006). From then, the Soviets launched a series of single-module space stations in order to compete with the US in the space race.

Launches were coordinated with the Soviet 5-year plans. Salyut 1 in 1971 (followed by Salyut 2, 3 and 4), then Salyut 5 in 1976 (followed by Salyut 6 and 7). The program consisted of the Almaz space stations (named Salyut 3 and 5), which were thought to have a military purpose and the Salyut space stations for civilian use.

An overview of the Salyut space station series is listed in the appendix.

For the Soviets the space station program was important. In 1978, President of the Soviet Union Leonid Brezhnev stated: "Mankind will not forever remain on Earth, but in the pursuit of light and space will first timidly emerge from the bounds of the atmosphere, and then advance until he has conquered the whole of circumsolar space. We believe that permanently manned crews will be mankind's pathway into the universe."
(OTA, 1983 p. 5)

Analysis will mainly be based on the last station of the Salyut series, the Salyut 7, if not stated otherwise.

Use
Soviet space stations (civilian and military)
Operational: 1971 - 1991
Inhabited: 1971 - 1986
Crew size: one resident crew (2-3) and visiting crews (2-3)

Architecture
Launch Vehicle: Proton
Crew Transportation: Soyuz
Elements: single element space stations with docking possibilities

Volume: 90 - 100m^3
Max. Diameter: 4.15m
Lengths: 13 - 15m
Mass: 19t

Selected Missions
Salyut 1: 1971, first space station

Salyut 3: 1974, one crew occupied the station for 15d

Salyut 4: 1974, two crews occupied the station for 28 and 62 days

Salyut 5: 1976, two crews occupied the station for 49 and 16 days

Salyut 6: 1977 - 1982, six occupancy crews (96, 140, 175, 185, 13, 74 days);
Ten visiting crews, first modular space station

Salyut 7: 1982- 1986, five occupancy crews (211, 150, 237, 169, 125 days);
Five visiting crews

*Sources are listed in the Appendix

3.3 SALYUT SPACE STATION

CONFIGURATION AND LAYOUT

The Salyut space stations were launched as a single module. The architecture was constantly improved and also cosmonauts' ideas and suggestions were taken into consideration for the space station design. (B.J.Bluth, 1987 p. I-120; Portree, 1995 p. 90, U.S. Congress, 1983 p.27)

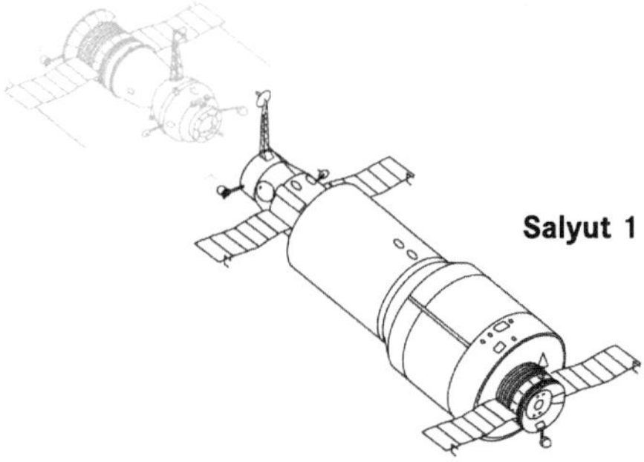

Salyut 1

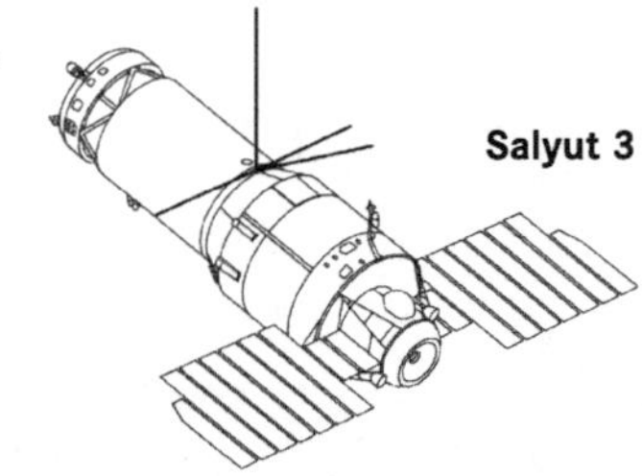

Salyut 3

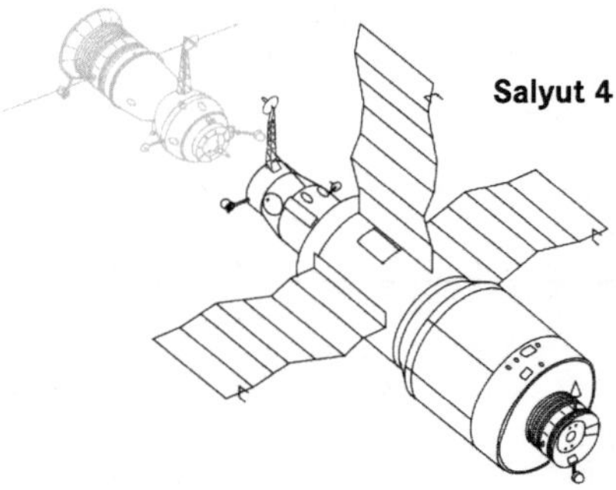

Salyut 4

A:

Salyut 7

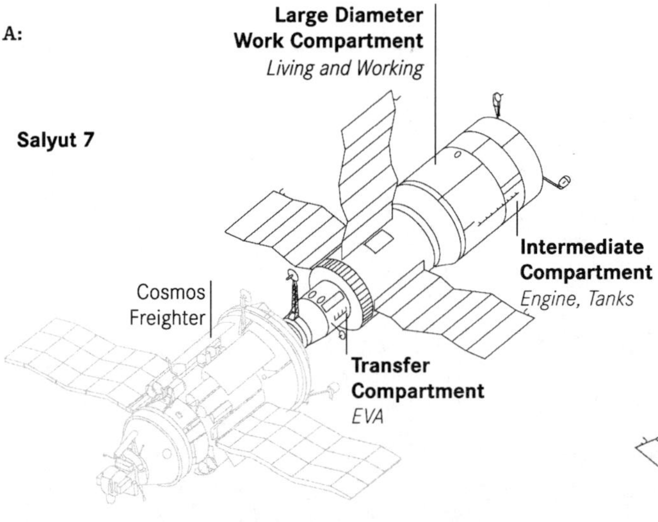

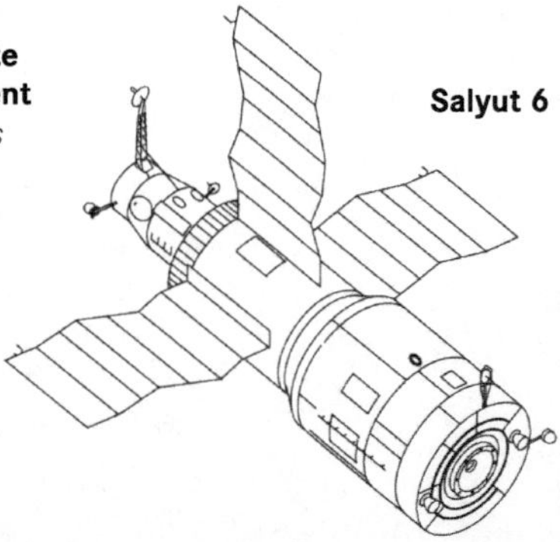

Salyut 6

A: Spacestation series: Salyut 1 - 6 and Salyut 7 with Cosmos Freighter

(credit A: NASA, source: Mir Harware Heritage, D.S.F. Portree, adapted by the author)

The **first-generation of Salyut space stations (Salyut 1 to 5)** were designed as 'Long-Duration-Stations' [russ. acronym DOS-type station]. They could not be refuelled and had only one docking port. (Portree, 1995)

In Salyut 1 the interior design used various colours (light and dark grey, apple green, light yellow) for supporting the astronauts' orientation in weightlessness (Portree, 1995).

Salyut 2 failed shortly after launch.

In Salyut 3 (Almaz) the floor was covered with Velcro to aid the cosmonauts in moving around. It had one standing bunk and one foldable bunk as well as a shower (Portree, 1995 p. 69).

Salyut 4 had Velcro carpeting, exercise machines and a shower (Harrison, 2001 p. 64; Caprara, 2000).

Salyut 5 (Almaz) consisted of a spherical transfer module with four hatches, a large diameter work compartment and a small diameter living compartment. The largest room was the working compartment, with a habitable volume of about 40m^3 (Portree, 1995 p. 73). A rapid fire cannon was to be mounted on the station for defence against the U.S. (Haeseler, 1998).

In addition to scientific experiments and the testing of hardware space stations of the first-generation were also used for military observations.

The **second-generation of Salyut space stations (Salyut 6 and 7)** provided two ports for docking of visiting expeditions (OTA, 1983 p. 15). The large-diameter compartment was longer than at the first-generation Salyut stations: 6.0m versus 4.1m (Portree, 1995 p. 75).

On Salyut 6 living conditions were further improved. A new propulsion system was installed that allowed refuelling in orbit. A water-regeneration device was integrated into the stations life-support system and new spacesuits were tested. In 1981 the module Cosmos 1267 docked with Salyut 6 to improve its functionality.

Salyut 7 was the most advanced and most comfortable space station of the Salyut series. A set of modifications to the interior made it more liveable. There were approximately 20 windows with shades on the Salyut 7. To protect the inside of the windows, they were covered with removable glass panels (B.J.Bluth, 1987 p. I-26). The colour scheme was improved and a refrigerator was installed (OTA, 1983 p. 27). The ceiling on the Salyut 7 was white; the left wall was apple green and the right one beige (B.J.Bluth, 1987 p. I-33).

The **Soyuz** is the Russian transportation and escape vehicle that first flew in 1967. It took cosmonauts and supplies to the space station Salyut and later to the Mir space station. An improved version, the Salyut-TMA is still being used today to bring astronauts to the ISS. It consists of three modules; a service module, an orbital module and a descent module.

The **Progress** is the Russian unmanned resupply spacecraft. It supplied the Russian space stations and is still used today to supply the International Space Station. It can be filled with waste from the space station to be de-orbited and destroyed on re-entry.

3.3

SALYUT SPACE STATION

CONFIGURATION AND LAYOUT

The space station Salyut 7 consisted of three sealable compartments; the Transfer Compartment, the Work Compartment, and the Intermediate Compartment (B.J.Bluth, 1987; OTA, 1983).

The **Transfer Compartment** had a docking port for the manned and unmanned freighters and contained the EVA hatch. It was used to store the spacesuits and for preparation of space walks.

The transfer compartment had seven portholes with scientific equipment arrayed circularly around the cylindrical part of the model (OTA, 1983 p. 51).

The **Work Compartment** consisted of two areas; a small-diameter work zone and a large-diameter work zone; it served as the main living and working area. The small-diameter work zone was equipped with the main control console, the

A:

A: Cosmonauts D. Prunariu and V. Savinikh on board of Salyut 6 space station, 14 - 22 May 1981
B: Graphic, Salyut 6 / 7 interior layout

(credit A: Dumitru-Dorin Prunariu; B: Author, based on B.J.Bluth, 1987)

food system, and the communication equipment. The large-diameter work zone contained the radio systems, the power supply control system, the sleeping facilities, the foldable shower, the exercise equipment, and the medical research equipment. It also contained two spherical chambers that were used for scientific experiments or for disposing trash. (B.J.Bluth, 1987)

The Transfer and Work Compartment could be separated by removable panels.

The **Intermediate Compartment** was the section between the Work Compartment and the second docking port. It had two portholes for visual observation.

B:

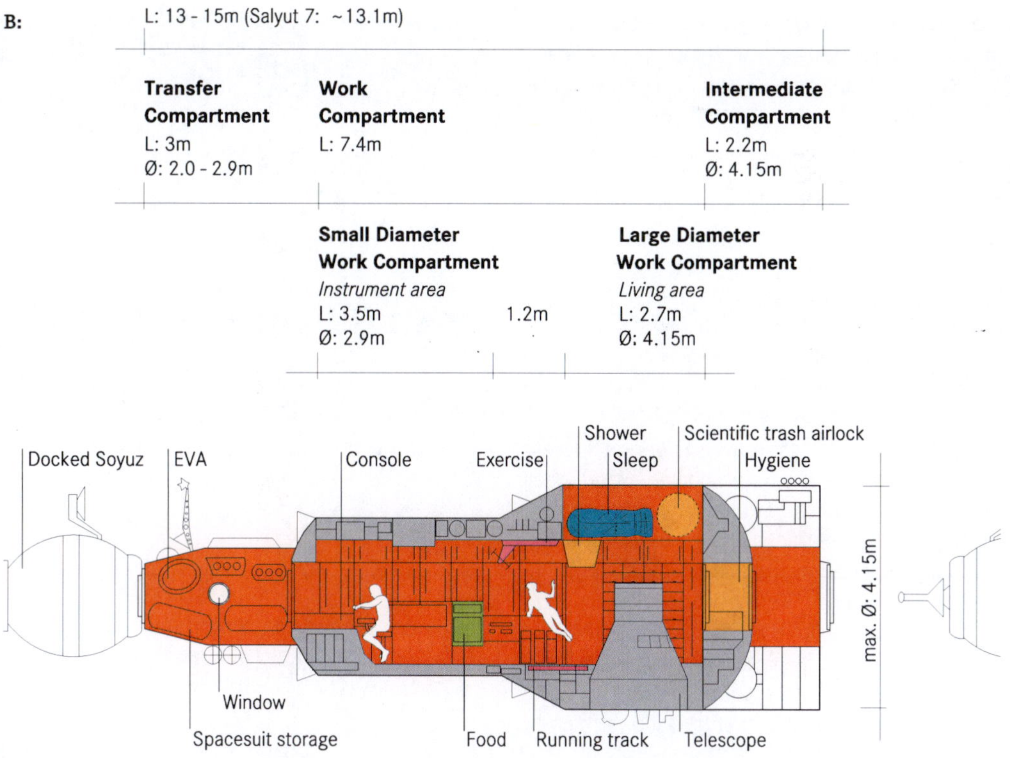

3.3 SALYUT SPACE STATION

TIME ALLOCATION OF HUMAN ACTIVITIES

The later Salyut missions were very long, up to 237 days. Earlier missions were not structured according to the habitual Earth circadian rhythm. Later the crew maintained a day of 24 hours of work and rest with an extended rest period. Over time work days were reduced in favour of rest days and the time dedicated to sleep was lengthened. Because communication was only possible via line-of-sight, sleeping was scheduled during those orbits that did not allow communication with Soviet mission control (U.S. Congress, 1983 p. 32).

The daily and weekly cycle was as follows (Toussaint):

Salyut 1 and 3: 7 workdays a week but the missions were short at first. The usual timetable foresaw 8h sleep, 8h work and 3h exercise and 1h for eating. On Salyut 4 and 5 cosmonauts were given 1 day of rest.

Salyut 6 and 7: 2 days rest a week.
Salyut 6:
8-9h sleep, 2h exercise, 1.5-2h eating, 1.-1.5h ground contacts, 2h free time, 8h experiments
Salyut 7:
9h sleep, 2h exercise, 8.5h work, 4.5h eating, rest, talk

In general cosmonauts took a shower once every ten days. They prepared the shower system after breakfast, as it was a daylong activity (B.J.Bluth, 1987 p. I-82). According to Bluth, work tasks included the observation of Earth (33%), technological experiments (33%), astronomical observations (16 %) and biological and medical experiments (16%). Other everyday activities included the control of onboard systems of the station, Soyuz and the Progress; maintenance work and conversations with mission control (B.J.Bluth, 1987 p. III-23).

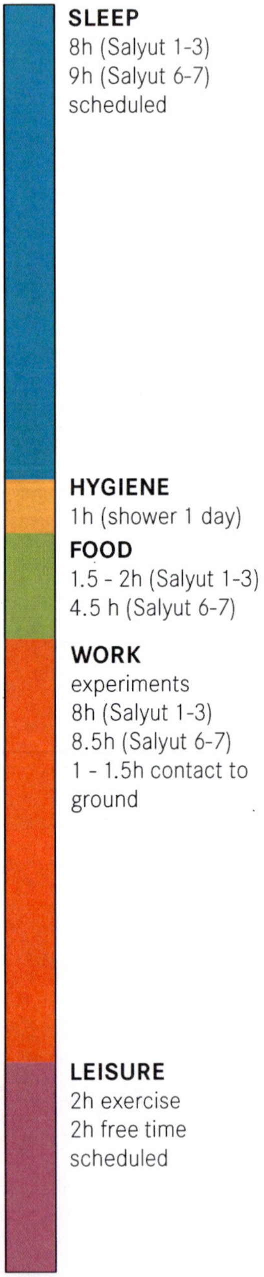

A:

A: Graphic, Time Allocation of Human Acitivites: Salyut
B: Graphic, Spatial Allocation of Human Activities: Salyut *(credit A, B: Author)*

SPATIAL ALLOCATION OF HUMAN ACTIVITIES

In the early Salyut 1-5 space stations, there was very little clear separation between functions because they were all in one main compartment. In the Salyut 6 and 7 stations, the module was structured in four main zones, each with different activities and functions assigned to them. On Salyut 7 the Work Compartment was divided into the instrument area and the living area.

The layout of the station was designed to provide convenient placement of scientific equipment and maximum comfort for the crew (Vasilyev, et al., 1974). The stations were designed with a mix of Earth and microgravity orientation. For example the exercise equipment and sleeping accommodations were located on the ceiling. To facilitate the cosmonaut's orientation, colours and light design was introduced. (B.J.Bluth, 1987 p. 5)

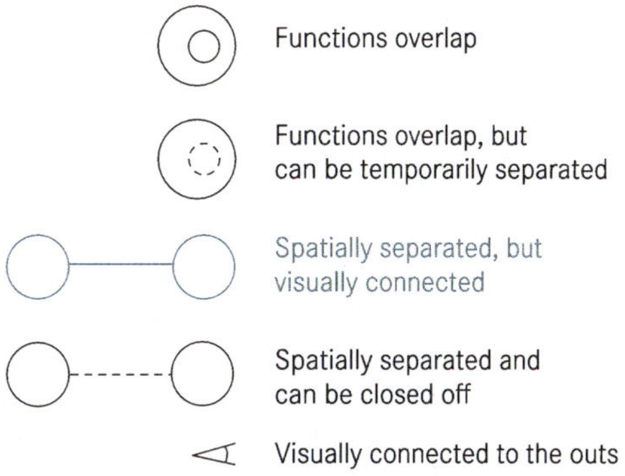

B:

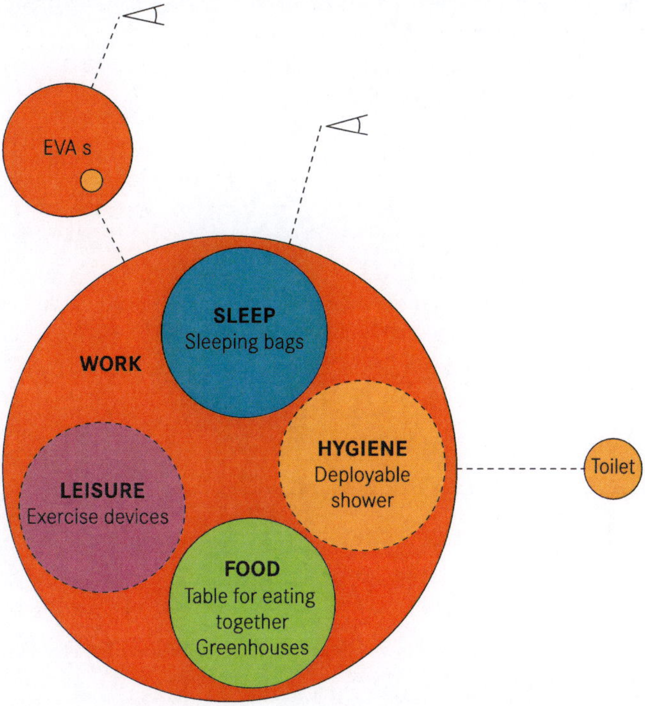

3.4 SKYLAB SPACE STATION

A: Skylab Space Station in Orbit
(credit: NASA)

OBJECTIVES

Skylab was the first US space station. It was launched on 14 April 1973 by a Saturn V rocket - up to now - the last rocket that was able to support human missions to the Moon and to LEO with such a big diameter and fairing capacity.

The overall mission objective of Skylab was to create and test a living and working environment for a long period of time that is self-sufficient and independent from frequent re-supply. The power had to be generated by onboard systems. These requirements lead to the development of a reliable Life Support System (LSS) that also forms the basis for today's ISS systems. (NASA [Skylab], 1977)

Skylab hosted three crews from 1973 – 74, lasting 28, 59 and 84 days. Then the space station was kept in a 'parking orbit' to be re-activated after the development of the Shuttle. Between 1975 and 1981 there was no launch vehicle left for another mission to Skylab. Due to delays in the development of the Space Shuttle the Skylab space station had to be given up. Re-entry of the station was in 1979. (NASA [Skylab], 1980)

Use
First American space station
Operational: 1973 – 1979
Inhabited: 1973 – 1974 (in total 171 days)
Crew size: three crews (3 crewmembers)

Architecture
Launch Vehicle: Saturn V
Crew Transportation Vehicle: Saturn V, Apollo CSM
Elements: single element space station

Volume: 320m^3 (OWS 270.2m^3)
Max. Diameter: 6.6m
Lengths: 26m

Missions
Skylab 2: 28 days
Skylab 3: 59 days
Skylab 4: 83 days

*Sources are listed in the Appendix

3.4 SKYLAB SPACE STATION

CONFIGURATION AND LAYOUT

The Skylab space station was built, using Saturn V and Apollo components and technology. The whole station was about 26m long and had a diameter of 6.6m. The Apollo Command Module docked to the Multiple Docking Adapter (MDA). In the MDA scientific and operational hardware was installed around the cylindrical module. The Saturn V hydrogen tank was converted into a big two-storey living and working space. It was the 14m Orbital Workshop Module, which had been the largest living and working space ever constructed for a space station.

The industrial designer Raymond Loewy gave a large input to the design of the orbital workstation. Eight habitability studies were conducted under his supervision. His positive influence on the interior design is expressed in a letter from 1974 by Dr. George Mueller, NASA's deputy administrator for manned space flight as follows:

A:

A: Skylab space station photographed by the Skylab 4 crew
B: Skylab elements

(credit A: NASA; B: NASA, adapted by the author, source: Skylab News Reference (March 1973))

"I do not believe, that it would have been possible for the Skylab crews to have lived in relative comfort, excellent spirits, and outstanding efficiency had it not been for your creative design, based on a deep understanding of human needs, of the interior environment of Skylab and the human engineering of the equipment and furnishings which the astronauts used." (Loewy, 1980 p. 206)

Figure B shows the main elements of the space station: the Orbital Workshop (OWS), the Multiple Docking Adapter (MDA), The Apollo Telescope Mount (ATM), solar wings and the docked Apollo Command and Service Module. The Apollo Command and Service Module was the logistic vehicle for the space station. (NASA [Skylab], 1977)

B:

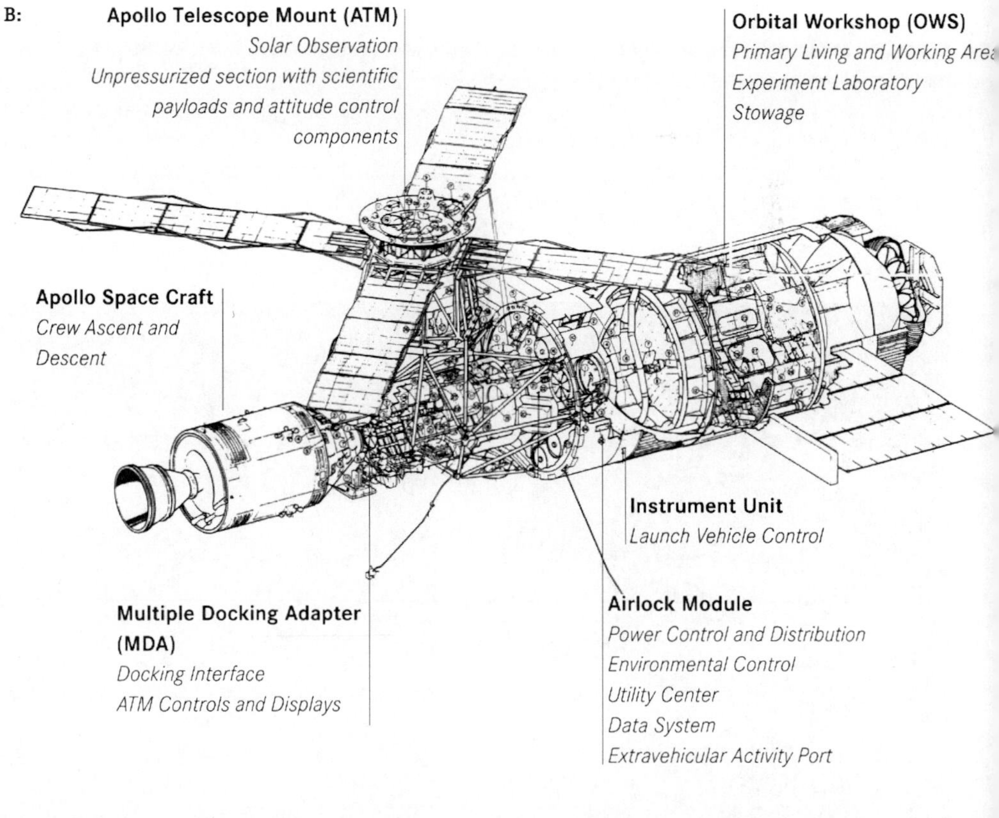

3.4 SKYLAB SPACE STATION

LAYOUT OF THE ORBITAL WORKSHOP

The primary living and working area was situated in the Orbital Workshop Module (OWS). The Forward Compartment of the Orbital Workshop was a half circle dome. Experiments were installed around the walls and on the floor. The lower floor included the sleep compartment, the wardroom, waste management compartment, and the Experiment Work Area. The adjacent oxygen tank served as the trash and the waste-disposal compartment.

The sleep compartment provided space for three astronauts and was next to the waste management compartment. In the wardroom, the astronauts prepared and ate their meals. The experiment compartment included the lower-body negative-pressure device, an ergometer and a foldable shower system.

To move around safely and to be held securely in place, the OWS had walls and ceilings made of a metal triangular grid-work *[Fig. B]*. The original design provided the astronauts with specially designed shoes with triangular plates fastened to the soles *[Fig. A]*, which fitted in the triangular openings of the grid. However, once on-orbit, the crew found them awkward to use; unnecessary as they learned to move and stabilize themselves in micro-G; and actually dangerous if one crew member should kick another with the metal shoe. They soon stopped using the triangle sole-plate shoes. The triangle grid proved tremendously useful for attaching and securing equipment. Other restraints which were provided, but not needed as soon as the crew was adjusted to move in microgravity were chair-type body restraints and the 'Fireman's pole'. It was attached between the dome hatch and the compartment floor hatch. (NASA [Bull.10], 1974; NASA [Bull.11], 1975)

Other restraints included handrails, handholds, straps, foot restraints and temporary restraints such as clips and snaps. (NASA [Skylab], 1977)

A:

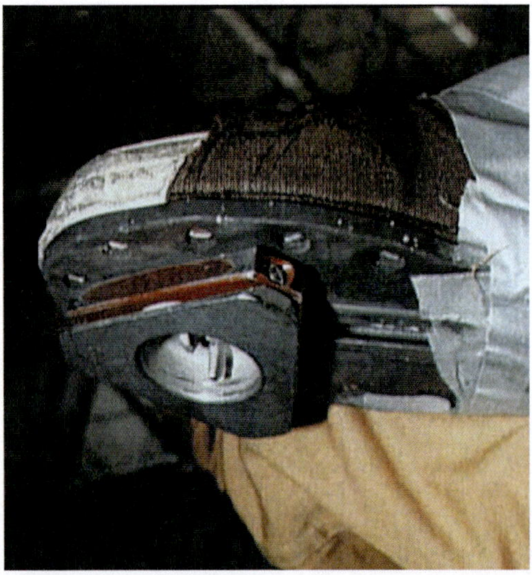

B:

A: View of triangle-shaped cleat taped on the bottom of an astronauts shoe
B: View of Orbital Workshop in Skylab
C: Graphic, Skylab Space Station interior layout

(credit A, B: NASA; C: Author, based on NASA documents)

C:

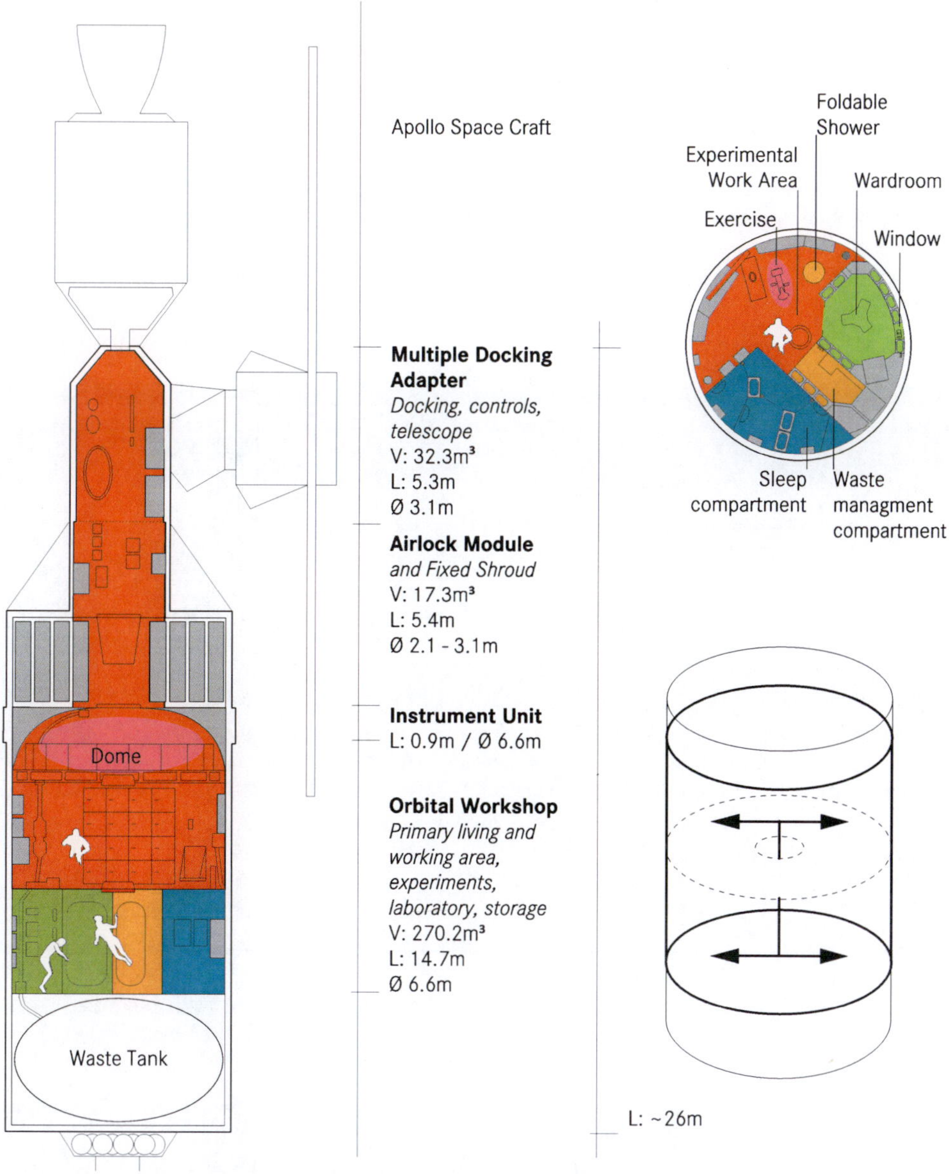

3.4 SKYLAB SPACE STATION

TIME ALLOCATION OF HUMAN ACTIVITIES

The daily and weekly cycle of the Skylab astronauts was as follows:

Skylab astronauts worked 8h a day and had about 9h for sleeping and body hygiene, 2h for eating, 2h for exercise and 2h for leisure activities. They had contact to mission control for about an hour every day. (NASA KSC, 2000)

All Skylab crewmembers had a scientific and piloting background. Joe Kerwin on the first crew was a medical doctor. All the others were military pilots.

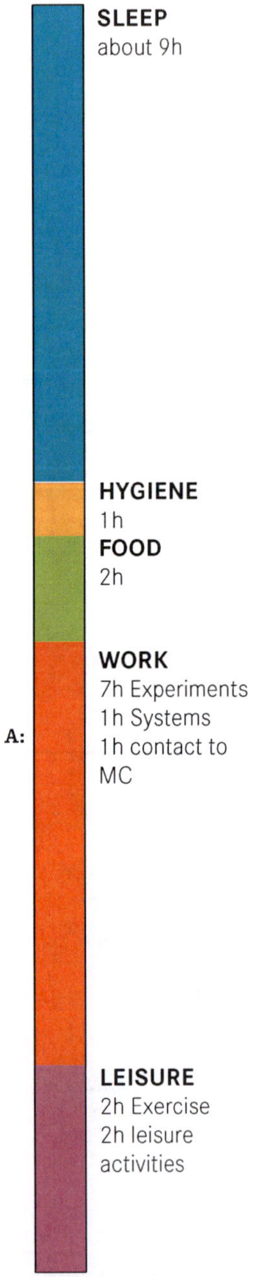

A: Graphic, Time Allocation of Human Acitivites: Skylab
B: Graphic, Spatial Allocation of Human Activities: Skylab *(credit A, B: Author)*

SPATIAL ALLOCATION OF HUMAN ACTIVITIES

Human activity areas were clearly separated. The upper, zone called the "dome area" was where most scientific and technical experiments and work was located. Below the dome area was a cylindrical zone that housed the crew living quarters between triangle grid floor/ceiling decks. A sleep compartment with three individual areas was provided. The waste management compartment with the toilet could be closed off. The astronauts prepared and ate their food in the wardroom which also had a window to look outside. The foldable shower and the exercise equipment were located in the experiment compartment.

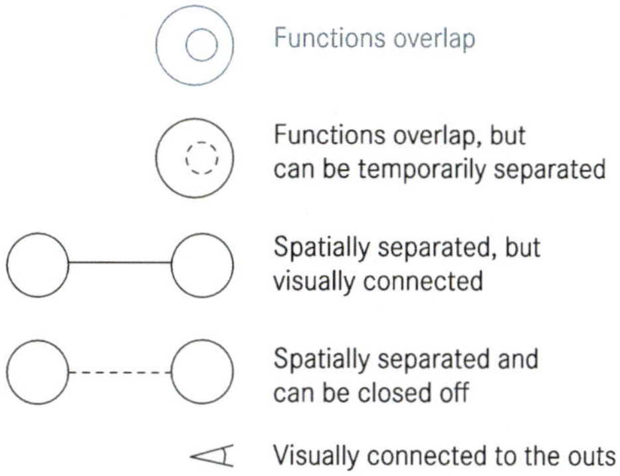

B:

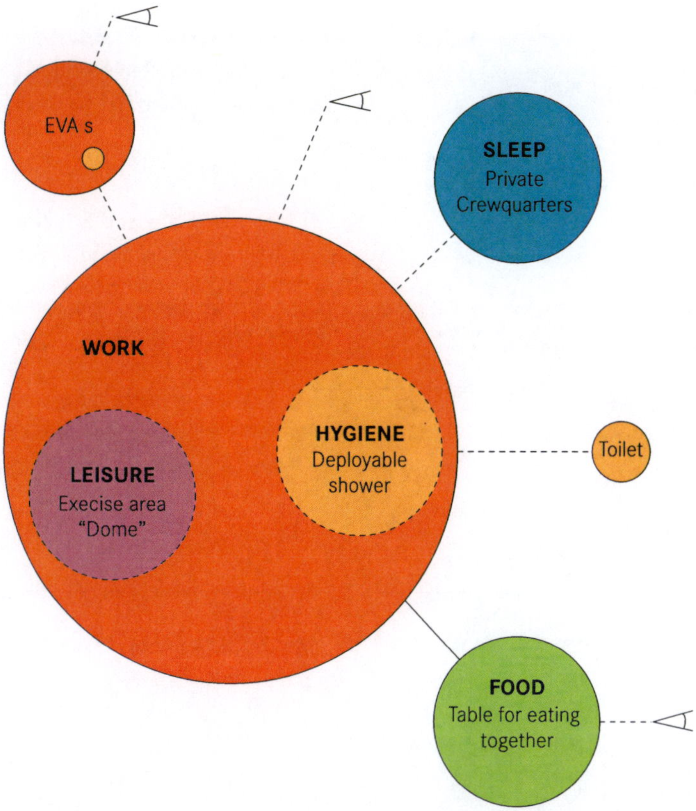

3.5 SPACE SHUTTLE

A: Space Shuttle Discovery approaches the International Space Station for docking. The Leonardo Multi-Purpose Logistics Module can be seen in the cargo bay *(credit: NASA)*

OBJECTIVES

The Space Transportation System (STS) with the Space Shuttle Orbiter is a reusable launch system that is launched with the Space Shuttle Main Engines augmented by two solid rocket boosters to space. It lands on a runway like an airplane. Its main function is to deliver payloads to Earth orbit. It can carry 7 people for 17+ days in nominal conditions. It has served to bring crews, equipment and modules to the Mir Space Station and to the ISS.

It has been used as a platform to build up the International Space Station and to transport people and equipment to the station. The Shuttle is also used to repair unmanned satellites in orbit. (Nasa [Shuttle], 2010)

Use
Transport Vehicle
Operational: since 1981
Crew size: 2 for test flights, 4 to 7 nominal (+3 rescue)

Architecture
Partially re-usable launch and fully reusable re-entry system

Elements: The Space Transportation System (STS) consists of the: External tank, Solid Rocket Boosters, and Orbiter Vehicle

Shuttles: Challenger and Columbia lost in accidents; operational: Discovery, Atlantis, and Endeavour

Habitable Volume: 66m^3
Middeck Floorplan Length: 4m
Middeck Floorplan Width: 2.7 (front) – 3.7m (rear)

Selected Missions
STS-1: 1981, first shuttle mission

STS-80: 1996
Longest flight with a Shuttle: 17d 15h 53m 18s

STS-134: 2011
Last shuttle mission

* Sources are listed in the Appendix

3.5 SPACE SHUTTLE

CONFIGURATION AND LAYOUT

Six Shuttles were built. The orbiter 'Enterprise' was an atmospheric test and training vehicle and never flew to orbit. 'Columbia' was the first shuttle in orbit, but was lost with its crew in 2003. 'Challenger' and its crew were lost in 1986. NASA built Endeavour to replace the Challenger. The shuttles Discovery, Atlantis and Endeavour remained operational. The latest version orbiter is the STS `Endeavour', which benefited from the Extended Duration Orbiter program that was underway in the 1980s. In principle, it could stay in orbit for up to one month, with additional consumables carried mainly in an augmentation module in the payload bay.

The Space Shuttle consists of three main units; the orbiter, the external tank and two solid rocket boosters. The orbiter is reusable and consists of the Forward Fuselage, the Mid Fuselage and the Aft Fuselage. The solid rocket boosters are also reusable and are recovered at sea after each launch and reconditioned for reuse.

A:

B:

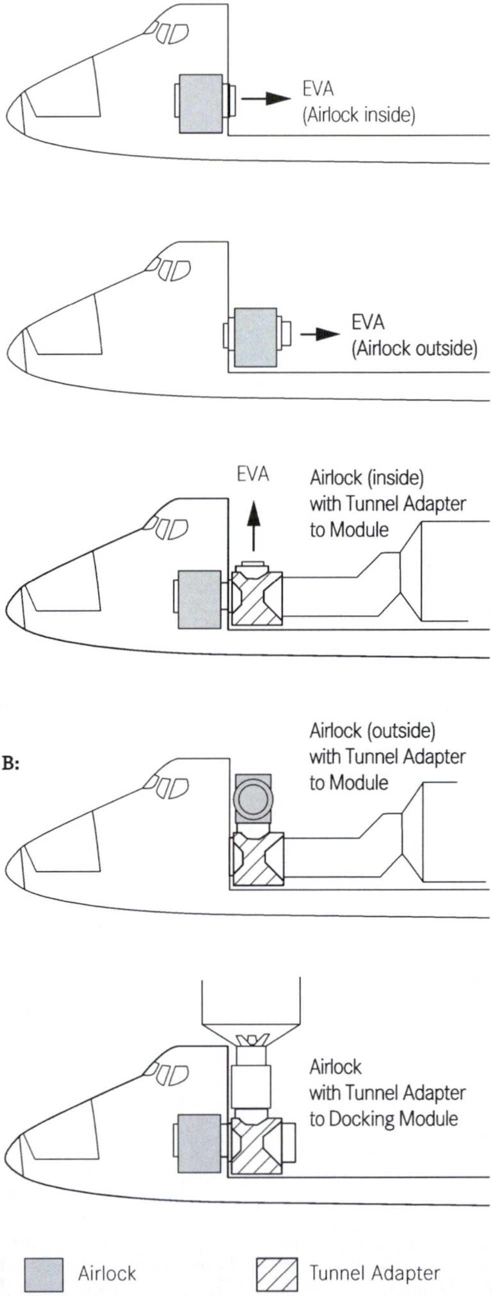

A: The docked Space Shuttle Discovery with the Leonardo Multi-Purpose Logistics Module in the payload bay
B: Space Shuttle Airlock configurations
C: Space Shuttle Orbiter

(credit A, C: NASA; B: NASA, adapted by the author from Space Shuttle News Reference)

Airlock

The airlock of the Shuttle can be used in a number of ways *[Fig. B]*. It can be located inside or outside the crew module to provide access to the payload bay. A tunnel adapter can be used to connect the airlock to other modules in the bay and for direct docking to the station a docking module can replace the airlock module. (NASA [Shuttle Reference], 1981 p. 3-11)

C:

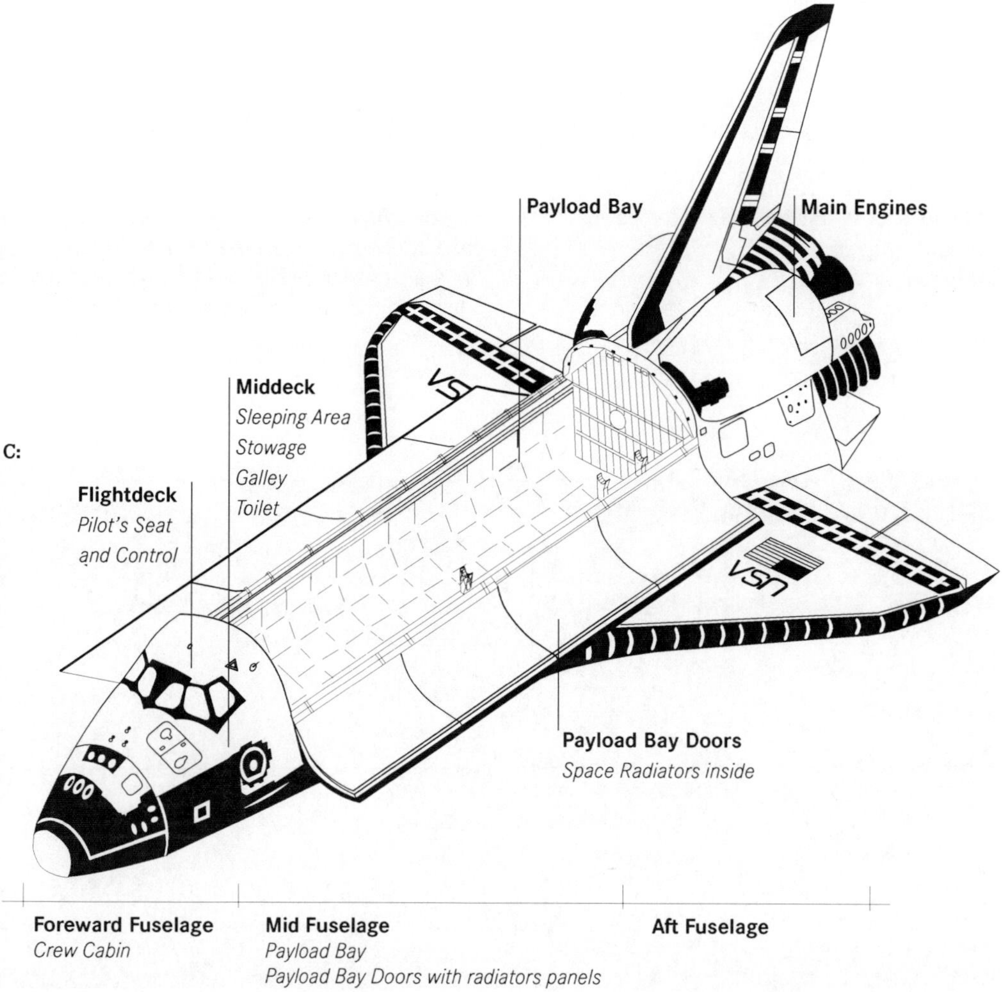

3.5 SPACE SHUTTLE

CONFIGURATION AND LAYOUT

The Forward Fuselage contains the crew cabin. Its pressurized volume is about $80m^3$ and its habitable volume is about $66m^3$. It has two crew levels consisting of the Flight deck and the Middeck plus the Avionics Bay, which contains the life support equipment. (Nasa [Shuttle], 2010)

Flight Deck

The flight deck contains the cockpit *[Fig. A]* with two pilot seats and two additional crew seats behind them on the aft flight deck for launch and landing the displays and controls to be piloted. It has six high optical quality windows and four aft-windows to observe the payload bay and to pilot the orbiter during rendezvous and docking. On orbit the cabin arrangement can be changed in order to allow for the handling of the payload.

Middeck

The Middeck provides the facilities for crew habitability. It contains the sleep stations for the astronauts, the waste management system, the personal hygiene station, the work and dining table, food preparation and stowage facilities. Standard Modular Stowage Lockers are installed next to the avionics bay (Boeing, 1997). One window is located in the crew entrance/exit hatch. (NASA [Shuttle Reference], 1981)

During launch and landing the Middeck provides three seats for astronauts, which are removed in orbit. Additional three contingency seats can be installed, instead of the sleeping provisions and stowage. A hatch provides access to the Airlock and to the Payload Bay. The Airlock can be used in various locations.

Payload Bay

The Payload Bay can be used to extend the Shuttle's pressurized volume through the installation and use of the Spacelab or the SpaceHab modules.

A:

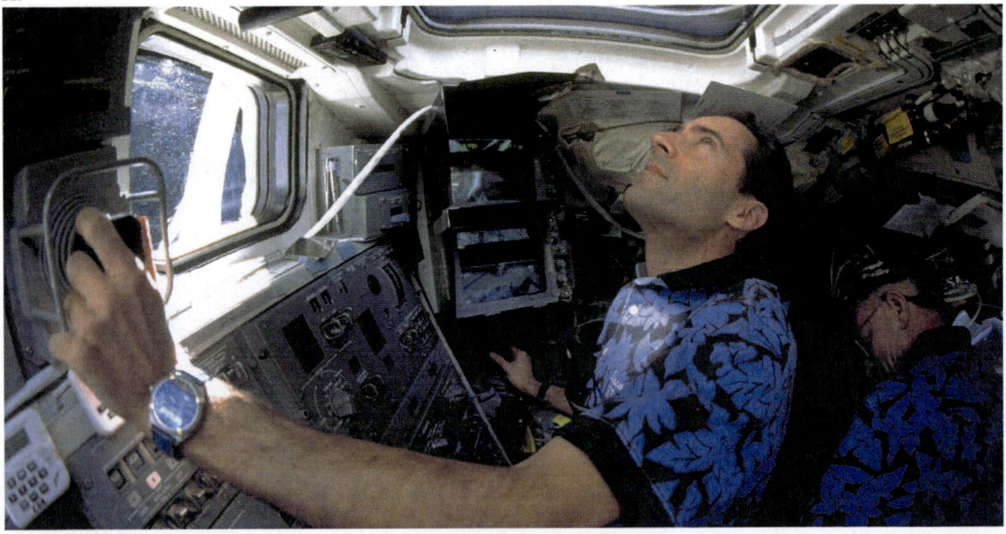

A: Jean-François Clervoy in the cockpit on the Shuttle's Flight Deck
B: Graphic, Space Shuttle interior layout

(credit A: NASA, Jean-François Clervoy; B: Author, based on NASA documents)

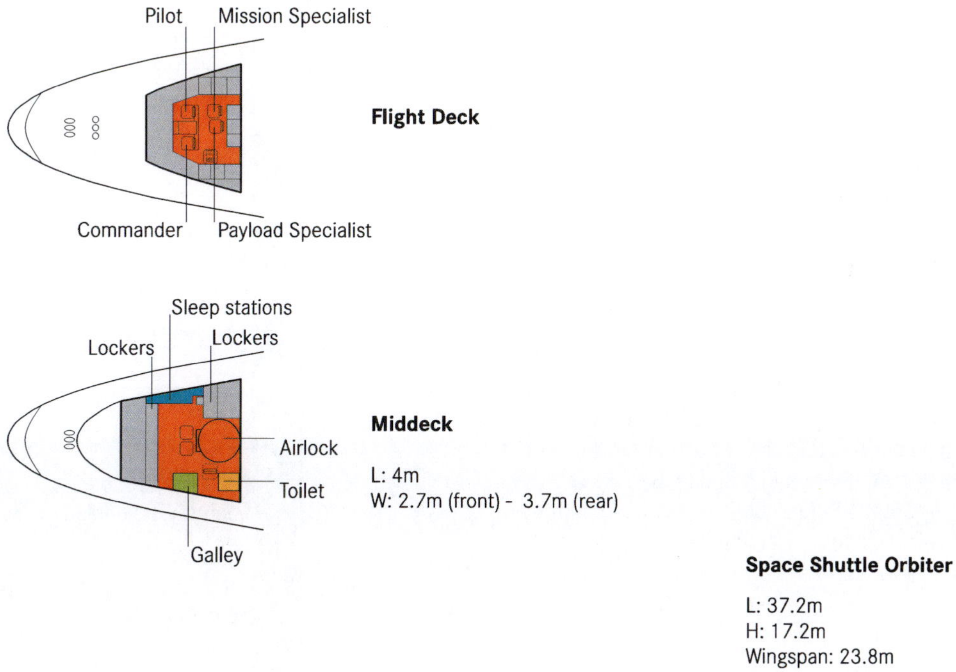

B:

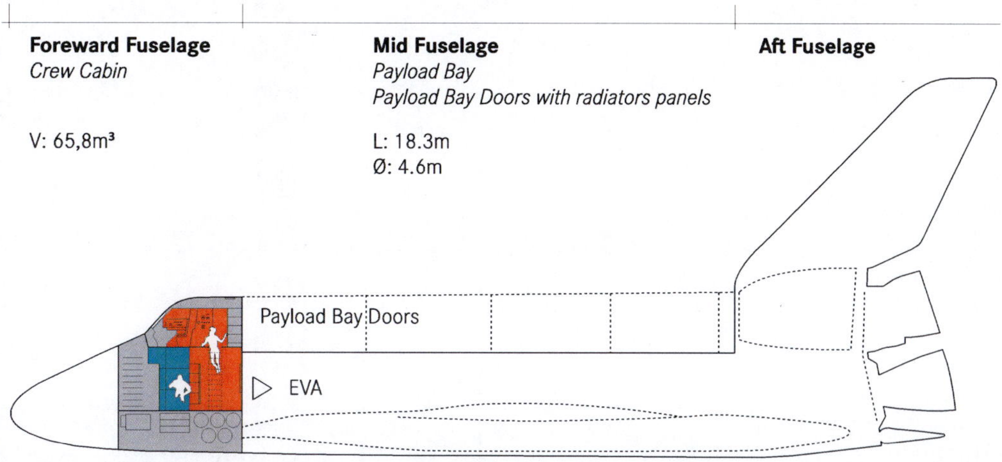

3.5 SPACE SHUTTLE

TIME ALLOCATION OF HUMAN ACTIVITIES

Shuttle flights nominally last about seven to ten days and can be extended to approximately 17 days. The nominal crew consists of seven astronauts (NASA [Facts], 2002).

The daily routine of astronauts depends on the mission, but each astronaut has to follow a detailed schedule each day (NASA [Shuttle Reference], 1981 p. 5-6). The average workday is 16 hours with about 8 hours sleep (NASA [Factsheet], 2006). Sometimes astronauts on Shuttle missions work on shifts.

Three 1-hour meal periods are scheduled into the work day for all, including preparation, eating and cleaning-up. In addition to time scheduled for sleep and meals, each crewmember has housekeeping tasks that require about 15 minutes a day. (NASA [Shuttle Reference], 1981 p. 5-6) Shuttle astronauts exercise 30 minutes each day.

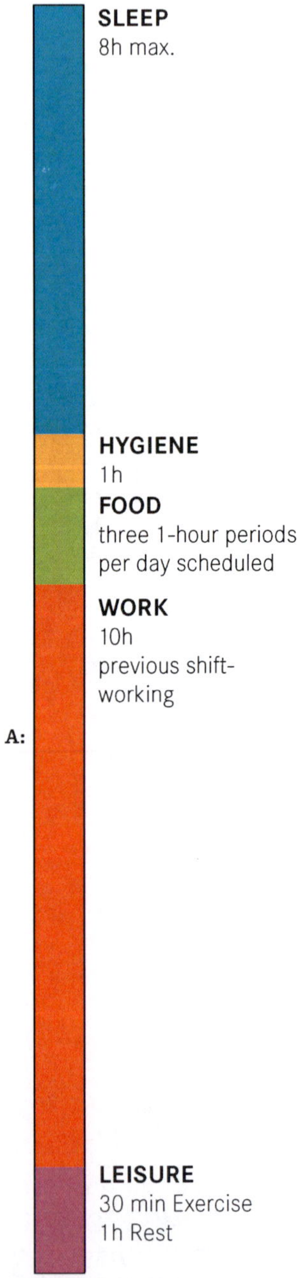

A: Graphic, Time Allocation of Human Acitivites: Shuttle
B: Graphic, Spatial Allocation of Human Activities: Shuttle *(credit A, B: Author)*

SPATIAL ALLOCATION OF HUMAN ACTIVITIES

The Middeck of the Space Shuttle is the living quarter; its overall length is about 4m, its overall width are about 2.7m at the front and 3.7m at the rear. The crew sleeps, eats and conducts hygiene activities here. It is equipped with a toilet, a washbasin, a galley with an oven, lockers, sleep stations or sleeping bags and an airlock to enter the cargo bay. (NASA [Factsheet], 2006)

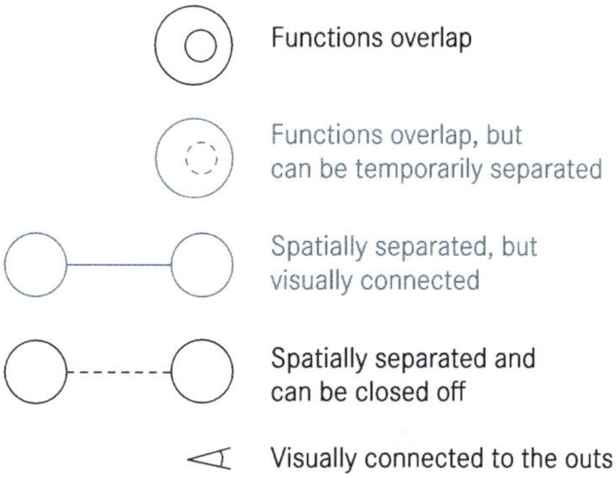

B:

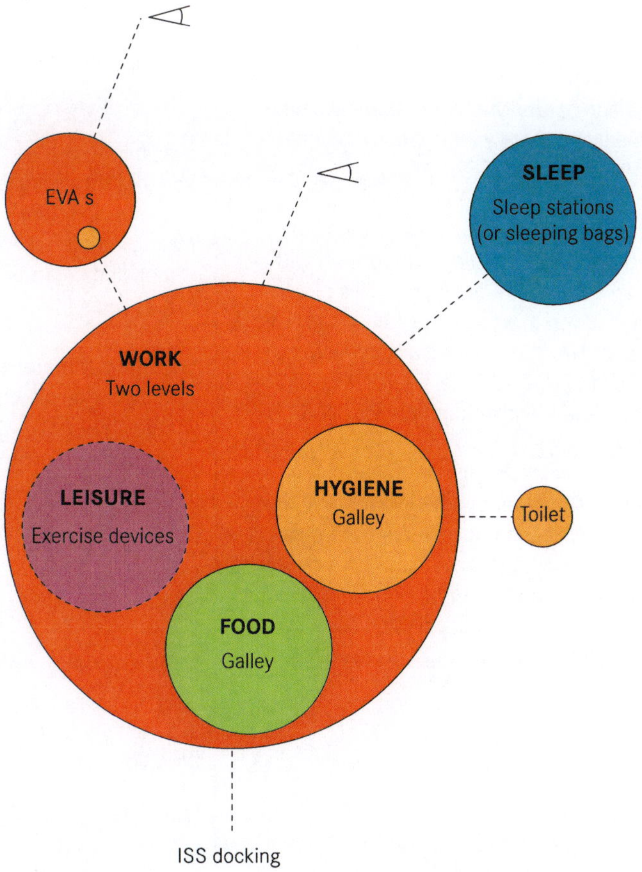

3.6 MIR SPACE STATION

A: Mir Space Station taken from STS-81 mission *(credit: NASA Marshall Space Flight Center)*

OBJECTIVES

The Mir Space Station was, even if unplanned, the first international space station. It was launched in 1986 and had 14 years of useful lifetime in orbit; Mir hosted 105 cosmonauts from 11 nationalities. The term 'Mir' is often translated, meaning peace or world (cf. Wikipedia). In 1996 Frank L. Culbertson Jr., Director of the Shuttle-Mir Program proposed another possible meaning of the word 'mir' at a Congress of The Association of Space Explorers. He refers to the original use of the word 'mir' meaning the concept of a village; where the local people lived close together to better share the limited resources. (Culbertson, 2004)

The Mir Space Station experiences built the basis for future space missions. Different docking techniques and resupply methods were evaluated, which helped lead to truly international cooperation on Mir and later on ISS.

In 1994 the Shuttle-Mir program started. It was based on the goals of learning how to work with international partners, reducing the risks of developing and assembling a space station, gaining experience for NASA on long-duration missions and to conducting life science, microgravity, and environmental research programs. (NASA [Shuttle-Mir], 2004)

Mir was only planned for two people for long duration missions up to about a year long. 15 years later up to 6 cosmonauts stayed on board over a short duration. During the Shuttle-Mir missions (1994-1998) Mir was also occupied by seven long-term NASA astronauts.

Use

Soviet/Russian civilian space station, first international space station, with an original design life of five years like its Salyut 6 and 7 predecessors
Operational: 1986 - 2001
Inhabited: 1986 - 2000
Crew size: one resident crew (2) and visiting crews (< 3); max. 6 not including NASA Space Shuttle missions

Architecture

Launch Vehicle: Proton
Crew Transportation: Soyuz, Space Shuttle
Elements: Mir core, plus the Kvant-1, Kvant-2, Kristall, Spektr, and Priroda laboratory modules, and the Docking Module;
Automatic docking of modules, automatic docking of Progress spacecraft with manual controls available
Volume: Mir complex $380m^3$
Core Module $90m^3$
Max. Diameter: 4.35m
Dimensions: 33m x 31m x 27,5m
Mass: 130 - 140t

Selected Missions

Constantly occupied from September 1989 - August 1999
29 occupancy crews and 3 Soyuz visiting crews and nine Shuttle visiting crews
EO-15/EO-16/E0-17
1994 January 8 - 1995 March 22
Valeri Polyakov spent 438 days in once (longest manned spaceflight)
EO-21/EO-22
1996 March 22 - 1996 September 26
Shannon Lucid stayed 188 days in orbit (longest woman and longest American flight duration to that time)

* Sources are listed in the Appendix

3.6 MIR SPACE STATION

CONFIGURATION AND LAYOUT

Mir is distinguished from its predecessors of the Salyut series by six docking ports that made the space station Mir the first inhabited space station with a truly modular concept (Salyuts 6 and 7 could accommodate one Cosmos Lab Module). All modules were launched unmanned and docked automatically to the core module.

The **Core module** of the space station Mir was launched in 1986, while the space station Salyut 7 was still in orbit. It had two docking ports and four berthing ports providing 90m^3 of habitable volume.

Throughout the orbital lifetime of Mir, the Soviets/Russians carried out a campaign of expansion by adding modules to the abundant docking ports. These modules included Kristall, KVANTs 1 and 2, Priroda and Spektr.

In March, 1987 the module **Kvant 1** was launched. It was used for astrophysics and attitude control. From 1989 on, **Kvant 2** provided the station with additional capabilities, such as an EVA airlock, a urine-regeneration system and a shower compartment. The module **Kristall**, launched in 1990 was used for material processing and Earth observation and was later used for the Shuttle docking. The Svet greenhouse was located in this module. In May 1995, the Geophysical Module **Spektr** that would be used for Earth observation and sampling and analysis of the Earth's atmosphere and the orbital environment was added. **Priroda**, the International Ecology Research Module was added in 1996, to gather data on the Earth and its atmosphere.

In 1994 the Shuttle-Mir program started. The first docking of the Space-Shuttle Atlantis *[Fig. A]* took place in June 1995 bringing the Mir 19 crew to the space station and taking the Mir 18 crew to Earth. For a short term the space station hosted 10 humans on board (McDonald, 1998 p. 9). The Docking Module was installed in November 1995 to preclude needing to move Kristall each time a Space Shuttle docks to another port.

The final architecture of Mir comprised of six modules. The spherical Transfer compartment of the Core Module provided four berthing ports for expansion modules and one docking port. The station had two docking possibilities for the Russian Progress or Soyuz freighters and one with a docking adapter for the American Space Shuttle.

In its final configuration the total habitable volume was about 380m^3. It consisted of modules that were launched with a Proton rocket that had a maximum diameter of 4.35m.

A:

A: STS-71, Space Shuttle Atlantis docked to Mir Space Station, photo taken by the Mir crew
B: Mir Space Station complex 1994

(credit A: NASA, B: NASA, Mir Hardware Heritage)

B:

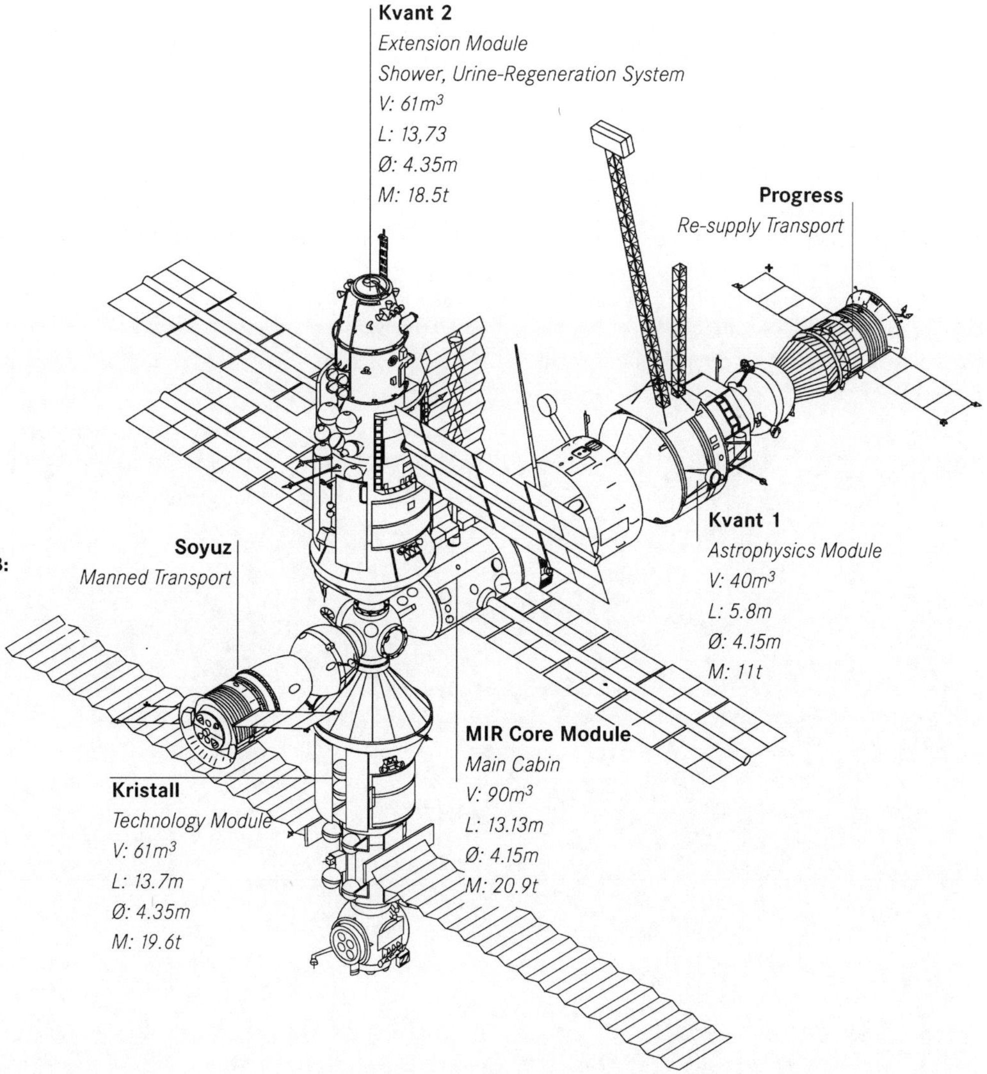

3.6 MIR SPACE STATION

CONFIGURATION AND LAYOUT

According to B. J. Bluth (1987), cosmonauts' ideas and suggestions were taken into consideration for improving the space station design. The architecture of the space station Mir's modules was based on already known technology, and was similar to the Salyut space stations. Due to the modular approach, the available space was increased. Soviet research had acknowledged that the lack of habitable volume can lead to negative physical and psychological effects (B.J.Bluth, 1987).

The **Mir Base Block or Core Module** *[Fig. B]* had a total volume of about 90m^3 and consisted of a large diameter section and a small diameter section. It was the main living module. Private crew cabins for two cosmonauts were provided and the crew prepared their food and ate here. The base block was also used for communication, maintenance and repair, while the experiments were conducted in the attached dedicated modules.

Unlike other modules, the Core Module used colours to mimic 'Earth orientation'. The small diameter section had a dark-green floor and light green walls. The large diameter section had a brown floor and yellow walls. Both sections had a white ceiling with fluorescent lighting (Portree, 1995; Burrough, 1999). The station complex had many windows for looking outside. They had shutters to protect them from the impacts of orbital debris (B.J.Bluth, 1987). The Core Module had windows on the 'floor' for Earth observations and one in each of the two crew cabins. In Kvant 2 was a large Earth observation window. The 'walls' had removable panels for storage.

A:

A: Austromir-91 crew in the Mir core module
B: Graphic, Mir space station interior layout

(credit A: Franz Viehböck, B: Author, based on NASA documents)

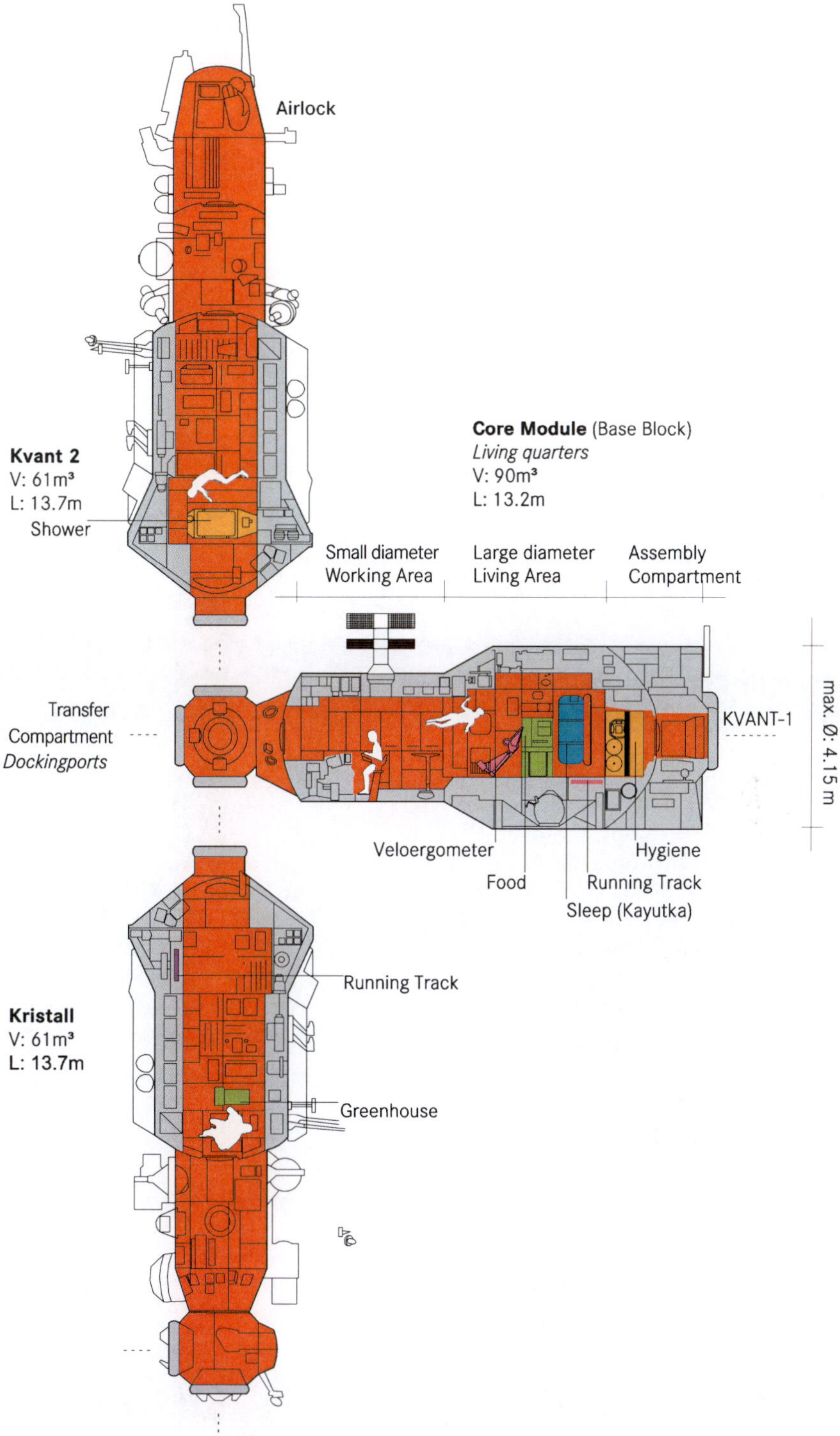

B:

3.6 MIR SPACE STATION

TIME ALLOCATION OF HUMAN ACTIVITIES

The space station Mir hosted two permanent crewmembers and visiting crews. Later, it would host a short-term crew up to seven. From 1989 until 1999 the Space Station Mir was permanently occupied.

The daily cycle on Mir was similar to the one on Salyut 7. Cosmonauts were scheduled 9h sleep, 2h exercise, 8.5h work and 4.5h for eating, rest and communication. The daily work schedule was based on the usual Earth cycle and biological rhythm. The usual day started at about 8:00 AM and ended about 11:00 PM for all cosmonauts – no split-shifts were used. Their weekly schedule foresaw 5 days work and 2 days rest (Toussaint; B.J.Bluth, 1987).

In his book 'Off the Planet', Jerry M. Linenger describes the beginning of a typical day as follows (Linenger, 2000 p. 91):

08:00 - 08:20	morning hygiene and toilet
08:20 - 09:50	activate space acceleration measurement system
09:50 - 10:40	exercise: treadmill
10.40 - 10.50	personal hygiene
10:50 - 13:00	spacesuit preparation
13:00 - 14:00	lunch
14:00 - 19:00	work
19:00 -	catch up on work tasks

The crew communicated with ground control several times a day, often between meal times. (Foale, 1999, 96)

During Linenger's mission, the Sunday was scheduled free. He also reported, that he had only one or two hours off during the week, because of work load.

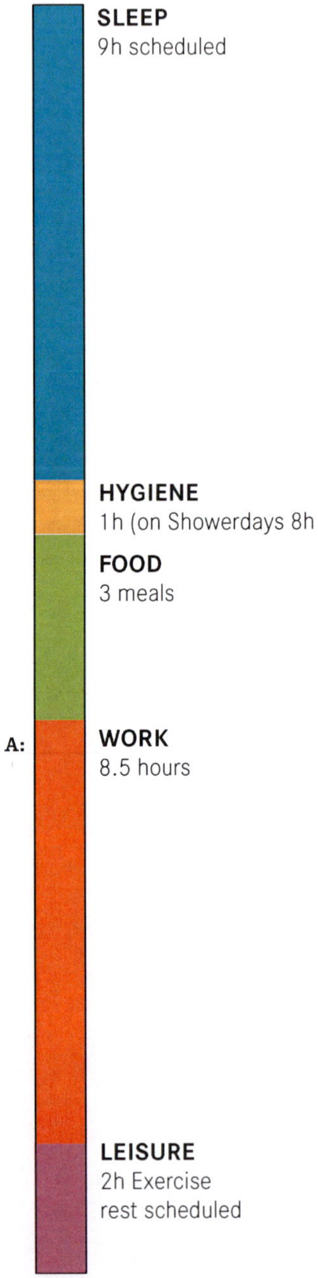

A: Graphic, Time Allocation of Human Acitivites: Mir
B: Graphic, Spatial Allocation of Human Activities: Mir *(credit A, B: Author)*

SPATIAL ALLOCATION OF HUMAN ACTIVITIES

The planning of the interior module foresaw separation of the activity zones work, rest, eating and sleeping (B.J.Bluth, 1987 p. 55). Each module housed a dedicated function or set of functions. The Core module provided the main living quarters and the other modules specific scientific functionality. The station provided permanent crew quarters (kayutkas) for two cosmonauts that could be closed off. Additional crewmembers slept in a scientific module. The activities food, hygiene and exercise were mainly arranged in the Core module.

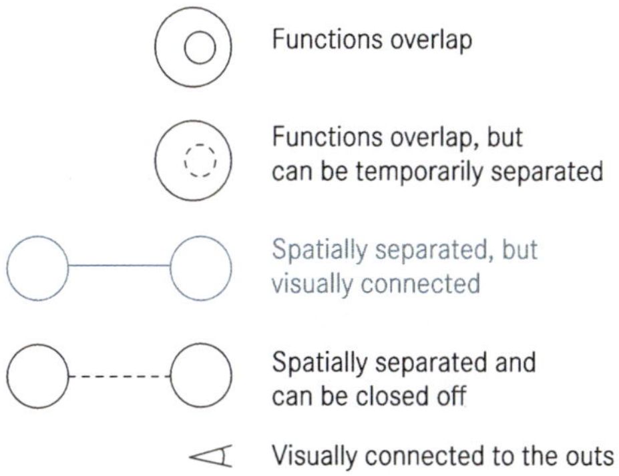

B:

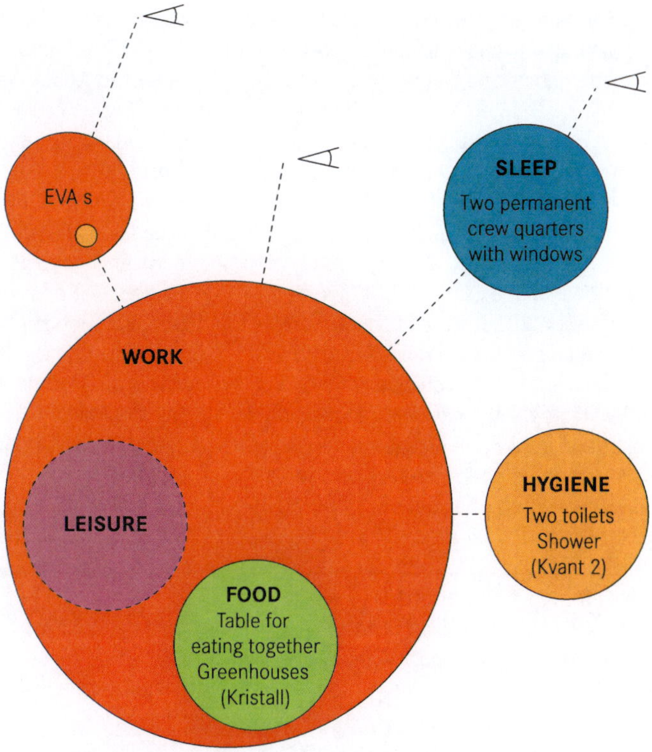

3.7 ISS SPACE STATION

A: International Space Station on orbit, 2009 *(credit : NASA)*

OBJECTIVE

The SKYLAB program started in 1970, as an extension of the Apollo Program, soon after the first two successful landings on the Moon. When it became clear in 1979 that the USA would need to abandon the Skylab because the Shuttle program was at least two years behind schedule, NASA started a new program. NASA formed the Space Station Task Force at NASA HQ in 1982 and started the design process in 1983 with the formation of the Space Station Concept Development Group. Then, with the CDG-1 configuration (Cohen, 1986) concept established, in 1984 US President Ronald Reagan announced, that "(...) I am directing NASA to develop a permanently manned space station and to do it within a decade" (NASA [Reagan], 2007). From the earliest days in 1983, NASA included ESA, Japan's NASDA, Canada's CSA and the European national space agencies (particularly Italy, the Netherlands, and Germany) in planning. The first plans for an international space station including the former Soviet Union emerged shortly after the Berlin Wall came down in 1989. On November 20th 1998, the first module 'Zarya' was launched from Baikonur, Kazahkstan.

The International Space Station (ISS) "serves as a habitat for its crew, a command post for orbital operations, and a port for the rendezvous and berthing of smaller orbiting vehicles. It functions as an orbital microgravity and life sciences laboratory, a test bed for new technologies in areas like life support and robotics, and a platform for astronomical and Earth observations" (NASA[ISS], 2006 p. 21).

The leading partners today are NASA, Canada (CSA), Japan (JAXA), the Russian Federation (Roscosmos) and the European Space Agency. (NASA [ISS], 2010)

Use
International space station
Operational: since 1998
Inhabited: since 2000
Crew size: one resident (6) and visiting crews

Architecture
Launch Vehicle: Proton, STS
Crew Transportation: Soyuz, Space Shuttle
Elements: Zarya, Unity Node 1, Zvezda, Truss elements, Destiny, Quest, Canadarm2, Pirs, Harmony Node2, Columbus, Kibo-JEM, Poisk, Node3

Volume: 935m^3 at assembly-complete
Max. Diameter: 4.5m
Length, overall: 110m
Mass: 419t

Selected Missions
Expedition 1: 2000 October 31 - November 2 permanent occupation begins; W. Shepherd, Y. Gidzenko, S. Krikalev
Expedition 2: D. Tito - the first space tourist
Expedition 16: 2007 October 10 - April 19; Y. Malenchenko, P. Whitson, Sh. Shukor, L. Eyharts, G. Reisman, D. Tani; The ESA Columbus laboratory module arrives
Expedition 17: 2008 April 8 - October 23; S. Volkov, O. Kononenko, G. Chamitoff; The first part of Japan's Kibo-Japan Experiment Module (JEM) lab arrives, Kibo means "hope"
Expedition 25: 2010 September 25 - November 30; D. Wheelock, O. Skripochka, A. Kaleri, Sh. Walker, F. Yurchikhin; The ISS becomes the longest habitable space outpost ever built.

* Sources are listed in the Appendix

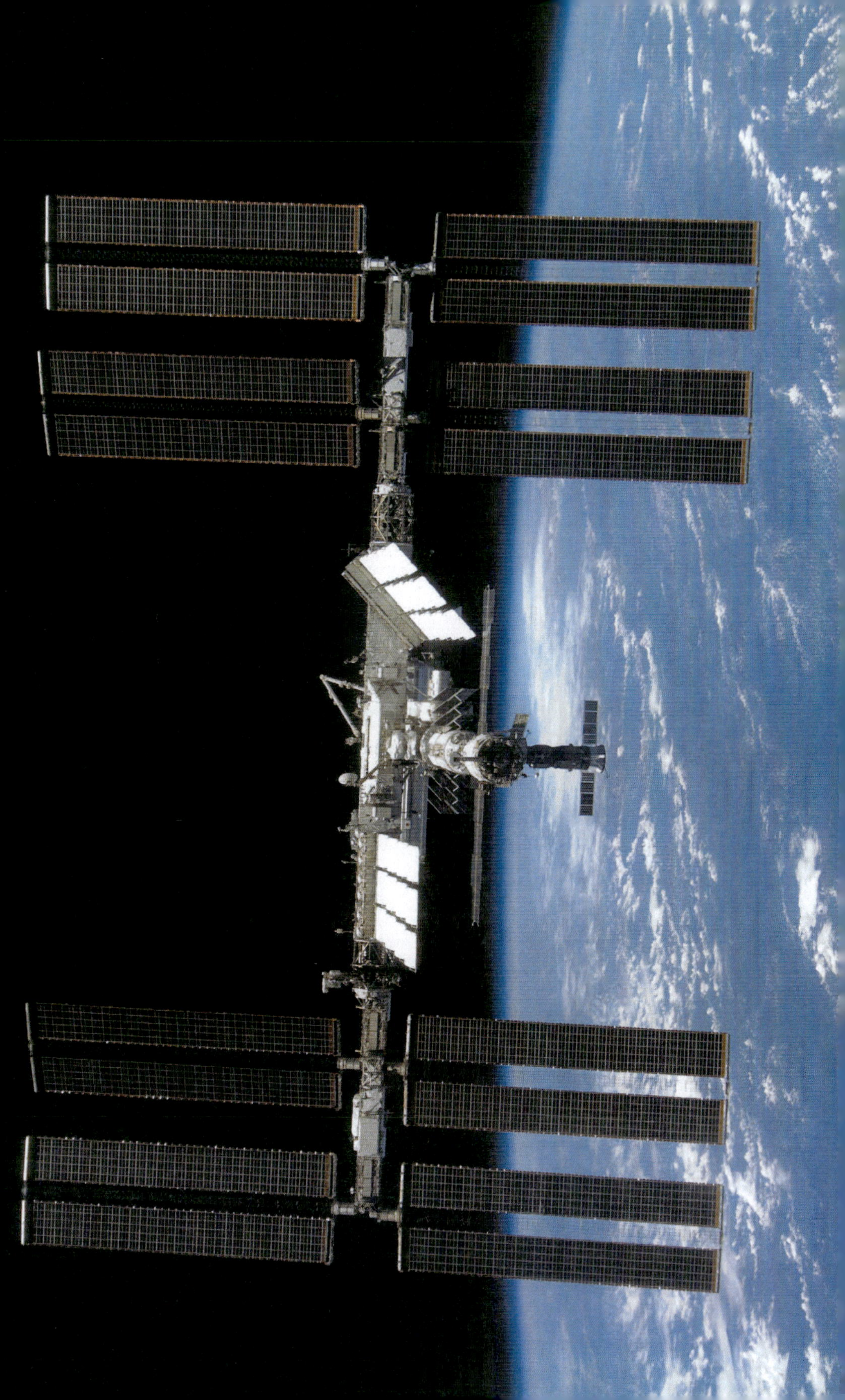

3.7 ISS SPACE STATION

CONFIGURATION AND LAYOUT

The International Space Station is based on a modular architectural concept. When completed, it will be the largest space station complex ever. It will have an internal pressurized volume of about 900m^3 incorporating a total of six modules and three nodes. *Figure A* shows the elements of the station.

The structural element of the ISS - the "backbone" is provided by a truss system. It serves attachment points for the modules and the technical equipment, such as radiators and batteries The habitable modules of the ISS are mainly pressurized cylindrical modules that are connected using the docking ports of the three Nodes. The size of the U.S., European and Japanese modules is based on the fairing capacity of the Space Shuttle.

The first module **Zarya** or **functional base block FGB** was launched in 1998 by the Russians. The FGB is a self-contained module with a pressurized volume of 71.5 m^3. It provided the main system functions, such as power, communication and altitude control, until the Service module was attached. Its architecture is based on the previous Mir modules. It is now used mainly for storage and as a backup system.

The **Service Module (Zvezda)** was derived from an improved Mir Core module design. It provided the early habitable space and is still the structural and functional centre of the Russian segment of the ISS. These functions included the life-support system electrical power distribution, data processing system, flight control system and a propulsion system. The Transfer Compartment can be used for EVA.

Following the Zarya and Zvezda, the ISS Partners began launching the laboratory modules and nodes. The **U.S. Lab (Destiny)** was the first science module on the ISS. It hosts 24 racks for accommodation and control of ISS systems in the U.S. segment and scientific research racks and has a large high quality optical window. Three **Nodes** with six docking ports connect the modules or extend the docking approach and access.

The **Japan Experiment Module (JEM)** provides a scientific laboratory for material processing and life science research. JEM includes an external, exposed research facility or "porch" and has a docking port to accommodate its own logistics module.

The ESA Research Laboratory **Columbus** is the European scientific and technological research module. In this module experiments in materials science, fluid physics, life science and technology are conducted. (NASA [ISS], 2010)

The **Cupola** is an attachment to Node 2 designed for observation of operations outside the ISS. It has six windows with protective shutters.

For the transportation of crew, modules and cargo the following space craft are used:
The **Soyuz** is the Russian crew transportation vehicle to the ISS. One Soyuz is always docked to the ISS and functions as a crew return vehicle in emergency situations. It is replaced every 6 months (NASA [ISS], 1998 p. 25).
The **Progress** is used as a resupply vehicle for cargo and propellant deliveries. The **Space Shuttle** is used to complete the assembly of the ISS and for crew transportation. The **Multi-Purpose Logistic Module** transfers payloads to the ISS on-board the Shuttle.
The European **Automated Transfer Vehicle (ATV)** is an expendable, unmanned resupply spacecraft. It supplies the ISS with propellant, water, air, payload and experiments and can re-boost the station into a higher orbit. In April 2008, ATV 'Jules Verne' docked to the ISS for the first time.

A: Assembly of the International Space Station *(credit A: ESA - D. Ducros)*

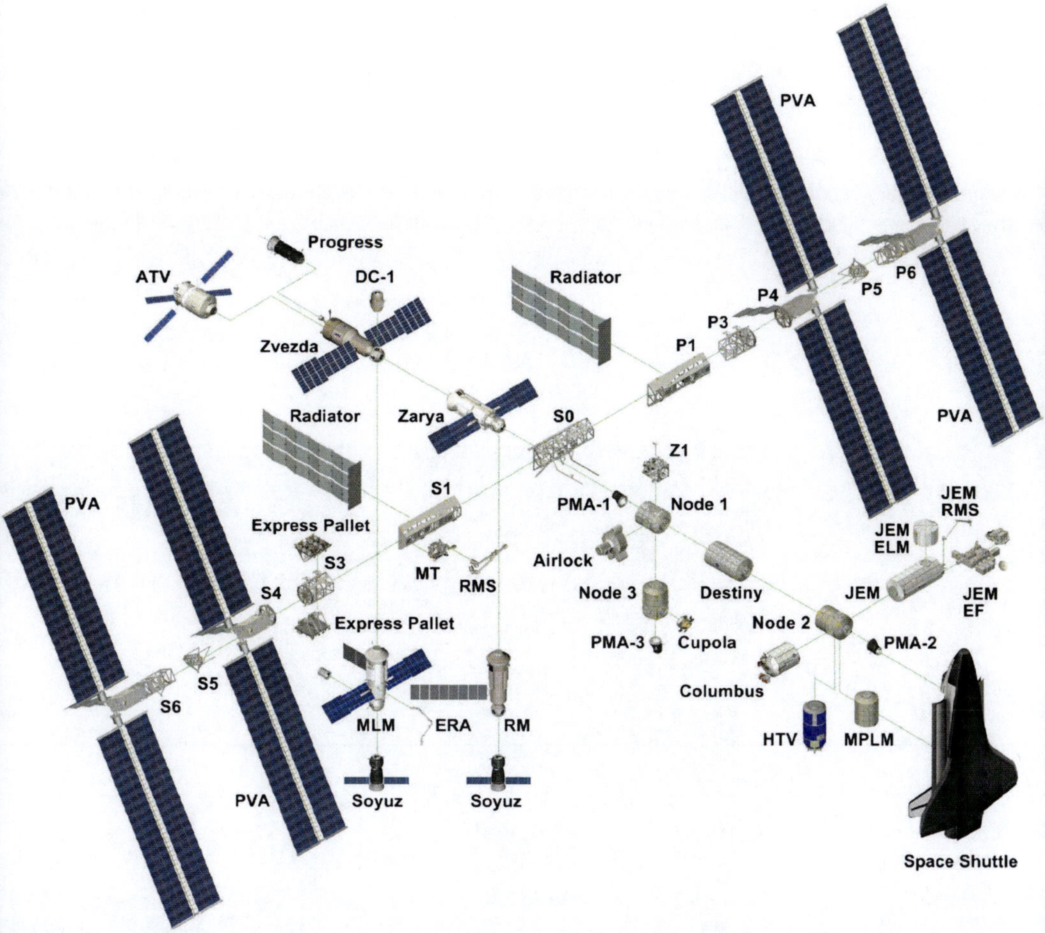

3.7 ISS SPACE STATION

CONFIGURATION AND LAYOUT

According to Kitmacher four main principles were applied for the design of the interiors of the ISS (NASA [ISS], 2006 p. 23):
- Modularity,
- Maintainability,
- Reconfigurability, and
- Accessibility

The interior layout of the U.S., Japanese and European modules is modularized at the rack level and is reconfigurable *[Fig. A]*. Racks consist of a rack structure, locker structure and stowage trays. The functional rack unit is the space in which to install a rack to the longitudinal stand-offs built into the NASA, ESA, and JAXA modules *[Fig. B]*. All the racks are 1.05m wide, but their depth and height can vary according to the module in which they will be installed. In the US Destiny Lab, the maximum dimension is 2.10m high by 0.9m deep. The space between the stand-offs in JEM and Columbus are shorter, so the "international" racks are also shorter.

The International Standard Payload Rack (ISPR) contains the research payloads. The Active Rack Isolation System (ARIS) is attached to an ISPR and isolates them from vibrations (NASA[ISS], 2006). The 'seat-track' derives from commercial airline cabin floors where they are used to attach the seats. On ISS, the seat track provides a line of anchor points on the front vertical columns to attach hardware to the front of U.S., European, and Japanese racks (NASA [ISS], 1998 p. 285).

In total 34 research racks are installed on the station (13 in the U.S. Laboratory, 11 in the Japa-

A:

A: Astronaut Greg Chamitoff, Expedition 17 relocates an experiment rack in the Kibo laboratory
B: International Space Station Rack Configuration *(credit A: NASA; B: Author, based on NASA documents)*

nese Experiment Module and 10 in the European Columbus Research Laboratory). Additional racks are used in the nodes to accommodate life support and other infrastructural functions.

The Russian modules provide stowage space behind the panels along the walls that are either hinged doors or screwed to the structure. The Russians have specially designed items, such as for food and clothes (NASA [ISS], 1998 p. 293).

External research accommodations and mounting spaces are provided on the U.S. Truss, the Russian Service Module, on the Japanese Experiment Module and on the Columbus modules. Payloads are accomodated on special racks (EXPRESS) that are mounted with the ISS's robotic arm.

B:

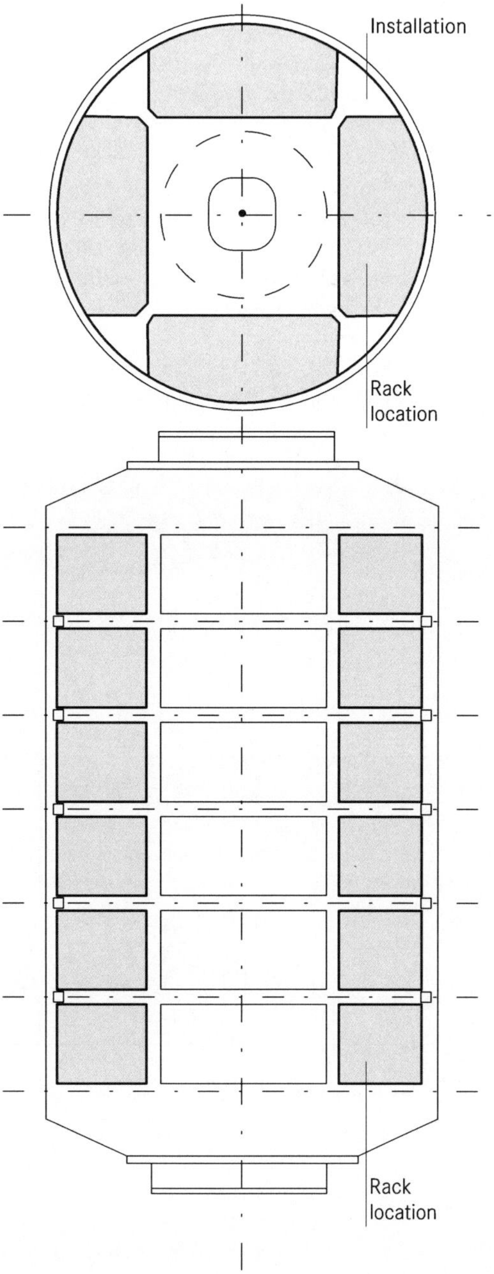

3.7 ISS SPACE STATION

CONFIGURATION AND LAYOUT

Zvezda Module

The Service Module (Zvezda) provided the early habitable space and is still the structural and functional centre of the Russian segment of the ISS. It provided the life-support system electrical power distribution, data processing system, flight control system and a propulsion system. Its architecture is similar to the Mir station's Core Module; it was manufactured in the same facility as the Mir modules. It provides two personal sleeping quarters, a toilet and hygiene facility, a galley with a table, and exercise facilities. Zvezda has about 14 windows (NASA [SM], 1999 p.2). The Transfer Compartment provides the spacewalk capability for the Russian Orlan-M spacesuits.

US Lab

The U.S. research laboratory, called Destiny is docked between Node 1 and Node 2. It is 8.5m long and provides 24 rack locations and additional crew support equipment. The cold storage unit 'Minus Eighty Degree Laboratory Freezer (MELF) keeps samples at cold temperatures. The module provides an optically pure telescope window with a diameter of about 50cm and an ergometer with vibration isolation.

A:

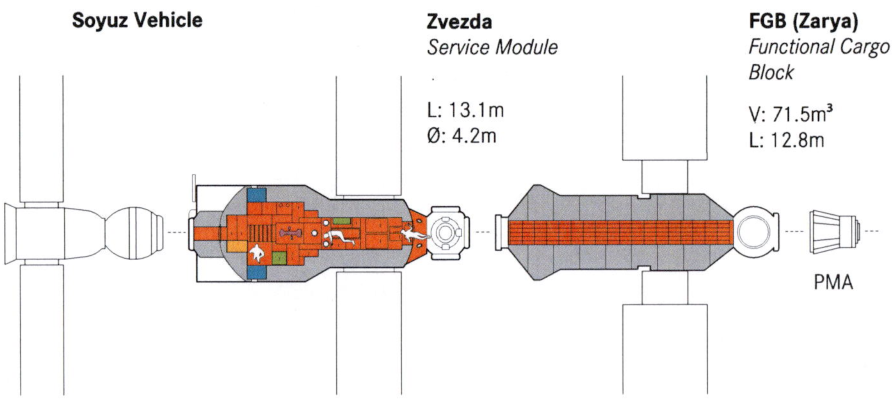

A: Graphic, ISS Modules: Zvezda, FGB, Node 1, US Lab Destiny, Columbus, JEM KIBO, Node 2, interior layout

(credit A: Author, based on NASA documents)

Columbus Module

The European research laboratory is docked to Node 2. It is 7m long and provides 16 rack locations. Science facilities include the Biolab, the European Physiology Module (EPM) and a Fluid Science Laboratory (FSL). It has an external payload facility (NASA [ISS], 2010).

Kibo

Kibo is the Japanese Experiment Module (JEM) that is docked to Node 2 on the opposite side from Columbus. It consists of the Pressurized Module (PM), the Experiment Logistics Module (ELM-ES), the Exposed Facility (EF) and the Remote Manipulator System. The pressurized module has a length of 11m with an inner diameter of 4.2m.

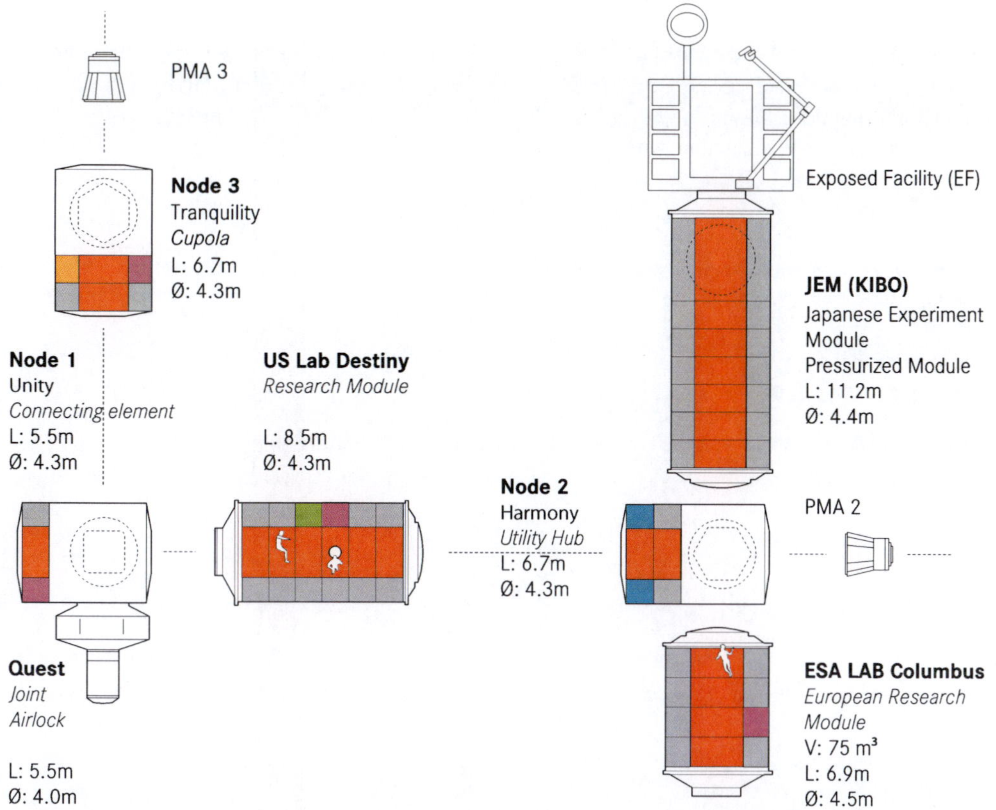

3.7 ISS SPACE STATION

TIME ALLOCATION OF HUMAN ACTIVITIES

The average workday of an astronaut is about 16 hours with a sleep period of 8 - 8.5 hours. An example schedule for astronauts at the ISS is as follows (CSA, 2006):

1.5 hours: Post Sleep
(Station inspection, hygiene, breakfast)

0.5 hours: Planning and coordination
(Daily planning conference and status report)

2.5 hours: Exercise
(Cardiovascular exercise with the cycle ergometer or treadmill; strengthening exercises, equipment set-up, personal hygiene)

1 hour: Lunch

1.5 hours: System operations
(Reports, emails, trash collection, work preparation)

6.5 hours: Work
(Assembly, maintenance, medical operations, on-board training, routine operations)

2 hours: Pre Sleep
(Evening meal, hygiene)

8.5 hours: Sleep

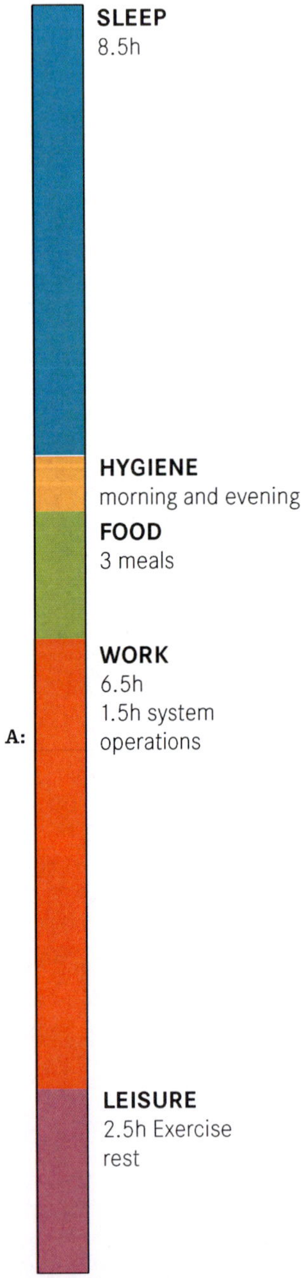

A: Graphic, Time Allocation of Human Acitivites: ISS
B: Graphic, Spatial Allocation of Human Activities: ISS *(credit A, B: Author)*

SPATIAL ALLOCATION OF HUMAN ACTIVITIES

Each module of the ISS has a specific function. The U.S. Lab, JEM, and Columbus module provide the scientific research facilities. The U.S. Module and the Zvezda Service Module provide the living facilities. Two sleeping compartments are provided in the Zvezda module. Four crewquarters are installed in Node 2. Two waste and hygiene compartments are provided, one in the Zvezda module and one in Node 3. Exercise equipment is located in Zvezda, Node 1, Node 3, the US Lab and the Columbus module.

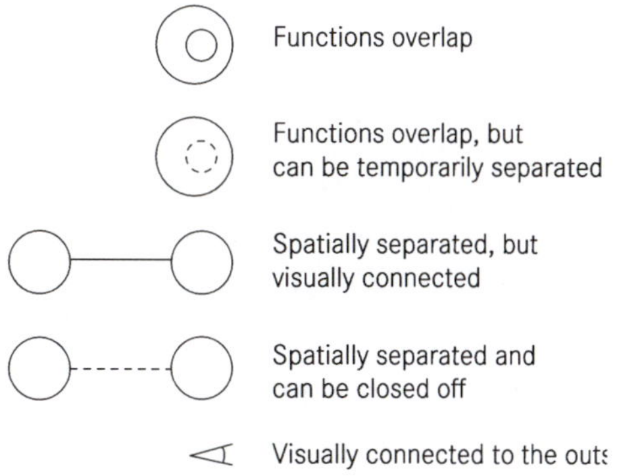

B:

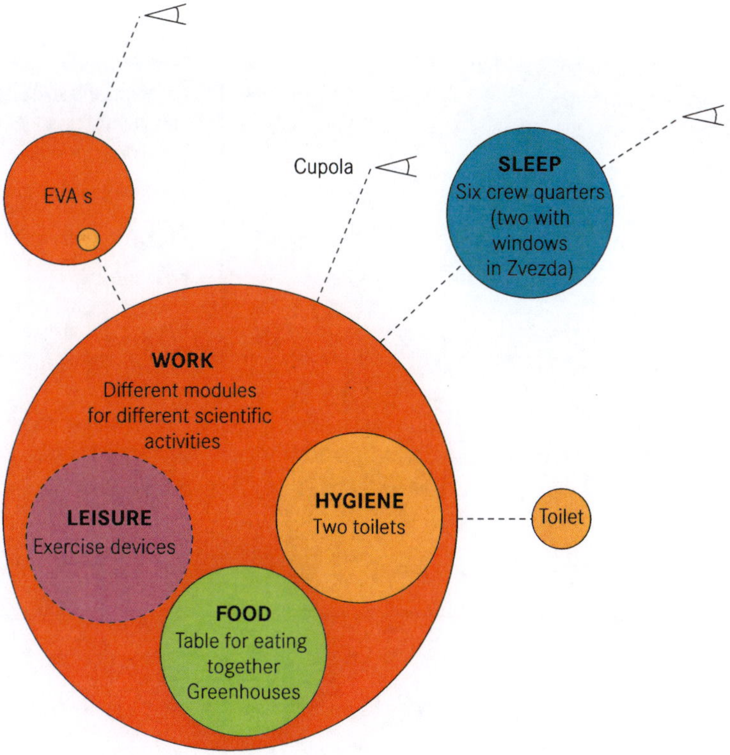

3.8 SUMMARY

A: Cupola view, image taken by the ISS Expedition 23 Crew *(credit : NASA, The Gateway to Astronaut Photography of Earth)*

Today we look back to a series of realized and inhabited space habitats: Apollo Spacecraft and Lunar Module, the Space Shuttle Orbiter, and the Space Stations; Salyut, Skylab and Mir. The International Space Station has been operational for more than ten years and it has become the largest habitable space outpost ever built. The experiences and lessons learned offer a great opportunity for the research of the relationship between humans and their inhabited space in an extreme environment.

Following the introduction of relevant Russian, American and International extra-terrestrial habitats, the tables on the following pages summarize selected issues.

The comparison **„Architectural Concept"** [3.8.1] summarizes issues related to:

- Architectural Configuration
- Spacecraft Autonomy
- Life-Cycle and Maintenance
- Crew Autonomy

The comparison **„Interior Layout"** [3.8.2] summarizes issues related to:

- Habitable Volume
- Mission Lengths
- Crew Size
- Spatial Orientation
- Restraints

The comparison **„Spatial Allocation of Human Activities"** [3.8.3] summarizes issues related to the human activity zones within the space habitat:

- Sleep
- Hygiene
- Food
- Work
- Leisure

3.8.1
ARCHITECTURAL CONCEPT COMPARISON

APOLLO	SALYUT	SKYLAB
Architectural configuration		
Single Volume; Command and Service Module; Lunar Lander; Lunar Rover	Single element space station with docking ports	Single element space station with multiple docking adapter; Automated launch and configuration
Spacecraft autonomy		
CO_2 chemically removed (with consumable) No O_2 generation CSM: Fuel cells provided electricity and drinking water	CO_2 removal Salyuts 1-5 no O_2 generation Salyuts 6-7, O_2 production Water Processing Solar panels and treadmill	CO_2 chemicallly removed (via regenerative source) No O_2 generation No Water Processing Solar panels and fuel cells
Separate LSS for CSM and LM		Partially closed loop system (2 canister molecular sieve)
Life cycle and Maintainance		
CSM and LM used once, no growth potential	Space station series; gradual change of hardware	Used once, limited growth potential
Operational: 1967-72 11 missions, 6 lunar landings Inhabited: 1968–72, 8–12 days	In orbit: 1971-91 Inhabited : 1971 - 1986 Salyut 7: > 800 days	In orbit: 1973–1979 Inhabited: 1973-74 28, 59 and 84 days
Low maintenance, with the exception of the Apollo 11 accident	High maintenance and extensive development of hardware	
Crew autonomy		
Almost constant communication (line of sight)	Limited communication with mission control	
Task-related schedule; Crew had some autonomy	Strict schedule	Strict schedule (was not always followed by the crew)

sources: Eckhart, 1996; Larson et al., 2000; NASA [ISS], 2010; NASA [ECLSS], 2009

SHUTTLE	MIR	ISS
Reusable launcher for use in 1G and microgravity	Modular space station; Automated launch and configuration	Modular space station with truss-backbone structure; Shuttle and robotic arm is needed for configuration in orbit
C02 chemically removed (with consumable) No 02 generation No Water Processing Fuel cells generate electricity Self-contained, short mission duration	C02 removal 02 production Water Processing Solar panels Recycling of moisture condensate, wash water and urine	C02 removal 02 recovery and generation Water processing Solar panels and cells to generate electricity Recycling of air (4 bed molecular sieve), wastewater, urine to produce drinking water; production of oxygen
Re-usable, limited growth potential (cargo bay) In orbit: since 1981 Inhabited: 7 - 17 days Low maintenance on orbit, maintained on Earth	Expandable, gradual change of hardware, multiple modules In orbit: 1986-2001; 15 years Inhabited: 1986 - 2001 4.594 days Extensive maintenance but long life cycle	Expandable, concept development of inflatable modules In orbit: since 1998 Inhabited: since 2000, since 2009 by a permanent crew of six
Almost constant communication to ground Low autonomy	Limited communication with mission control Strict schedule (was not always followed by the crew)	Constant communication and monitoring via telemetry Strict experiment schedule, many experiments are conducted with ground support or from ground; Low autonomy

3.8.2 INTERIOR LAYOUT COMPARISON

	APOLLO	SALYUT	SKYLAB
Habitable Volume	CM: $6m^3$ LM: $4.5m^3$	$90 - 100m^3$	$320m^3$ OWS: $270m^3$
Mission Lenghts	Longest mission: 12d 17h 12m (Apollo 15) Longest mission on Lunar Surface: 3d 02 h 59m 40 s (Apollo 17)	Longest mission: 237 d (Salyut 7)	Three missions 28, 59, and 84 days Longest mission: 83d (SL-4)
Crew Size	CM: 3 (LM: 2)	2-3, + visiting crews	3
Spatial Orientation	CM: horizontal and vertical LM: horizontal Restraints: seats in the CM and sleeping bags	Mainly horizontal: floor, wall, ceiling; Exceptions: exercise equipment, windows; Use of colour code to support spatial orientation Restraints: early stations had seats	OWS: horizontal, defined floor, wall and ceiling; Upper deck: free space; Exceptions: Toilet; was located on the wall Restraints: triangular grid system (floor, wall), shower restraints, table restraints, foot restraints

sources are listed in the Appendix

SHUTTLE	MIR	ISS
$66m^3$	$380m^3$ Core Module: $90m^3$	At assembly complete: $935m^3$ (pressurized volume)
7 - 17d		Varies, 6 months
Longest mission: 17d 15h 53m 18s (STS-80)	Longest mission: 438d (Polyakov on EO-3/EO-4)	Record for the longest uninterrupted human presence in space
7 (typical) + 3 (emergency)	Resident crew of 2, + visiting crews, later up to 6 (not including Space Shuttle missions)	Initial crew of 3, since 2009 nominal permanent crew of 6, + visiting crews
Horizontal, two levels; For use in 1G and microgravity	Similar to the Salyut space stations, mainly horizontal; Use of colour code	Mainly horizontal
Restraints: multiple restraints (sleep, toilet, food, work, exercise)	Restraints: Multiple restraints (sleep, toilet, food, work, exercise)	Restraints: Multiple restraints (sleep, toilet, food, work, exercise); Cupola crew restraints

3.8.3 ALLOCATION ACTIVITIES COMPARISON

APOLLO

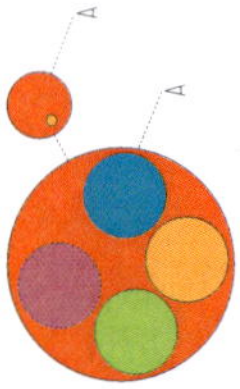

SALYUT

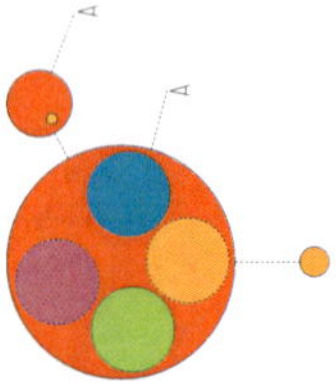

SKYLAB

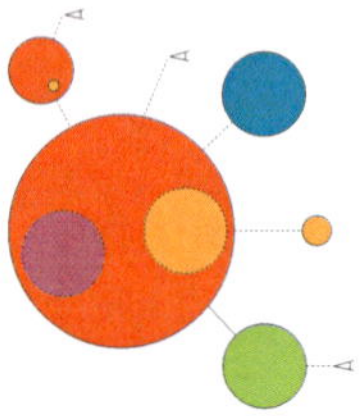

	APOLLO	SALYUT	SKYLAB
Sleep	In main module	In main module	Spatially separated in private crew quarters
Hygiene	In main module	Deployable shower and personal hygiene in work area; Toilet close to work area	Collapsible shower in work area; Spatially separated hygiene area
Food	In main module	Wardroom with table in work area	Spatially separated wardroom for preparation and eating of food
Work	In main module; outside on the lunar surface (EVAs)	Instrument area could be partitioned from living area; EVAs	Experiment area and Dome in the OWS; EVAs
Leisure	In main module; outside on the lunar surface	Exercise and recreation in main module	Dedicated are for exercise in work area, private crew quarters

SHUTTLE

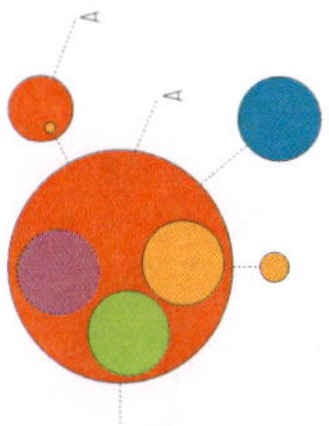

MIR

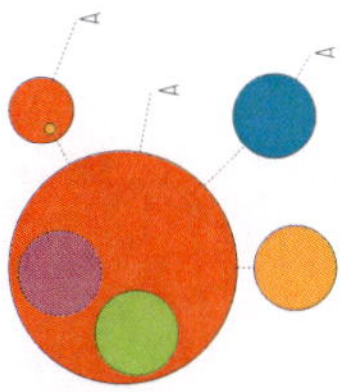

ISS

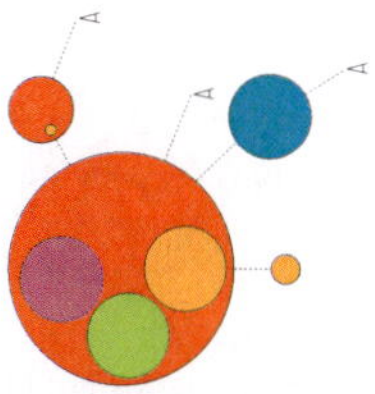

SHUTTLE	MIR	ISS
Depending on mission; in main module (sleeping bags) or dedicated spatially separated area (sleep boxes)	Spatially separated in individual cabins	Spatially separated in individual crew quarters
No dedicated area for advanced personal hygiene; Separate toilet area	Permanent shower in Kosmos; Toilet with curtain in core module	No shower; Two toilet compartments
No dedicated area to eat; Galley rack for food preparation	Food cabinet with table in work area	Food cabinet with table for all astronauts
In main volume; EVAs	In core module and dedicated science modules; EVAs	Dedicated modules and rack system; EVAs
Exercise and recreation in work area (Middeck, Flight Deck)	Exercise in work area, but in different modules; Recreation in individual cabins; other modules	Exercise In work area, but in different modules; Recreation in crew quarters

3.9 REFERENCES

Additional references on Space Station Facts are listed in the Appendix.

B.J.Bluth, M. Helppie. 1987. Soviet Space Stations as Analogs. NASA Headquarters Washington, USA : NASA Grant, 1987. NAGW-659.

Boeing. 1997. Middeck Interface Definition Document (Space Flight Operations Contract). Houston, Texas, USA : NASA Johnson Space Center, Subcontractor Boeing North American, Inc., 1997. NSTS-21000-IDD-MDK.

Burrough, Bryan. 1999. Dragonfly - NASA and the crisis aboard Mir. London, GB : Fourth Estate Limited, 1999. ISBN 1 84115 087 8.

Caprara, Giovanni. 2000. Living in Space - From Science Fiction to the International Space Station. Milan, Italy : Firefly Books Ltd. 2000. ISBN 1-55209-549-5

Cohen, Marc M. 1986 August. Lightweight Structures in Space Station Configurations. In, Proceedings of the First International Conference on Lightweight Structures in Architecture (vol. 1, p. 507-541). University of New South Wales, Australia, 24-29 August 1986: Unisearch, Ltd.

CSA. 2006. Living in Space - An Astronaut's Schedule. [Online] Canadian Space Agency, 18 08 2006. [Cited: 02 10 2009.], http://www.asc-csa.gc.ca/eng/astronauts/living_relaxing.asp#workday.

Culbertson, Frank L. Jr. Director, Shuttle-Mir (Phase 1) Program. 2004. Spaceflight Nasa - The Meanings of Mir. [Online] 4 April 2004. [Cited: 28 March 2008.], http://spaceflight.nasa.gov/history/shuttle-mir/references/r-documents-mirmeanings.htm.

DLR. 2006. From Cold War to international cooperation - the story of the ISS. [Online] DLR, 24 July 2006. [Cited: 12 December 2009.], http://www.dlr.de/iss/en/desktopdefault.aspx/tabid-1945/2746_read-4182/.

Eckhart, Peter. 1996. Spaceflight Life Support And Biospherics. U.S.A. : Space Technology Library, 1996. ISBN 1-881883-04-3.

Harrison, Albert A. 2001. Spacefaring - The Human Dimension. Berkeley/Los Angeles/London : University of California Press, 2001. 0-520-22451-1.

Haeseler, Dietrich. 1998. Almaz Space Station - Inside Views of a Full-Scale Training Model. SPACEFLIGHT. Vol. 40, 1998, Vol. July.

Jones, Eric M. 1995. Apollo 12 - Post-EVA Activities in the LM.http://history.nasa.gov/alsj/a12/a12.posteva1.html : Apollo 12 - Lunar Surface Journal, 1995.

Kennedy, John F. May 25, 1961. May 25, 1961.

Larson, J. Wiley and Pranke, Linda K. 1999. Human Spaceflight - Mission Analysis and Design. New York, USA : Mc Graw Hill, 1999. ISBN 0-07-236811-X.

Linenger, Jerry M. 2000. Off The Planet - Surviving five periolous months aboard the Space Station MIR. U.S.A. : McGraw-Hill, 2000. ISBN 0-07-136112-X.

Loewy, Raymond. 1980. Industrial Design. London, GB : British Library Cataloguing in Publication Data, 1980. ISBN 0-947795-38-3.

McDonald, Sue. 1998. Mir Mission Chronicle, November 1994 - August 1996. Houston Texas, USA : NASA, Lyndon B. Johnson Space Center, 1998. NASA TP-1998-208920.

NASA [Bull.10]. 1974. Skylab Experience Bulletin No. 10: Body Restraint Systems. Houston, Texas, USA : NASA, Lyndon B. Johnson Space Center, 1974. JSC-09544.

NASA [Bull.11]. 1975. Skylab Experience Bulletin No. 11: Personal Mobility Aids. Houston, Texas, USA : NASA, Lyndon B. Johnson Space Center, 1975. JSC-09545.

NASA [ECLSS]. 2009. Environmental Control and Life Support System (ECLSS), System Engineering Workshop, Presenter: L. Peterson : Life Sciences Department. ISU SSP 2009, Ames Research Center, USA

NASA [Facts]. 2002. Kennedy Space Center - Space Shuttle Facts & Statistics. [Online] November 2002. http://www-pao.ksc.nasa.gov/kscpao/factoids/shfacts.htm#edo.

NASA [Factsheet]. 2006. Living and Working in Space - Factsheet. Houston, Texas, USA : NASA Lyndon B. Johnson Space Center, 2006. FS-2006-05-019-JSC.

NASA [ISS]. 1998. International Space Station Familarization. Houston Texas, USA : Lyndon B. Johnson Space Center, 1998. TD9702A.

NASA [ISS]. 2010. Reference Guide to the International Space Station. Assembly Complete Edition. Washington, DC, USA : Library of Congress Cataloging-in-Publication Data, 2010. NP-2010-09-682-HQ

NASA [ISS]. 2006. Reference Guide to the International Space Station. [ed.] Gary, editor Kitmacher. Washington,

DC, USA : Library of Congress Cataloging-in-Publication Data, 2006. ISBN 0-9710327-2-6.

NASA [Reagan]. 2007. History. [Online] NASA, 8 August 2007. [Cited: 5 July 2009.] http://history.nasa.gov/printFriendly/reagan84.htm.

NASA [Shuttle]. 2010. Space Shuttle: News, Reference Information, and detailed mission profiles for all missions. [Online] 2010, http://www.nasa.gov/mission_pages/shuttle/main/index.html.

NASA [Shuttle Reference]. 1981. Space Shuttle News Reference. U.S. Government Printing Office, 1981. 723-000/562.

NASA [Shuttle-Mir]. 2004. Shuttle-Mir Programm, Goals. [Online] Curator: Kim Dismukes, 04 04 2004. http://spaceflight.nasa.gov/history/shuttle-mir/welcome/goals.htm.

NASA [Skylab]. 1980. Skylab Reactivation Mission Report. Systems Analysis and Integration Laboratory, Science and Engineering Directorate. Alabama: George C. Marshall Space Flight Center, 1980. NASA TM-78267.

NASA [Skylab]. 1977. Skylab, our first space station. Washington D.C., US : NASA, Scientific and Technical Information Office, 1977. Prepared by George C. Marshall Space Flight Center, Edited by Leland Belew. NASA-SP-400.

NASA [SM]. 1999. International Space Station: The Service Module: A Cornerstone of Russian International Space Station Modules. Houston, Texas, US: Lyndon B. Johnson Space Center, 1999. IS-1999-09-ISS019JSC.

NASA. 1971. Apollo Operations Handbook - Extra Vehicular Mobility Unit. Houston, Texas, USA : Manned Spacecraft Center, 1971. MSC-01372-1.

NASA KSC. 2000. Project Skylab. [Online] 25 November 2000. [Cited: 15 October 2008.], http://science.ksc.nasa.gov/history/skylab/skylab.html.

OTA. 1983. SALYUT - Soviet Steps Toward Permanent Human Presence in Space. Washington D.C. : U.S. Congress, Office of Technology Assessment, 1983. Andelin, John; Naismith, Nancy; Mills, William; Rogers, Thomas F.; Chandler, Philip P.. OTA-TM-STI-14.

Portree, David S.F. 1995. MIR Hardware Heritage. Houston, Texas, USA : NASA, Lyndon B. Johnson Space Center, Information Services Division, 1995. NASA RP 1357.

Schmitt, Harrison. 2009. Transcript not published. [interv.] Sandra Häuplik-Meusburger. Vienna, Austria, 2009.

Toussaint, Marc. Russian Manned Space Programme & Russian Space Stations Evolutions from Salyut 1 to MIR. ESA-HQ Paris. ESA-HQ Pars : date not specified.

U.S. Congress. 1983. Salyut: Soviet Steps Toward Permanent Human Presence in Space - A Technical Memorandum. Office of Technology Assessment. Washington, D.C., USA : Library of Congress Catalog, 1983. OTA-TM-STI-14.

Vasilyev, M. P., et al. 1974. Salyut Space Station in Orbit, Translation of ‚Salyut na Orbite', Mashinostroyeniye Press, Moscow, 1973, 160 pages. Washington, D.C., USA : NASA, National Aeronautics and Space Administration, 1974. NASA TT F-15,450.

Wikipedia [Mir]. 2010. Mir Space Station. [Online] Wikipedia, 10 October 2010. http://en.wikipedia.org/wiki/Mir.

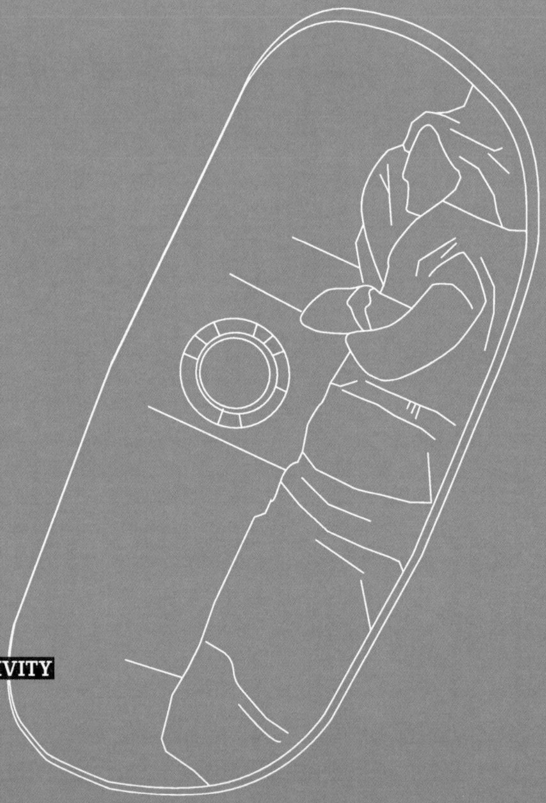

PART 4: HUMAN ACTIVITY

SLEEP

The human activity category ‘Sleep’ includes the sub-activities rest; preparation for sleep, relaxation, sleep as well as associated translation paths and stowage areas. Sleep concepts applied within the selected habitats range from sleeping bags, to hammocks and boxes to private crew quarters.

4.1 SLEEP APOLLO

CONCEPT Sleeping bags, "curled up", couches (CM), hammocks (LM)
LOCATION Command Module/Lunar Module
SPECIFIC Not a priority

RÉSUMÉ

The **Command Module** was equipped with sleeping bags underneath the couches. The centre couch could be moved out of the way during flight to provide space for three astronauts next to each other.

The **Lunar Module** used for Apollo 11 had no sleep provisions. But they were not needed, because according to Jones (1995 pp. 24-25), Neil Armstrong got no sleep at all and Buzz Aldrin had about two hours "curled up in corners". Besides excitement, reasons for the bad sleeping conditions were the high level of noise, low temperature and the light, according to Armstrong "light came through those window shades like crazy" (NASA [Debriefing A11], 1969 p. 81).

Starting with the Apollo 12 mission, the lunar module was equipped with two hammocks and an additional heater to increase the cabin temperature for the astronauts. Before sleeping, the hammocks had to be deployed one to be hung up close to the ceiling, the other one beneath it across the cabin. They slept on top of each other in a stretched out position *[Fig. C]*. The hammock base was insulated and the astronauts used additional blankets. Apollo 12 astronaut Bean reported a problem with tightening the Beta cloth covering his hammock (NASA [Debriefing A12], 1969).

During sleeping periods, astronauts tried to minimize noise with the help of earplugs and by turning the environmental control system down (NASA [Report A15], 1971). Apollo 11 and 12 astronauts took off their shoes while sleeping, but left their space suit and helmet on, not only because of safety reasons. With the helmet on it was quieter and breathing-in of hazardous lunar dust was prevented (NASA [Debriefing A11], 1969). After the extra-vehicular activities their suits and shoes were covered with dust (NASA [Report A15], 1971).

A:

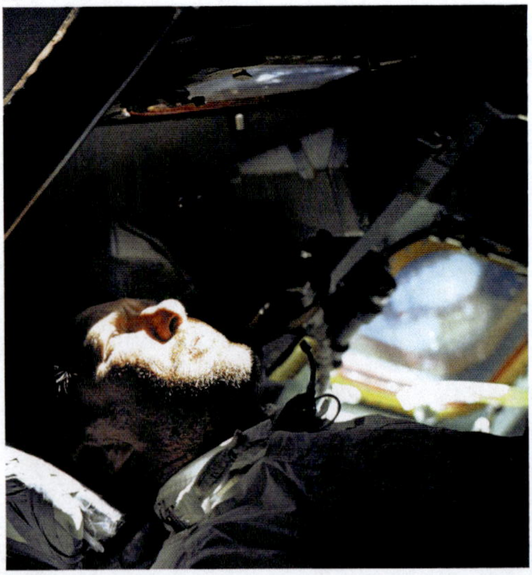

B:

A: Apollo Astronaut Walter M. Schirra looks out of the rendezvous window
B: Three astronauts inside CM Simulator
C: Sleep positions in the Lunar Module
D: Eugene Cernan sleeping aboard Apollo 17 spacecraft

(credit A, D: NASA; B: NASA Johnson Space Center; C: Original Image from NASA, adapted by the Author)

ASTRONAUTS' EXPERIENCES

USABILITY

Alan Bean (Apollo 12): *"I felt like I was tired, towards the end of the second EVA and I felt like it wasn't from the physical effort. It was from the* ***lack of good sleep****. I didn't take the pill because it was not a macho thing to do. But, looking at it, I would do it now. In the future, people ought to do everything they can - including that - to get some good sleep, so they've got their maximum energy the next day. I don't feel like I had it. I felt like I was really running out of gas. I think I didn't sleep well because I was just nervous and excited. I think it was just being hyper and being worried about it going well."* (Jones, 1995 p. 25)

LIVABILITY

Michael Collins (Apollo 11): *"Speaking of window shades (...) I like to have my sleeping accommodations dark, as dark as I can get them (...) They [window covers] were quite difficult to install (...) They were tight, but not nearly as tight as they were on flight".* (NASA [Debriefing A11], 1969 p. 50)

Gene Cernan (Apollo 17): ***"What a waste of time****. My mind whirls as I lay in the hammock, wide awake. I was mentally and physically whipped, but felt I should not be loafing around in my underwear while there was a whole Moon to explore just beyond that little hatchway. We only had about sixty hours left, and time had warped. When we were outside, the hours just galloped away, but inside the spacecraft, the clock didn't seem to move at all, and our rest period passed with agonizing slowness. Eventually, we slept."* (Cernan, et al., 1999 p. 330)

FLEXIBILITY

Alan Shepard (Apollo 14): *"We did find that we had to take the boots off because there's so much dust in your overshoes that we did take those off before we went to bed."* (NASA [Debriefing A14], 1971 p. 172)

Harrison Schmitt (Apollo 17): ***"Sleep is a very individual thing."*** (Schmitt, 2009)

C:

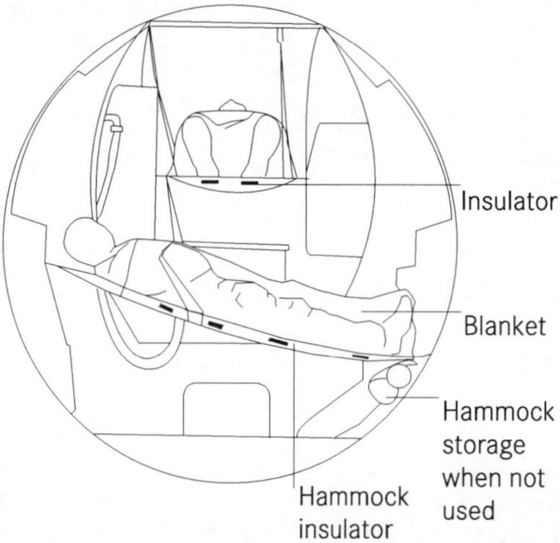

D:

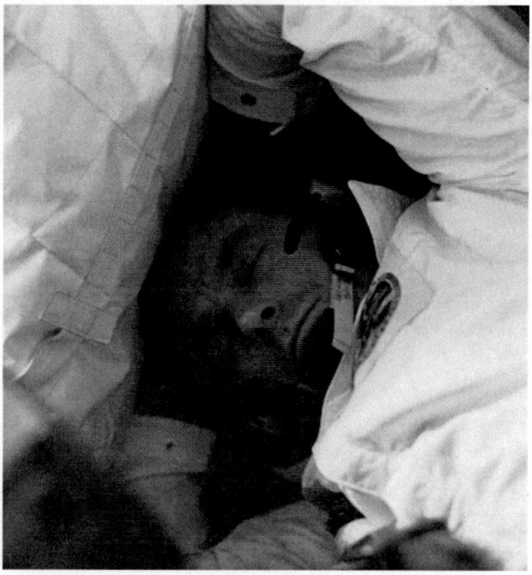

4.2 SLEEP SALYUT

CONCEPT Sleeping bags with sheet inserts, bunks
LOCATION Large-diameter Work Compartment, on the wall; Orbital compartment
SPECIFIC Screening of sleeping area

RÉSUMÉ

Two sleeping areas were located port and starboard on the ceiling in the large-diameter area of the Work Compartment. In addition cosmonauts could sleep in the orbital compartment of the Soyuz spacecraft. The location of the sleeping place was also designed to have a good overview of the station (Vasilyev, et al., 1974).

The cosmonauts slept in insulated sleeping bags with bed-sheet inserts. The sleeping bags were equipped with air vents and snapped fasters on all sides that could be opened if a cosmonaut felt hot. For sleeping, cosmonauts took their clothes off, but put on winter boots, because weightlessness prevents normal blood circulation and their feet got cold (B.J. Bluth, 1987 p. III-37; Lebedev, 1990). Sometimes sleeping aids were taken.

On **Salyut 1** cosmonauts slept in sleeping bags attached to the walls of the large-diameter compartment or in the orbital module of the docked Soyuz. (Portree, 1995 p. 67)

On **Salyut 3** cosmonauts slept using one standing bunk and one foldayway bunk. (Portree, 1995 p. 69)

On **Salyut 6**, Ryumin's body height caused problems. “He had to strap his bed to the floor and tuck his arms in, because he was unable to squeeze into his allotted sleeping area on the ceiling.” (B.J. Bluth, 1987 p. III-37)

On **Salyut 7**, blinds to close off the sleeping areas from the living quarters were installed. This screening was similar to mosquito netting, to avoid breathing in dust or mechanical particles during sleep. (B.J. Bluth, 1987 p. III-37)

A:

A: Cosmonauts Kovalionok, Prunariu and Savinykh in their sleeping bags on board Salyut 6 *(credit A: Dumitru-Dorin Prunariu)*

ASTRONAUTS' EXPERIENCES

USABILITY

Valentin Lebedev (Salyut 7): *"Two or three years ago I heard some proposals that each crew member should have a separate cabin. I don't agree, at least not with only four crew members.* ***You can't separate people from each other****. The closeness during long term missions is crucial. Crew members can act faster in emergencies."* (Lebedev, 1990 p. 257)

LIVABILITY

Vladislav Volkov (Salyut 1): *"Our sleeping areas resemble a beehive (in the forest) where bees are flying in and out. There are also small openings into which we crawl when sleep time comes and emerge when we hear the command to wake up (this means the duty officer awakens you by touching your shoulder and sometimes your head)."* (Vasilyev, et al., 1974 p. 139)

Valentin Lebedev (Salyut 7): *"Ten p.m. and I haven't slept at all. Constant thoughts. Thoughts about my house, my flight, my work, and my friends. I need to sleep at least a couple of hours, but I can't."* (Lebedev, 1990 p. 148)

Mikhailovich Grechko (Salyut 6, 7): *"Because there are too many new prospects in space, so much is interesting that* ***I don't like eating or sleeping****. I like experiments and my duty was engineering (...)"* (Grechko, 1993)

FLEXIBILITY

Valentin Lebedev (Salyut 7): *"Sleeping on board is fun.* ***You can sleep here in any position****: standing, upside down, or on the ceiling. Since you don't feel any gravity, even if you lie on the side the whole night long, your side won't go to sleep. You sleep without tossing and turning."* (Lebedev, 1990 p. 112)

Dumitru-Dorin Prunariu (Salyut 6): "Everyone could choose any place inside to sleep. You have the sleeping bag and you just anchor yourself so you don't float. (Prunariu, 2011)

4.2
SLEEP SKYLAB

CONCEPT Private crew quarters
LOCATION OWS, crew quarters
SPECIFIC Vertical, against-the-wall
0.7m x 0.96m x 1.98m (~1.3m³)

RÉSUMÉ

Skylab was the first space-station that provided private crew quarters. The sleep compartments were situated in the Orbital Workshop (OWS) Crew Deck, next to the waste compartment, the wardrobe and the experiment work area and provided accommodation for three astronauts. The compartment had a corridor which lead to three separated sleep areas *[Fig. A]*. (NASA [Skylab LL], 1974; NASA [Bull.3], 1974

A key feature of Raymond Loewy's design for the sleep compartments was that the floor plan for each of the three was different to create a sense of individual identity for each compartment. Elements of the crew quarters included sleep restraints, storage lockers, privacy partitions, lighting, a light baffle, privacy curtains, mirrors, towel holders and a communication box. A desk was not provided. All materials used were non-flammable and many materials were specially developed and used in Skylab for the first time (NASA [Skylab], 1977; NASA [CS], 1974).

The sleep restraint *[Fig. C]* consisted of a metal 'sleep restraint frame' that was attached to the floor and ceiling grids in the crew quarter. It could also be used in other locations, such as in the Dome or the MDA. The 'thermal back assembly' provided thermal insulation. The 'comfort restraint' and 'top blanket' was similar to a sleeping bag. In addition, a headrest and body straps were provided. (NASA [TM], 1974)

Astronauts were sleeping in an up-right against-the-wall orientation, which posed no difficulties in zero gravity (NASA [Skylab LL], 1974). Sleeping in a vertical position was acceptable, but some astronauts preferred to sleep upside down "to keep air from blowing up their noses" (NASA [Skylab LL], 1974 p. SLL2_48). Astronaut Weitz did not sleep in the crew quarters; he slept in the OWS Forward Dome (Weitz, 2000).

The sleep areas were used for sleeping and relaxation during off-duty times. Sometimes they were used for interim storage of trash bags.

A:

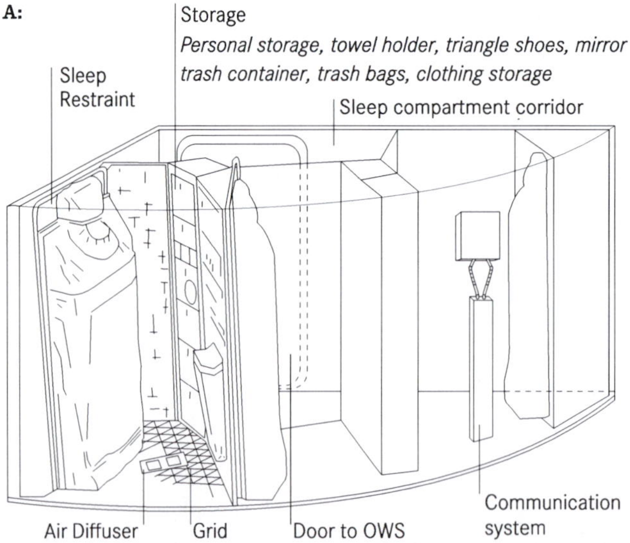

B:

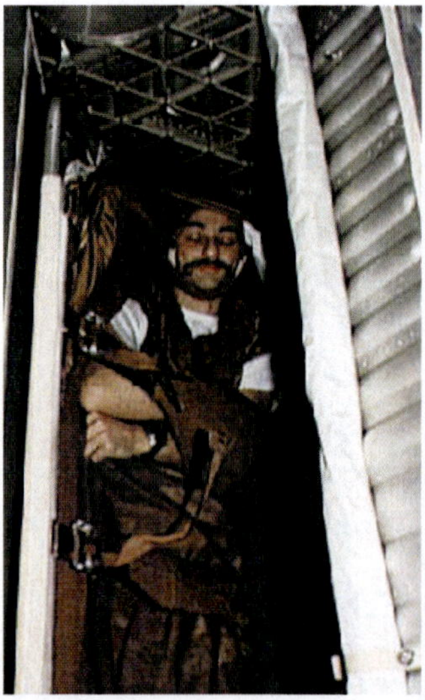

A: Crew Quarters in Skylab
B: Skylab 3 astronaut Owen Garriott in his sleep restraints in the crew quarters of the OWS
C: Skylab Sleep restraint

(credit A, C: Original Image from NASA, adapted by the Author; B: NASA)

ASTRONAUTS' EXPERIENCES

USABILITY

Jack R. Lousma (Skylab 3): *"We'd usually be awakened after about six hours of sleep or so, sometimes with music and other treatments by the ground, appropriate for the day, of course, and always exciting to understand what that was or look forward to what it was going to be. So we'd glide out of our sleeping bags and sort of get our clothes on and rendezvous somewhere there, usually around the kitchen table, and start having breakfast."* (Lousma, 2001 p. 55)

LIVABILITY

Paul Weitz (Skylab 2): *"But as far as my space, your space, we didn't do that. We didn't have that. The sleep compartments were, of course, separated, segregated (...)."* (Weitz, 2000 p. 28)

Charles Conrad (Skylab 2): *"I really had the sensation I was sleeping on a bed on the ground."* (NASA [Bull.3, 1974 p. 14)

FLEXIBILITY

Paul Weitz (Skylab 2): *"So every night I'd unbuckle my sleep restraint, which had a metal frame, and I'd take it out and I'd take it up above in the big open area and I'd stretch it across the place so I was sleeping horizontally."* (Weitz, 2000 p. 28)

Gerald Carr (Skylab 4): *"**It's got to be a place that can be modified** in the way any individual desires."* (NASA [Bull.3], 1974 p. 75)

C:

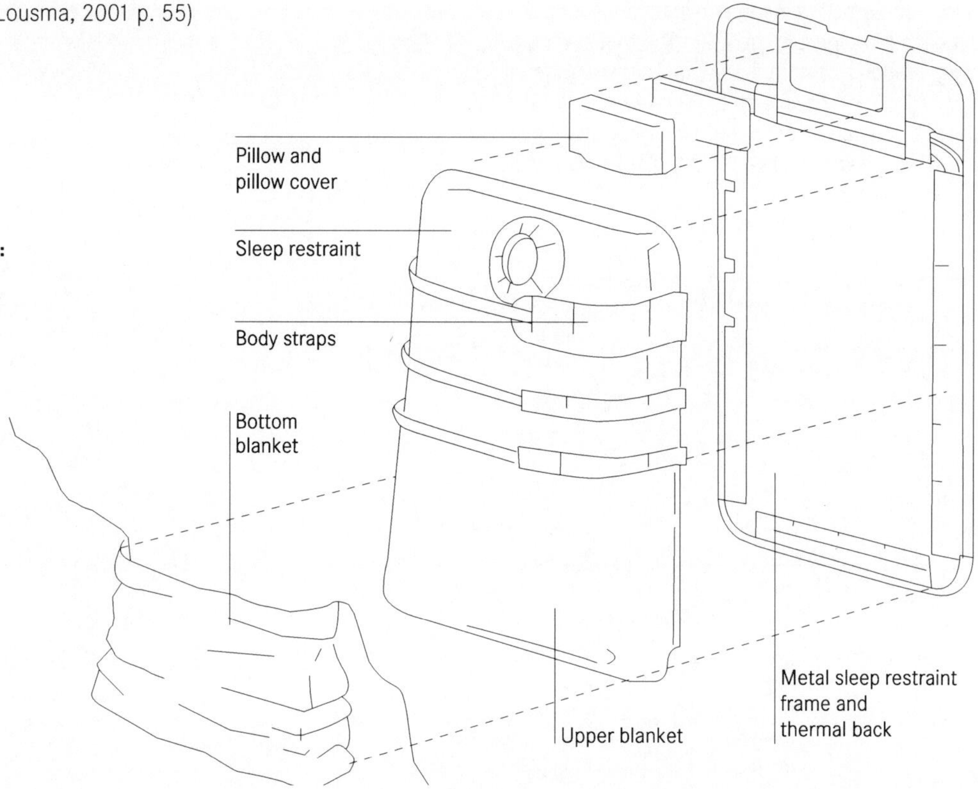

4.4 SLEEP SHUTTLE

CONCEPT Sleeping bags, rigid sleep stations, sleep restraints
LOCATION Middeck
SPECIFIC Different sleeping configurations

RÉSUMÉ

Depending on the mission, different sleeping configurations can be used *[Fig. A]*. Sleeping bags are attached to the Starboard Middeck wall. They are equipped with adjustable restraining straps, a pillow and a head restraint. Sleeping bags are made of perforated Beta material or Nomex material. They can be attached to the Middeck locker either horizontally or vertically. (NASA [Shuttle Reference], 1981) Astronauts can also sleep in the Flight Deck or in the Airlock.

If astronauts have to work shifts, the rigid sleep station compartments are installed, which are located on the starboard side of the middeck. Two types of sleep stations can be used. The three-tier rigid sleep station accommodates three crewmembers. It is made of plastic honeycomb panels and weighs about 93kg. The four-tier rigid sleep station accommodates four crewmembers and is made of metal with a weight of about 79kg. These contain a sleeping bag, personal stowage provisions, a light and ventilation (NASA [Crew Compartment], 2002).

A:

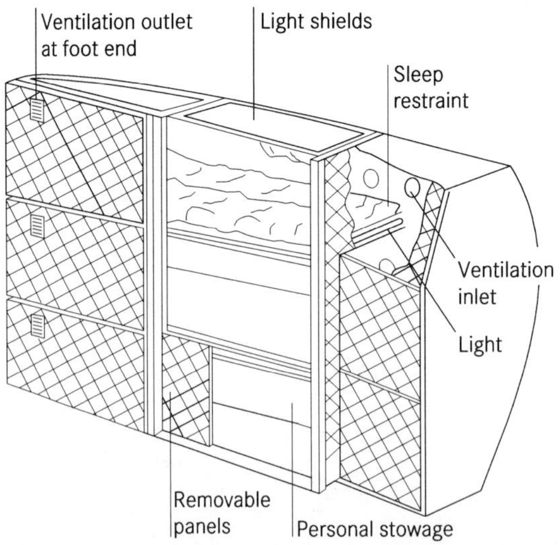

Each crewmember is supplied with eye covers and ear plugs as well as special harnesses for sleeping in microgravity.

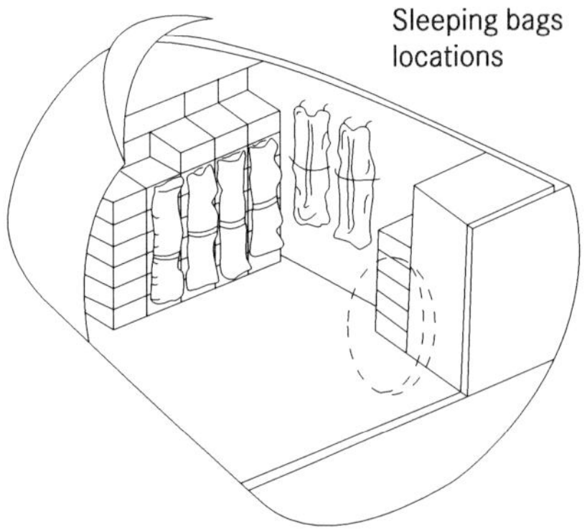

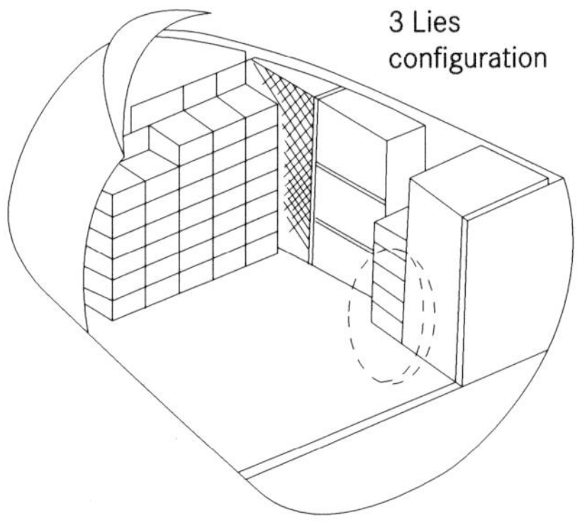

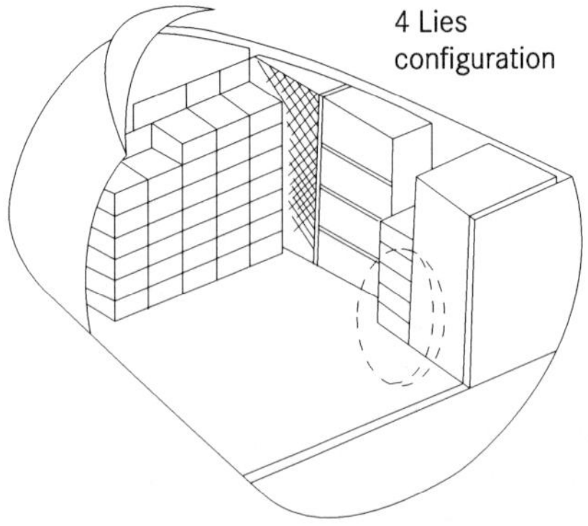

A: Sleeping configurations on the Space Shuttle
B: Jean-François Clervoy in his sleeping bag
C: STS-107 crewmembers in their sleep station on the middeck of the Space Shuttle Columbia

(credit A: Original Image from NASA, adapted by the Author, based on NASA documents; B: NASA, Jean-François Clervoy; C: NASA)

ASTRONAUTS' EXPERIENCES

USABILITY

Paul Weitz (NASA): *"(...) And I was going to sleep when I heard this little tinkling noise (...) and it was Story [Musgrave], and he's in the airlock fooling around with the suits (...) and I would doze off and then I'd wake up (...) "Story, if you're going to play all night, you're going to have to do it outside!"* (Lenehan, 2004 p. 190)

LIVABILITY

Jean-François Clervoy (ESA): *"And if you think, you are in a box, you feel like you are in a box underground ... but if you close your eyes and just feel the weightlessness, you can think, you are in an enormous volume (...)* ***you can imagine you are in a big room****. It's ok. It's comfortable."* (Clervoy, 2009)

FLEXIBILITY

Greg Chamitoff (NASA): *"Your body just naturally moves into a completely* ***natural position****, your arms gonna float up (...) where the natural position is (...) it is very comfortable."* (ReelNASA, 2008)

Gerhard Thiele (ESA): *"The reason that one cannot set up ones sleeping bag anywhere (in microgravity) is that care must be taken that one does not end up in a corner with insufficient air circulation."* (Thiele, 2010)

B:

C:

4.5 SLEEP MIR

CONCEPT Private crew cabin (kayutkas), sleeping bag
LOCATION MIR base block, Modules
SPECIFIC Vertical orientation, 2 private cabins with windows

RÉSUMÉ

The space station Mir provided two individual crew quarters (cabins, kayutkas), located in the large-diameter compartment in the Mir base block. Each crew quarter could be closed off from the work area. Later when Mir was occupied by more than two astronauts (Interkosmos and Shuttle-Mir program) other astronauts had to sleep in their sleeping bag 'somewhere' in the station. They often slept in other modules, such as Kristall, Priroda or Spektr. The commander usually slept in one of the kayutkas in order to act fast in an emergency (cf. Burrough, 1989).

Each cabin was equipped with a porthole window, a "vertical" sleeping bag, personal storage space, an integrated fold-out writing surface, intercom and a mirror (B. J. Bluth, 1987). The crew quarters allowed for personalization and cosmonauts put up personal items or pictures 'around their beds'.

A:

B:

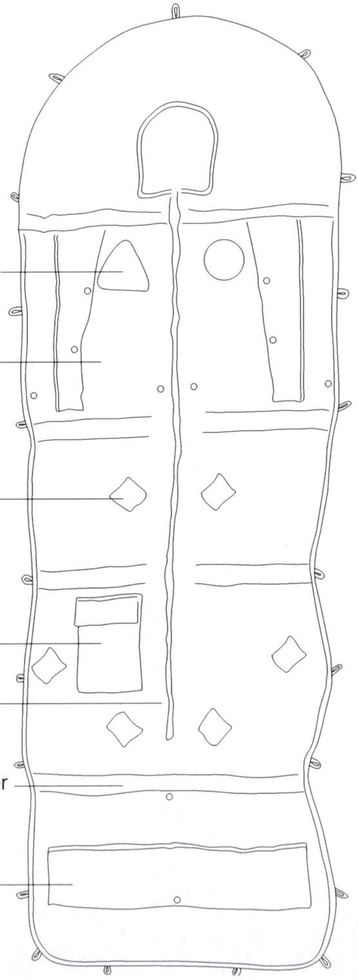

A: Astronaut Norman Thagard in the Mir crew quarters in 1995
B: Cosmonauts' sleeping bag
C: Reinhold Ewald's sleeping place in the Kristall Module *(credit A: NASA; B: Author; C: DLR/Reinhold Ewald)*

ASTRONAUTS' EXPERIENCES

USABILITY

Michel Tognini (ESA, Mir-Antares): *"You don't need a bedroom for a short mission. I don't need a bedroom."* (Tognini, 2009)

Jerry Linenger (Mir NASA 3): *"Unstrapping myself from the wall I was sleeping on and twisted to an upright position - I always slept upside down on the wall."* (Linenger, 2000 p. 90)

LIVABILITY

Mike Foale (NASA): *"At eleven pm all windows would be 'closed' (by operating external shutters) so that the regular orbit periods of daylight would not disturb their 'night's' slumber. The internal lights, always on for their work period, would be switched off."* (Foale, 1999 p. 98)

FLEXIBILITY

Sergei Avdeyev (Russian): "Sleeping in a weightless environment is very comfortable once you get used to it because any position - face up or down, on your left side or on your right - feels the same. In your head you tell yourself where is up, so you can pretend to be sleeping lying down, with a pillow under your head. But ***the concept of what is the ceiling and what is the floor is all in the mind.***" (Avdeyev, 2001)

Mike Foale (NASA): "but such was the success of the ceiling/ wall/floor configuration on Sasha's [Kaleri's] psychology he was unable to sleep 'Vertically' on the wall, so instead he anchored his sleeping bag in Kristall, which had no such orientation." (Foale, 1999 p.96)

C:

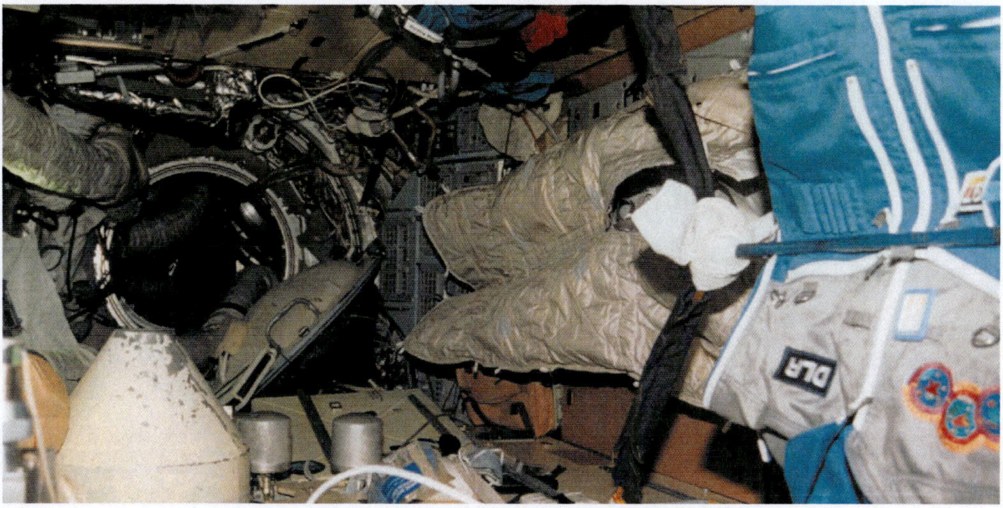

4.6 SLEEP ISS

CONCEPT	Private crew quarters, sleeping bags
LOCATION	Zvezda Service Module (2), modular rack-sized crew quarters (4)
SPECIFIC	vertical orientation; Russian kajutka with windows, TeSS Rack size 110x200x90cm ~ 2,00m^3

RÉSUMÉ

Two sleeping compartments are located in the 'Russian side' of the ISS, in the Zvezda Service Module *[Fig. C]*. Each crew cabin (kayutka) is equipped with a sleeping bag, personal stowage capacities, a laptop computer, audio devices, ventilation, individual lighting and a porthole window to look out into space. They provide space to sleep, change clothes, do personal hygiene, and can be individually decorated.

Until the end of 2008, only two permanent sleeping places were provided for three astronauts. The third astronaut could sleep wherever she or he wanted. Astronaut Susan Helms slept in the Destiny Laboratory Module, which was far from her other crewmates, who slept in the Zvezda Service Module (NASA [Sleep], 2003).

Already in the late 1990s deployable crew quarters were developed and resulted in the prototype of a Temporary Sleep Station (TeSS) which was installed in US Laboratory Module Destiny in 2001 (Broyan, et al., 2008) *[Fig. A]*. It provided enhanced acoustic noise mitigation, integrated radiation reduction material, controllable airflow, communication equipment, redundant electrical systems, redundant caution and warning systems as well as multiple crewmember restraints, adjustable lighting, controllable ventilation, and interfaces for personalization of the crew quarter (NASA [STS-128], 2009 p. 34).

In November 2008 two full functional rack-sized crew quarters were installed in the Harmony Node 2 (NASA [STS-126], 2008). Two other crew quarters were installed in the Japanese Experiment Module (JEM). The four new crew quarters can be installed in any rack-sized location; for Expedition 26 they were installed in Node 2.

Currently, the ISS provides sleep facilities for the permanent crew of six. Visitors to the station use sleeping bags.

A:

A: Expedition 26 flight engineer Catherine Coleman peeks out of her sleeping quarters, 2010
B: Sleeping Rack
C: Astronaut Paul W. Richards in front of one sleep station in the Zvezda service module, 2001

(credit A, C: NASA; B: Author, based on NASA documents)

ASTRONAUTS' EXPERIENCES

USABILITY

Tom Jones (NASA): *"I set up my bunk on Destiny's starboard wall, clipping in parallel to the deck. Roman and Beamer soon settled nearby (...) leaving us in the dim glow from the FGB a couple of dozen feet after. A few hours later I woke, chilled by the Lab's efficient air conditioner, too much even for my sweater and sleeping bag. Still zipped in the bag, I unclipped the top (...) tugged myself past my sleeping friends, (...) into the darkened shuttle middeck. It was warmer there, and I was soon asleep."* (Jones, 2006 p. 318)

LIVABILITY

Frank De Winne (ESA): *"(...) It was a little bit cramped, we had to sleep in small places;* ***it was a little bit like camping*** *and, of course, for eight days on board the ISS you can for sure do that (...)."* (NASA [Interview], 2009)

FLEXIBILITY

Pedro Duque (ESA): *"Lets add another factor:* ***it's not possible to go home to sleep****. As the lab is in a remote region, one has to sleep inside it and eat inside it - pre-cooked food. There are even those who stay in the lab for up to a year. Every now and then one feels like going out, see something else, have a walk, open the window to get some fresh air, but that is not allowed either. In this lab the windows can never be opened, and the air gets recycled through filters."* (Duque, 2003)

B:

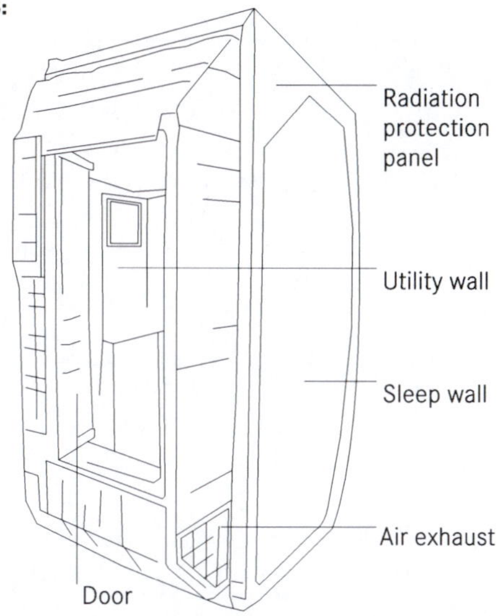

C:

4.7 SUMMARY OBSERVATION

	APOLLO	SALYUT	SKYLAB
Concept	Sleeping bag Hammocks	Sleeping bags attached to the wall	Permanent private crew quarters with sleeping bags in vertical position, private storage and communication
Review	"What a waste of time" (Cernan, Apollo 17)	Netting to close off areas were installed, but total separation of crewmembers was rejected due to the possibility of emergencies	Crew quarters were basically satisfactory; Astronauts requested flexibility on the restraint system and in blanket arrangements
Potential	Hammocks may be an option for additional sleeping possibilities in off-nominal situations or for short term missions (eg. Rover missions)	Cosmonauts wanted to be close on short missions. Could also be a topic during future exploration missions	Individual thermal control and flexibility in blanket arrangement should be integrated

Diagram

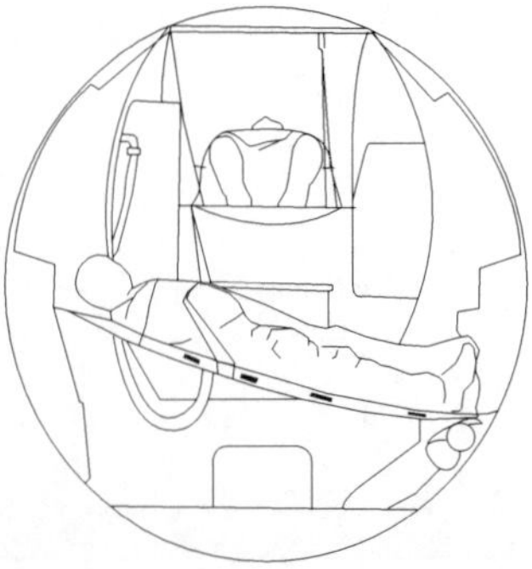

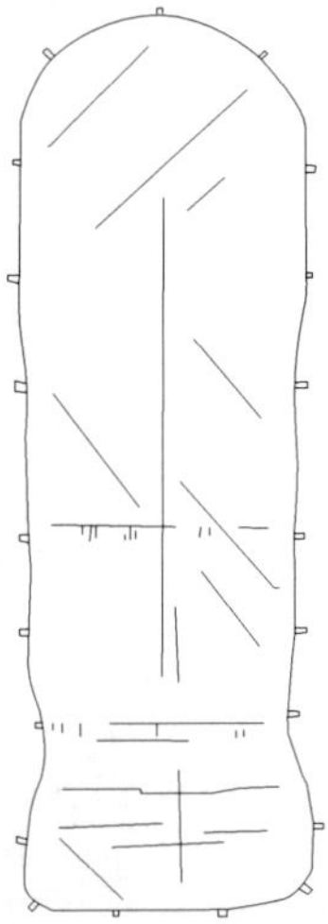

SHUTTLE	MIR	ISS
Sleeping bags; private sleeping boxes	Permanent private crew quarter (kajutkas) with sleeping bag, desk, intercom and a porthole	Two permanent private crew quarters with sleeping bags; four soft flexible temporary crew quarters
Different sleeping configurations are used, depending on the mission schedule (boxes, sleeping bags)	Some cosmonauts preferred to sleep in the attached modules, such as Kristall in order to have more privacy and/or better radiation protection	Some astronauts chose an individual location, such as the Node to place their sleeping bag and private storage
Adaptability in sleep configuration depending upon 'scenarios' seems useful	Windows are important, but not necessarily in the crew quarters. A dedicated private space is mandatory for long term mission	Sleeping provisions could be flexible in location; Virtual windows can replace real windows in certain cases

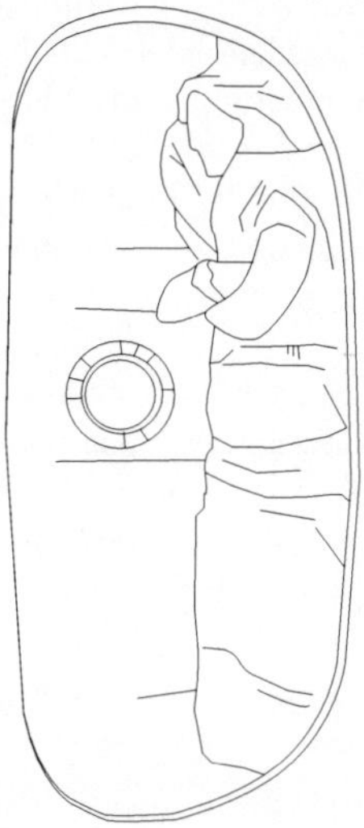

4.7 SUMMARY OBSERVATION

Throughout history, astronauts have reported sleep disturbances, but have also confirmed that good sleep is necessary for good work performance (Jones, 1995; Stuster, 1986). Similar observations have been made in analogue environments on Earth among crewmembers of polar, submarine, and spaceflight missions. Problems were associated with noise, temperature, work schedules, and stress amongst others (Connors, et al., 1999 p. 11).

For most of the Apollo Astronauts, scheduled sleeping time was more a nuisance, than a desired activity. (B.J. Bluth, 1987 p. 330). Apollo 17 astronaut Harrison Schmitt reported that he slept well on the Moon, but not continuously. "I found, that I slept for two, three hours and then would wake up and spend some time checking things out and then would go back sleep" (Schmitt, 2009).

Undisturbed Sleep

One design requirement in order to provide undisturbed sleep is insulation from outside light and noise.

During the Apollo missions available space was too small,.to separate the sleep area from other functions. Salyut stations had dedicated areas for sleep, but they were not spatially divided from the work area. Only a sort of net was provided around the sleep areas to prevent the cosmonauts breathing in dust and floating small parts (B.J.Bluth, 1987).

Skylab was the first space station that provided spatially and visually separated crew quarters. During sleep periods all of the spacecraft lights were turned off. One of the crew complaints was that the adjacency to the waste management compartment was a problem because the waste management compartment was very noisy and if one crewmember used it, it woke up the other two. They also complained that movements of other crewmembers disturbed sleeping. (NASA [Bull.3], 1974 p.15]

On Mir, individual cabins were provided for two cosmonauts, but they were located in the core module which was very loud. According to Jean-Pierre Haigneré the Kvant-2 module was one of the quietest places, as it had very few ventilators. The disadvantage of this module was the low radiation protection, because of the thin walls (Haigneré, 2009).

The International Space Station has only been recently equipped with crew quarters for the permanent crew of six. Two are permanently installed in the Zvezda module and four are mobile rack-sleep stations. Within the spatially separated areas, astronauts are screened from outside light and noise as well as protected from floating particles and objects. The new crew quarters have integrated radiation reduction material and caution and warning systems, thus offering an advanced protection while astronauts sleep.

Sleeping in Microgravity and Restraints

Since the Apollo missions, sleeping facilities included sleep restraints to fix the body and head to the preferred sleeping position and also to prevent the astronaut from floating over delicate equipment.

Sleeping in a micro-gravity environment is different than sleeping on Earth. The absence of gravity has severe impact on the human body, the natural and habitual body position changes to the neutral body posture, thus also affecting sleeping. One example is that astronauts report experiencing lower back pain when sleeping in a micro-gravity environment. On Earth people also experience back pain, but is associated with heavy spinal load – as a consequence of gravity. Astronaut Jean-

François Clervoy reports that when relaxing in the absence of gravity, one may experience back pain when the dorsal muscles are stronger than the abdominal muscles, because they tend to curve the lower spine (Clervoy, 2009). Research on low back pain in crewmembers is still going on and scientists believe the loss of calcium and mass of muscles to be the cause for weakening the muscle corset (ESA [Mus], 2003-09).

Salyut cosmonauts restrained their sleeping bag with strings and knots to the 'ceiling'. To ensure good air movement they could use portable ventilators. (Prunariu, 2011) Some Skylab astronauts requested more restraints to hold them against the firm back. The restraint straps they used worked work well in principle, but did not cover a sufficient portion of the body (NASA [Skylab LL], 1974 p. SLL2-48). Edward Gibson stated that "all you really needed was a blanket which would go across you, and three straps." (NASA [Bull.4], 1974 p. 40)

Today, multiple restraints can be used during Shuttle missions or on-board the ISS to help the astronauts maintain their sleeping positions and to help ensure a good sleep. A special harness is available that maintains the knees close to the chest, similar to a foetal position. The sleeping bag is equipped with special belts that can be tightened on the outside, to maintain the body inside the sleeping bag and to secure the desired body position in order not to have back pain. Astronauts can also use a "pillow", maintained with a special strap on the forehead, to have the feeling that the head is pressing on a pillow (Clervoy, 2009).

Anecdotal evidence shows that some astronauts prefer the feeling of micro-gravity with minimal restraints.

Privacy and Territory

Most of the interviewed astronauts reported that they had little requirement for privacy during their short-term missions. But there are many anecdotal references, especially from long-term astronauts, that privacy - as well as social life - is important to crewmembers. Likewise, research from analogue environments show that under prolonged isolation and confinement the need for private space increases (Stuster, 1996, p. 90; Kanas, et al., 2003 p. 164; Connors, et al., 1999 p. 101).

During the Apollo missions, astronauts slept next to each other in one volume and had no privacy, but since "most pilots are not used to privacy (...) it was not a problem" (Schmitt, 2009). The Salyut cosmonauts slept on the 'ceiling' of the large-diameter work compartment. They did not have private crew quarters, thus having a visual overview of the station from their sleep area, which seemed to be important at that time.

Although the Salyut cosmonaut Lebedev argued against the provision of separate cabins for astronauts in 1982 (Lebedev, 1990) because of safety reasons, the cosmonauts have since favoured the installation of private crew quarters. The requirement to separate sleep and work with moveable partitions was integrated early on. Skylab was the first space station that provided spatially separated and private crew quarters. On the space station Mir, individual cabins were provided for two cosmonauts, additional crewmembers had to sleep in their sleeping bag somewhere else in the stations.

Connors et al (1999 p. 86) cites John Archea (1977): "visual access and visual exposure are the two key aspects of privacy regulation." A space that on the one hand accommodates all crew members and on the other hand offers closure to provide individual private areas is important

4.7 SUMMARY OBSERVATION

for the functioning of a group. This arrangement is also the preference of former long-term Mir astronaut Jean-Pierre Haigneré (2009). For him a personal place for thinking, concentration and personal activities needs to be integrated into the design of any habitable space.

Although private crew quarters are provided for the permanent crew of six at the ISS, astronaut Hans Wilhelm Schlegel (2009) reported that a crew quarter on ISS provides the absolute minimum of privacy, "more would be better."

Object Management and Storage
Another requirement for private crew quarters is the provision of storage for personal items and clothes, requested by the Apollo astronauts and repeatedly an issue on Salyut and Mir space stations.

On Skylab some general equipment items were stowed in the crew quarters. This lead to some complaints, because "it was felt to be a personal area" (NASA [Bull.3, 1974 p. 13). On the other side, there was not enough private temporary stowage areas provided. One astronaut taped two trash bags to the wall close to his bed and used them as dresser drawers. (NASA [Bull.3, 1974 p. 4]

Anecdotal references show that astronauts have put their private belongings next to the place where they slept. Michel Tognini (2009) slept in the module Kristall during his stay on the Mir space station. A private area seemed not to be a priority for him; more important was to have a place, where he could put his camera or paperwork. He put his personal objects close to his sleeping bag. A place where astronauts put and secure hardware and items and "nobody disturbs anything" was also very important for Jean-Pierre Haigneré (2009). He stored his personal items next to his sleeping bag in the module Priroda.

Relations to the Outside
Apollo astronauts had the shutters from their windows open at all times; only when they wanted to sleep did they decide to close the shutters to prevent incoming light. They had a marvellous view of planet Earth as well as of the lunar surface.

The Salyut and Mir space stations had multiple windows and the Mir crew quarters provided a view outside even from the personal crew quarters. Although there was a big discussion concerning the inclusion of a big window on Skylab, it was finally installed (Compton, et al., 1983), but Skylab astronauts had no windows in the crew quarters.

Today on ISS, only the crew quarters in the Zvezda module allow a direct view to space. The other crew quarters have no windows to the outside. All crew quarters have doors that can be opened or closed while sleeping.

Perception of Space

Salyut cosmonauts often slept in the Orbital Work Module of the docked Soyuz. It's almost spherical shape made it more spacious in comparison to that of the Apollo Command Module (U.S. Congress, 1983 p. 11).

The design of the Soviet stations integrated a colour system to increase the feeling of spaciousness and also to improve mood. The interior outfitting of sleeping areas were painted in muted and cool colour tones (OTA, 1983 p. I-111).

Despite the small volume, the potential of a microgravity environment is described by Jean François-Clervoy. He reported that even the crew quarter [of the Space Shuttle] looks like a "sarcophagi", you can think, you are inside an enormous volume, because you float in the middle" (Clervoy, 2009).

Changing Mission and Crew Objectives

The Mir space station was initially designed for a permanent crew of two, but simultaneous occupation increased during the lifetime. This multiple occupation meant, that later only two crewmembers would be provided a crew quarter, the other cosmonauts and astronauts had to sleep somewhere else in the station.

References show that on the space stations Skylab and Mir the sleep quarters were used not only for sleeping. They were used during the day to relax or to take a short break from work or from the other crewmembers. Sometimes, as reported from Skylab and Mir astronauts, the crew quarters were even used for interim storage (NASA [Skylab LL], 1974).

The Space Shuttle is the only spacecraft where - depending on its mission - sleeping provisions can be configured on Earth. When the crew must work split shifts, sleep stations are provided. If a sleep station is not installed, astronauts use sleeping bags. However, the locations where a sleeping bag can be placed are predetermined to a certain extent. "The reason that one cannot set up one's sleeping bag anywhere (in microgravity) is that care must be taken that one does not end up in a corner with insufficient air circulation"(Thiele, 2010). Also locations for the sleeping bags depend on the type of mission, as different requirements mean changing interior [storage] arrangements.

Individual Preferences of Astronauts

Throughout the history of spaceflight, astronauts have chosen their individual sleeping position in the habitat, some developing their "own style" of sleeping positions.

In an interview, Skylab astronaut Paul Weitz, mentioned that he had difficulties sleeping "hanging from the wall". Thus, every night he would unbuckle his sleeping restraint from the metal frame and take it up to the big open area in the Orbital Workshop. There he stretched it across the module and slept horizontally (Weitz, 2000). Some of the other Skylab astronauts turned their sleep restraint or adjusted the netting and the bag (NASA [Bull.3], 1974 p. 19, 26, 51; NASA [Bull.4], 1974).

American Mir astronaut John Blaha slept on the floor of the module Spektr (Burrough, 1999 p. 114). Jerry Linenger (2000 pp. 90, 182) slept upside down on the wall in the module Spektr, to be next to an installed fan on the opposite floor, as there is no convection in microgravity and fans are the only means of moving air, thus providing fresh oxygen. Another reason for sleeping on the wall was that he did not have to "stow and unstow his sleeping bag every day" (Burrough, 1999 p.114).

4.7 SUMMARY OBSERVATION

To avoid floating free, he was using a bungee cord or a piece of Velcro.

Jean-Pierre Haigneré (2009) slept in the module Priroda, which he shared with Sergei Awdejew. For him it was a discreet place and he could look at Earth from "one of the nicest windows". Reinhold Ewald slept in the module Kristall. It was one of the more radiation protected modules of Mir, because of the technological heavy equipment on the walls. Ewald had his sleeping bag loosely fixed to ropes running through the station that his colleagues used as movement aid. When sleeping he had his head rested on - and his arms folded around these ropes (Ewald, 2009).

According to Shuttle astronaut Gerhard Thiele (2010), the commander and pilot often sleep in the flight deck. The reason is in order to be ready should a call come in from Mission Control during the sleep phase and something needs doing immediately.

In addition to the astronauts' individual preferences on the location, once 'settled-in', they have personalized their 'private corner'. Already on the Salyut stations, cosmonauts put up pictures and personal items around the area, where they slept (B.J.Bluth, 1987 p. I-77). Today, the crew quarters on the ISS have integrated interfaces for personalization, as well as individually adjustable airflow controls and lighting, which add to habitability.

Astronaut Jean-François Clervoy ponders about future spacecraft design "OK. Today I will change completely the layout of my bedroom, because I am fed up with, seeing that the bed is always there (...) if sometimes you decide, ok, my sleeping station for the next month will be there. Would be nice. Yes that is nice." (Clervoy, 2009)

A: Adaptable crew quarter concept for a Lunar Base *(credit A: Author)*

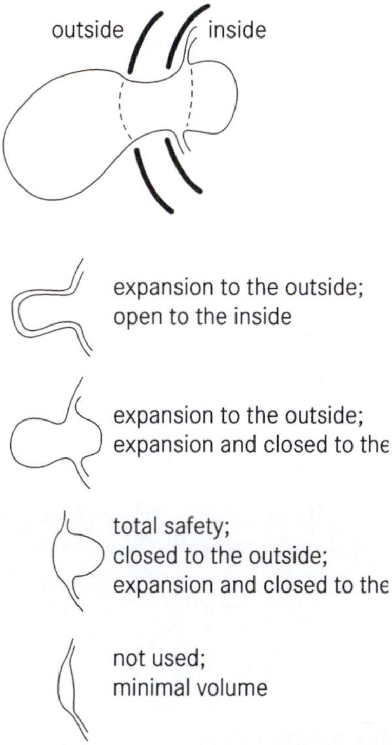

A:

4.8 COMPARISON USABILITY MATRIX

	APOLLO	SALYUT	SKYLAB
Availability and Equipment	CM: sleeping bags and couches for three crewmembers; LM: hammocks with blankets for two crewmembers; Astronauts slept in space suit on first missions	Sleeping bags for all cosmonauts; Sleeping bags with air vents and napped fasters	Private crew quarters for each astronaut; Included sleep restraints, private storage lockers, privacy curtains and communication system
Spatial Arrangement	CM: crewmembers slept next to each other on the launch and flight "couches"; LM: hammocks across the module	In the large-diameter Work Compartment; Sleep locations on the ceiling, next to the food supplies; Some slept in the Orbital module of the docked Soyuz	Private sleep compartment in the Orbital Workshop with a corridor; Some slept in the OWS forward dome
Object Management	No personal stowage	No personal stowage	Personal stowage in crew compartment; Triangular grid for fasten equipment
Ergonomic Safety	Individual fit of astronaut couches, sleeping bag = sleeping restraint	Special netting around sleeping area - to avoid breathing in small parts; Straps to restrain bed to the allotted sleeping area	Materials used were non-flammable and especially developed for Skylab; Grid system for moving around; Individual and modular restraints

SHUTTLE	MIR	ISS
Different configurations depending on the mission: three-tier or four-tier rigid sleep stations or just sleeping bags; Additional Equipment: eye covers, ear plugs, 0g-harness	Two private crew quarters for two crewmembers, sleeping bags for the other; Cabins were equipped with a port hole window, sleeping bag, desk and mirror	Two private crew quarters for two crewmembers, sleeping bags for the other; Cabins were equipped with a port hole window, sleeping bag, desk and mirror
Astronauts sleep in the Middeck; Sleeping bags are on the starboard wall; Rigid sleep stations are on the starboard side	Crew quarters were located in the large-diameter Compartment in the Mir Base Block; Some slept in Kristall, Kvant 2, Priroda or Spektr	Two crew quarters are integrated in the Zvezda module, four modular rack-sized sleep compartment are currently in Node 2
Personal stowage lockers, personal stowage in sleep station	Personal stowage in crew cabin (only for two)	Personal stowage in crew quarters
Durable, non-flammable and non-off-gassing materials, no sharp corners; Ventilation system in sleep stations, a variety of sleeping restraints	Less radiation protection in crew quarters (window)	Durable, non-flammable and non-offgassing materials, additional radiation protection and warning systems in the new crew quarters; A variety of sleeping restraints

4.9 COMPARISON LIVABILITY MATRIX

	APOLLO	SALYUT	SKYLAB
Territoriality and Privacy	Each crewmember had his couch/hammock; No privateness	Each crewmember had his sleeping bag; No spatial separation; Limited privateness Closeness was important for crew	Personal crew quarters with privacy curtains; Spatially separated from the work area but personal crew quarters next to each other
Sensory perception	Leaking window shades bothered sleeping; Music was used for relaxation; music wake-up call	High noise level; Slow air movement in certain places	Privacy curtains screened light from outside, but provided no sound protection
External relations	Small windows to Earth and Lunar surface;	Salyut had many windows, none especially for the sleep area; Window shutters were closed during sleeping periods	Private communication box in crew quarters; no windows
Internal relations	Astronaut slept at the same time; all activities in same module	Cosmonauts had a good overview of the station; View to the greenhouse	Spatially separated from other crewmembers and group domain

SHUTTLE	MIR	ISS
Rigid sleeping stations provide personal space for each crewmember	Two crewmembers had a personal area that allowed personalization; others chose adjacent modules	Six crew quarters provide privacy for the permanent crew and allow personalization
Rigid sleep stations are screened from outside light and noise	Crew quarters situated in the 'loud' work compartment, visual disclosure possible	Improved acoustic noise mitigation in new crew quarters, visual disclosure
Sleeping stations have no windows	Crew quarters had ‚personal' window to the outside	Two crew quarters in Zvezda module have windows to the outside; the other crew quarters have communication equipment
Rigid sleep stations are spatially separated from work area; Sleeping bags in group domain	Good overview of the main module from crew quarters	If closed, no internal spatial relations only via communication system

4.10 COMPARISON FLEXIBILITY MATRIX

	APOLLO	SALYUT	SKYLAB
Spatial Flexibility	CM: fixed sleep area (couches); LM: dedicated sleep area (hammocks) or in sleeping bag	Dedicated place on the ceiling; Sleeping bag positioning was flexible	Integrated crew quarters; Sleep restraint was mobile and two astronauts slept somewhere else
Object Flexibility	Deployable hammocks, window covers, adjustable sleeping bags	Netting, sleeping bags, mobile ventilators	Adjustable and mobile sleep restraint
Individual Flexibility	"Sleeping is individual"	Individual sleeping positions, sleeping pills	Crew quarters allowed personalization, flexibility in blanket arrangement, adjustable restraints

SHUTTLE	MIR	ISS
Shuttle allows different configurations: Fixed sleep stations or sleeping bags; Dedicated place for sleeping bags but positioning is flexible	Two integrated crew quarters; Some cosmonauts slept in different modules	Two integrated crew quarters and four modular crew quarters
Sleep stations have removable panels, adjustable sleeping bags	Crew quarters can be personalized; Third sleeping place (sleeping bag) 'somewhere' in the station	Four crew quarters can be relocated, two are fixed Sleeping bags can be located anywhere
Use of sleep restraints is optional	Crew quarters allowed personalization, foldable table, individual sleeping position	Crew quarters allow personalization and mounting for electrical assemblies (integrated utility wall)

4.11 DESIGN DIRECTIONS

The most important findings related to the human activity 'SLEEP' are summarized and structured in the form of design directions. These directions summarize the main points in a short sentence. The paragraph following refers to the empirical background of experiences made in extraterrestrial habitats.

USABILITY

Provide sleep facilities for all crewmembers

Sleep facilities need to be provided for each crewmember. The use of 'hammocks', sleeping bags or other temporary sleeping provisions are a good option for short term missions, when additional sleeping provisions are temporarily needed for visitors and in emergency situations. For long-term missions sleep facilities need to offer more functionality than just sleeping (personal recreation and work, communication, personal stowage).

Provide accessibilty

Sleep areas need to be accessible at all times to allow undisturbed sleep when required or in case one crewmember gets sick.

Separate sleep from other activities

Sleep areas shall be separated from the main work area and other activity zones in order not to disturb sleeping crewmembers.

Provide private storage

An exclusively private stowage area for clothes, laptops and other private items should be provided. In view of future lunar and Mars exploration scenarios (NASA [Mars 5.0], 2009; ESA [HMM], 2004) with several habitats and pressurized rovers, portable stowage units similar to the ISS cargo bags will ease handling and transfer from one place to another.

Protect sleep areas

Sleep areas need protection from floating particles, incoming radiation, and provide adequate air movement. They also need protection from outside light and noise.

Provide sleep restraints

Sleep restraints to fix the body and head to the preferred sleeping positions and also to prevent the astronaut from floating over delicate equipment shall be provided in microgravity. Individual and modular restraints are needed.

Integrate communication systems

Communication and warning systems need to be integrated close to the place where astronauts sleep to enable the crewmembers to act quickly in case of emergency.

LIVABILITY

Provide a private territory

A space where astronauts can retreat from the others is of importance for long-term missions. Due to volumetric limitations, large dedicated areas for privacy will probably not be feasible in the near future. The installation of personal crew quarters offers an easy possibility for providing astronauts with a private and individual place.

Increase spaciousness by design

Spaciousness can be increased by design and geometry. In 'Quantitative Modelling of Human Spatial Habitability', Jim Wise (1988 p. 14) describes spaciousness as the perceived extension of an enclosure. Edward Hall (1966 p. 51) writes about Frank Lloyd Wright's ability to successfully enhance the individual experience of space through careful use of textures and surfaces and the Japanese art of combining visual and kinaesthetic senses to 'make the most of small spaces'.

Integrate media facilities
If crew quarters are the only private areas they should provide for private communication to friends and family. Integrated media facilities can also substitute for the natural view to the outside (window).

Integrate visual access
In addition to user controlled visual disclosure sleep areas should provide visual access to the group domain, either through doors or openings.

FLEXIBILITY

Think mobile
Independent sleep restraints that can be used anywhere in the station improve usability. Crewmembers themselves can decide whether to sleep 'attached to the wall' or 'float in the middle'. Mobile ventilators support free choice of location.

Allow reconfiguration
Once a space habitat is in orbit, it is difficult to make major reconfigurations. For future-long term missions an option to reconfigure the habitat needs to be implemented. Crew quarters can provide flexibility and diversification in spatial configurations, and allow different individual scenarios depending upon the crewmembers preferences and mission concept. Flexible sleeping station concepts can personalize the mission crews' "home" and thus reflect the community's identity and living preferences.

Design for multiple use
Sleep areas shall offer functionality during the day for individual working, private communication, donning and doffing clothes or to 'back out' during work hours.

Integrate multipurpose furnishing
Flexible partitions can further enhance usability, to temporarily close off work and sleep areas.
A flexible interior design ensures that, the individual astronaut has direct physical and visual access to the group domain, but can also choose its current level of privacy by screening from the group (Häuplik, et al., 2003).

Allow individual adjustment
Different users have different preferences. "Sleeping is individual, just like here on Earth", said Apollo 17 astronaut Harrison Schmidt about sleeping, "and you would almost have to talk to everybody to get the full spectrum" (Schmitt, 2009). Lighting and temperature needs to be individually adjustable. In addition the place where astronauts sleep is a reflection of the user's personality, thus it shall allow individual outfitting and decoration. There is a wide range of individual variety.

Integrate feed-back-design
Comparable data on experiences in relation to the interior is important for the design and implementation of future sleep stations. An integrated 'feed-back-design' for the research on ergonomics could be an option in view of planetary missions, because data and preferences for sleeping in such conditions still need to be evaluated.

4.12 REFERENCES

Archea, John. 1977. The Place of Architectural Factors in Behavioral Theories of Privacy. J. Social Issues. Vol. 33, 1977, Vol. No. 3, pp. 177-181.

Avdeyev, Sergei. 2001. My life on Mir. [interv.] Amelia Gentleman.The Guardian, 14 March 2001. http://www.guardian.co.uk/science/2001/mar/14/spaceexploration.education

B.J.Bluth, M. Helppie. 1986. Soviet Space Stations as Analogs. Washington D.C., US : NASA Headquarters, 1986. NASA Grant NAGW-659.

Broyan, James Lee Jr., Borrego, Melissa Ann and Bahr, Juergen F. 2008. International Space Station USOS Crew Quarters Development. San Francisco, CA, USA : Proceedings of the 38th International Conference on Environmental Systems, 2008. 08 ICES-0222.

Burrough, Bryan. 1999. Dragonfly - NASA and the crisis aboard Mir. London, GB : Fourth Estate Limited, 1999. ISBN 1 84115 087 8.

Cernan, Eugene and Don, Davis. 1999. The last man on the Moon. USA : St. Martin's Griffin, 1999. ISBN 0-312-19906-6.

Clervoy, Jean-François. 2009. Transcript not published [interv.] Sandra Häuplik-Meusburger. Paris, France, June 2009.

Compton, W. David and Charles D., Benson. 1983. Living and Working in Space, A History of Skylab. Washington, US. : National Aeronautics and Space Administration, The NASA History Series, 1983. NASA SP-4208.

Connors, Mary M., Harrison, Albert A. and Akins, Faren R. 1999. Living Aloft, Human Requirements for Extended Spaceflight. Washington, D.C. : NASA, Ames Research Center, 1999. SP-483/ISBN: 1-4102-1983-6.

Duque, Pedro. 2003. Pen, Paper and Weightlessness: Astronaut Pedro Duque's Soyuz/ISS Diary. International Space Station : s.n., 23 October 2003. Diary from Space. http://www.space.com/missionlaunches/duque_soyuz_diary.html.

ESA [HMM]. 2004. Human Mission to Mars: Overall Architecture Assessment. Noordwijk, NL : ESTEC Concurrent Design Facility (CDF), 2004. CDF Study Report. CDF-20 (A).

ESA [Mus]. 2003-09. Study of Low Back Pain in Crewmembers During Space Flight (Mus). [Online] 2003-09. [Cited: 25 7 2009.], http://www.nasa.gov/mission_pages/station/science/experiments/Mus.html.

Ewald, Reinhold. 2009. Transcript not published. [interv.] Sandra Häuplik-Meusburger. Cologne, Germany, July 2009.

Foale, Colin. 1999. Waystation to the Stars - The Story of Mir, Michael and Me. London, GB : Headline Book Publishing, 1999. ISBN 0 7472 6331 0.

Grechko, Mikhailovich. 1993. Historical Reminiscences of Cosmonaut Georgi Mikhailovich Grechko, Part 1. [interv.] Dennis Newkirk and Jim Plaxco. http://www.astrodigital.org/space/intgrechko1.html, May 1993.

Haigneré, Jean-Pierre. 2009. Transcript not published. [interv.] Sandra Häuplik-Meusburger. Paris, France, June 2009.

Hall, Edward T. 1966. The Hidden Dimension. N.Y., USA : Anchor Books, 1966. ISBN 0-385-08476-5.

Häuplik, Sandra, Springer, Markus and Lorenz, Susanne. 2003. Tycho Lunar Base. Hochbau 2, TU Wien, ESTEC NL and TU Vienna : 2003.

Jones, Eric M. 1995. Apollo 12 - Post-EVA Activities in the LM. http://history.nasa.gov/alsj/a12/a12.posteva1.html : Apollo 12 - Lunar Surface Journal, 1995.

Jones, Tom. 2006. Skywalking, an Astronaut's memoir. U.S.A. : HarperCollins Publisher, 2006. ISBN 0-06-088436-3.

Kanas, Nick M.D. und Manzey, Dietrich Ph.D. 2003. Space Psychology and Psychiatry. Dordrecht, The Netherlands : Springer/ Microcosm Press, 2003. ISBN 978-1-4020-6769-3.

Lebedev, Valentin. 1990. Diary of a Cosmonaut - 211 Days in Space. United States and Canada : Bantam Air & Space Series, 1990. Vol. Vol.4. ISBN: 0-553-28778-8.

Lenehan, Anne E. 2004. Story, The Way of Water. USA : The Communications Agency, 2004. ISBN 0 9752286 0 9.

Linenger, Jerry M. 2000. Off The Planet - Surviving five periolous months aboard the Space Station MIR. U.S.A. : McGraw-Hill, 2000. ISBN 0-07-136112-X.

Lousma, Jack R. 2001. Oral History Transcript - Jack Lousma. [interv.] Carol L. Butler. Houston, Texas, 7 March 2001.

NASA [Bull.3]. 1974. Skylab Experience Bulletin No. 3: Architectural Evaluation for Sleeping Quarters. Houston, Texas, US : NASA, Lyndon B. Johnson Space Center, 1974. JSC-09537.

NASA [Bull.4]. 1974. Skylab Experience Bulletin No. 4: Design Characteristics of the Sleep Restraint. Houston, Texas, US : NASA, Lyndon B. Johnson Space Center, 1974. JSC-09538.

NASA [Crew Compartment]. 2002. Human Space Flight - Crew Compartment. [Online] Curator: Kim Dismukes, 4 July 2002. [Cited: 5 August 2007.] http://spaceflight.nasa.gov/shuttle/reference/shutref/structure/crew.html.

NASA [Skylab CS]. 1974. MSFC Skylab Crew Systems Mission Evaluation - Technical Memorandum. Alabama : George C. Marshall Space Flight Center, 1974. Technical Memorandum. NASA TM X-64825.

NASA [Debriefing A11]. 1969. Apollo 11 - Technical Crew Debriefing. Houston, Texas, USA : NASA, Manned Spacecraft Center, 1969. 276.

NASA [Debriefing A12]. 1969. Apollo 12 - Technical Crew Debriefing. Mission Operations Branch - Flight Crew Support Division. Houston, Texas, USA : Manned Spacecraft Center, 1969. p. 308.

NASA [Debriefing A14]. 1971. Apollo 14 Technical Crew Debriefing (U). prepared by: Mission Operations Branch Flight Crew Support Division. Houston, Texas, USA : Manned Spacecraft Center, 1971. February 17, 1971.

NASA [Interview]. 2009. International Space Station - Preflight Interview Frank De Winne. [Online] 05 June 2009. [Cited: 15 August 2009.] http://www.nasa.gov/mission_pages/station/expeditions/expedition20/de_winne_interview.html.

NASA [Mars 5.0]. 2009. Human Exploration of Mars: Design Reference Architecture 5.0. Mars Architecture Steering Group, NASA Headquarters. Houston, Texas, US : Editor: Bret G. Drake, NASA Johnson Space Center, 2009. NASA-SP-2009-566.

NASA [Report A15]. 1971. Apollo 15 Mission Report. Houston, Texas, USA : Manned Spacecraft Center, 1971. MSC-05161.

NASA [Skylab]. 1977. Skylab, our first space station. Washington D.C., US : NASA, Scientific and Technical Information Office, 1977. Prepared by George C. Marshall Space Flight Center, Edited by Leland Belew. NASA-SP-400.

NASA [Skylab LL]. 1974. Lessons Learned on the Skylab Program. Houston, Texas, USA : NASA, Lyndon B. Johnson Space Center, 1974. JSC-09096.

NASA [Sleep]. 2003. Space Sleep. [Online] 08 11 2003. [Cited: 23 5 2007.] http://spaceflight.nasa.gov/living/spacesleep/index.html.

NASA [Shuttle Reference]. 1981. Space Shuttle News Reference. Houston, Texas. : U.S. Government Printing Office, 1981. 723-000/562.

NASA [STS-126]. 2008. STS-126 'Extreme Home Improvements' - Space Shuttle Press Kit. USA : NASA, 2008.

NASA [STS-128]. 2009. STS-128 ‚Racking Up New Science' - Space Shuttle Press Kit. [Online] 2009. http://www.nasa.gov/pdf/379392main_STS-128_Press_Kit.pdf.

OTA. 1983. SALYUT - Soviet Steps Toward Permanent Human Presence in Space. Washington D.C. : U.S. Congress, Office of Technology Assessment, 1983. Andelin, John; Naismith, Nancy; Mills, William; Rogers, Thomas F.; Chandler, Philip P.. OTA-TM-STI-14.

Portree, David S.F. 1995. MIR Hardware Heritage. Houston, Texas : NASA, Lyndon B. Johnson Space Center, Information Services Division, 1995. NASA RP 1357.

Prunariu, Dumitru-Dorin. 2011. Transcript not published. [interv.] Sandra Häuplik-Meusburger. Vienna, Austria, 2011.

ReelNASA. 2008. Ask Astronaaut Greg Chamitoff: In Your Dreams. YouTube : Movie 2:33 , 2008. http://www.youtube.com/watch?v=AutZQ-4KpEQ.

Schlegel, Hans Wilhelm. 2009. Transcript not published. [interv.] Sandra Häuplik-Meusburger. Cologne, Germany, 2009.

Schmitt, Harrison. 2009. Transcript not published. [interv.] Sandra Häuplik-Meusburger. Vienna, Austria, 2009.

Stuster, Jack W. 1986. Space Station Habitability - Recommendations Based on a Systematic Comparative Analysis of Analogous Conditions. Anacapa Sciences, Inc. Santa Barbara, California : NASA, 1986. NAS2-11690, NASA Contractor Report 3943.

Stuster, Jack. 1996. Bold Endeavors: Lessons from Polar and Space Exploration. US Naval Institute Press, 1996. ISBN-10: 155750749X.

4.12 REFERENCES

Thiele, Gerhard. 2010. Transcript not published. [interv.] Sandra Häuplik-Meusburger. Vienna, Austria, December 2010.

Tognini, Michel-Ange Charles. 2009. Transcript not published. [interv.] Sandra Häuplik-Meusburger. Cologne,Germany, July 2009.

U.S. Congress. 1983. Salyut: Soviet Steps Toward Permanent Human Presence in Space - A Technical Memorandum. Office of Technology Assessment. Washington, D.C. : Library of Congress Catalog, 1983. OTA-TM-STI-14.

Vasilyev, M. P., et al. 1974. Salyut Space Station in Orbit, Translation of ‚Salyut na Orbite`, Mashinostroyeniye Press, Moscow, 1973, 160 pages. Washington, D.C. : NASA, National Aeronautics and Space Administration, 1974. NASA TT F-15,450.

Weitz, Paul J. 2000. Oral History Transcript - Interview with Paul J. Weitz. [interv.] Rebecca Wright. 26 March 2000.

Wise, James. 1988. The Quantitative Modelling of Human spatial Habitability. Ames Reseach Center : NASA Contractor Report, 1988. 177501.

PART 5: HUMAN ACTIVITY

HYGIENE

The category 'Hygiene' was divided into three sub-categories.

1. **Personal hygiene** (full and part body cleansing and change clothes)
2. **Toilet** (collect, store and process waste) and
3. **Housekeeping** (cleaning and maintaining a certain hygiene standard in the station.

5.1 HYGIENE APOLLO

CONCEPT Minimal personal hygiene, urine transfer and receptacle system, 'Apollo bags'
LOCATION CM / LM
SPECIFIC All resources brought from Earth

RÉSUMÉ PERSONAL HYGIENE

Anecdotal references indicate that personal hygiene and "getting clean" was very important for the Apollo astronauts. Apollo 12 astronauts shaved three to four times during their missions. For that, mirrors for the astronauts were provided in the Lunar Module and Command Module. Apparently they never used the mirrors in the Lunar Module, but only the ones in the Command Module during flight (NASA [Debriefing A12], 1969).

Apollo astronauts had only one set of EVA undergarments with the Liquid Cooling and Ventilation Garment (LCVG), incorporating the urine collection and transfer system *[Fig. B]*. In the Technical Crew Debriefing, astronaut Dick Gordon reported that "after a few days it got so clogged with urine and so dirty, that you just hated to put the thing back on" (NASA [Debriefing A12], 1969).

The hygiene kit consisted of a shaver, blades, shaving cream, a comb, toothbrush, toothpaste, and later dental floss. The astronauts could select individual products, which were bought "off-the-shelf" (NASA [Bull.14], 1975 p. 2). According to Gene Cernan (2007 p. 14), astronauts used chewing gum instead of brushing their teeth with toothpaste.

A:

B:

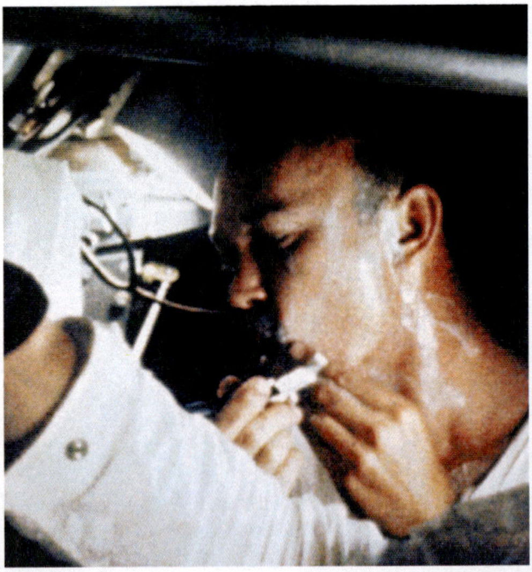

A: Apollo 10 Commander Stafford shaves in the CM
B: Apollo 11 Command Module Pilot, Astronaut Michael Collins decides to shave his beard after almost 8 days in the space flight
C: Neil Armstrong's space suit assembly (Extravehicular Mobility Unit - EMU): Urine Collection and Transfer Assembly, Liquid Cooling and Ventilation Garment, Helmet Assembly, Multi-layered pressure suit, Glove Assembly and Boots
(credit A, B, C: NASA)

ASTRONAUTS' EXPERIENCES

USABILITY

Gene Cernan (Apollo 17): *"We didn't have any hot water on the surface of the Moon. We had it in the Command Module. When we got back in there, we took our suits off, and stripped naked and got hot water and a hot washrag and gave ourselves a "bath". I cannot tell you how good that feels. Don't take a bath for 14 days and tell me how you feel.* ***We looked like coalminers.****"* (Cernan, 2007)

LIVABILITY

Neil Armstrong (Apollo 11): *"Of course, that's [gases] a big odour problem in the spacecraft."* (NASA [Debriefing A11], 1969 p. 118)

FLEXIBILITY

Harrison Schmitt (Apollo 17): *"[Shaving]* ***some did, some didn't,*** *and we had the equipment. It was not much different, than shaving here [on Earth] the surface tension keeps the shaving cream on your face, and just run the razor, it's very much the same."* (Schmitt, 2009)

C:

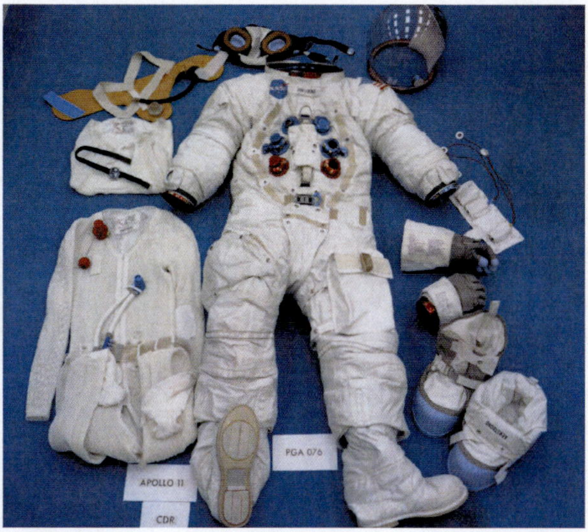

5.1

HYGIENE APOLLO

RÉSUMÉ TOILET

The Apollo waste management system consisted of a urine subsystem and a faecal subsystem (Sauer, et al., 1975; NASA, 1972):

Three devices were used for collecting and transferring urine: the urine transfer system (UTS), the urine receptacle assembly (URA), and the urine collection and transfer assembly (UCTA).

The **urine transfer system** *[Fig. A]* consisted of a roll-on rubber cuff- that was used like a condom with a flexible collection bag.

Either urine was collected using the bag or it was directly dumped by connecting with a flexible hose to the waste management panel, integrated in the cabin wall of the spacecraft to the exterior of the spacecraft. This system was also used in a slight variation to collect urine samples. Each astronaut had his personal UTS. The cuffs were colour-coded for each astronaut and replaced every day. This system was used up to and including Apollo 12, later it was used as a back-up system for following Apollo missions.

The **urine receptacle assembly** was the main system used on the Apollo missions and was used mostly during waking hours and flight. It had no collection function and urine was dumped by the waste management panel to the exterior using the same system as the UTS.

In general the URA system "seemed easy and convenient" for the astronauts to use. Reported problems included the astronaut's hands getting covered with urine every time the system was used. There was a constant film of urine on the inside of the cover and the Teflon top, which had to be cleaned with a towel. (NASA [Debriefing A12], 1969)

A:

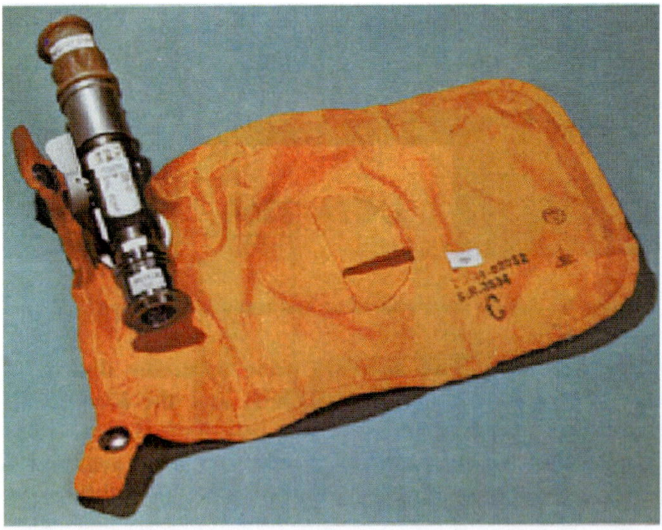

B:

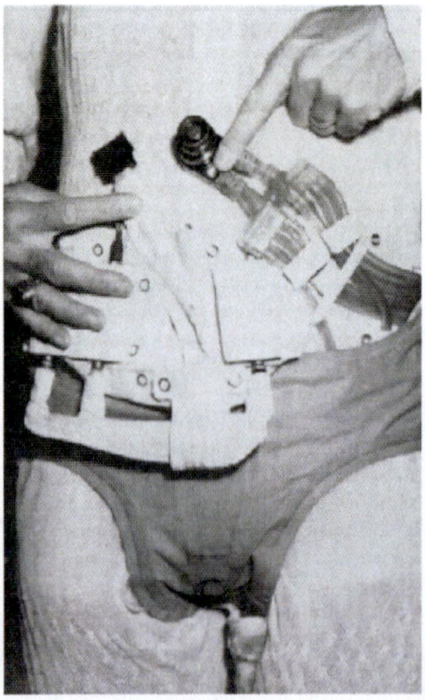

A: Apollo urine transfer system (UTS) with roll-on cuff
B: Urine collection and transfer assembly worn over the liquid cooling garment
C: Urine collection and transfer assembly (UCTA)

(credit A, B, C: NASA)

Procedure
(Sauer, et al., 1975; NASA, 1972)

1. The cup was removed from a cylindrical container.
2. Before urination it had to be connected to the waste management panel and the valve had to be turned to the 'dump' position in order to allow direct dumping.
3. Then the astronaut had to direct his urine stream to the cylindrical receiver, which contained a honeycomb cell insert that acted as a bundle of capillary tubes.
4. The urine went by the urine transfer hose to the waste management panel to the outside.
5. Then the hose had to be cleaned and the dump valve turned off. After use the URA was stored.

The **urine collection and transfer assembly** *[Fig. B, C]* was used by space-suited astronauts (during launch, EVAs and emergency situations). It was worn like a belt over the liquid cooling garment and consisted of a roll-on cuff and a collection bag. It could also be used after the suit was removed, by connecting the urine transfer hose to the waste management panel.

C:

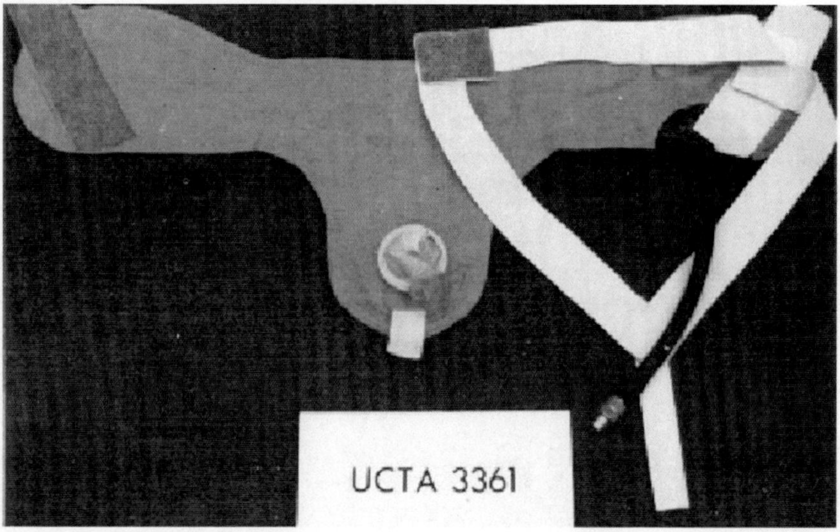

5.1 HYGIENE APOLLO

RÉSUMÉ TOILET

The faecal-collection assembly was used to collect, inactivate and stow faeces on the Command module. It consisted of a faecal bag ('Apollo bag') with germicides and a waste-stowage-vent system to discharge faecal odours outside the crew cabin. The so-called defecation-collection device was similar to a diaper. The 'Apollo bag' system *[Fig. A]* or waste stowage bag was a transparent plastic bag with an opening, a finger pouch and a pouch for tissue wipes and germicides. The faecal procedure was distasteful for the astronauts, therefore low residue foods and laxatives were generally taken before launch. Apollo astronauts also took drugs to reduce intestinal motility. (Sauer, et al., 1975 p. 2)

Faecal Procedure
(Sauer, et al., 1975; NASA, 1972)

1. The astronaut put his finger in the pouch to position the bag to the anus.
2. The bag was taped to the buttocks for collection of the faeces.
3. After "business", the astronaut used the 'finger pouch' to push the faecal to the bottom of the bag.
4. He then cleaned his buttocks with the tissues and sealed the bag.
5. When sealed, the astronaut had to knead the content and mix it with an integrated germicide.
6. Sometimes a probe was taken for biological examination on Earth.
7. Before being placed in the waste stowage compartment, the bag volume had to be minimized by coiling it up.

The **faecal and emesis collection device** (FCS) was similar to a diaper *[Fig. B]*. It was used when astronauts wore spacesuits and was worn under the liquid cooling garment.

A:

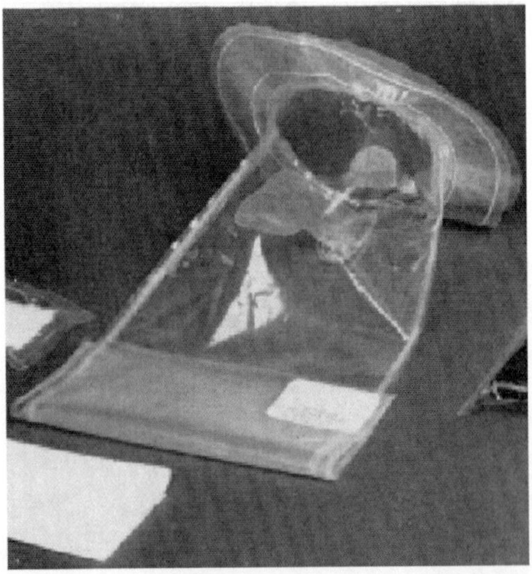

A: Faecal bag 'Apollo bag'
B: Faecal containment system for use during extravehicular activity *(credit A, B: NASA)*

ASTRONAUTS' EXPERIENCES

USABILITY
Harrison Schmitt (Apollo 17):
"As long as you have a normal bowel movement it's not a problem. Diarrhoea is a problem (...) because of surface tension being a dominant force. The fluid just migrates on the body, rather than going into the faeces bag." (Schmitt, 2009)

"I think the LM faeces bags are superior to the CSM's in that they have a goodly quantity of tissues cut to size and are quite good. I see no reason why those couldn't be the same kind of blue bags in the CSM." (NASA [Debriefing A17], 1973)

LIVABILITY
Gene Cernan (Apollo 17):
"Undesirable odors - I don't think there were any undesirable food odors (...) ***They were overwhelmed by the urine and feces odors*** *and the gas odors in the spacecraft."* (NASA [Debriefing A17], 1973)

B:

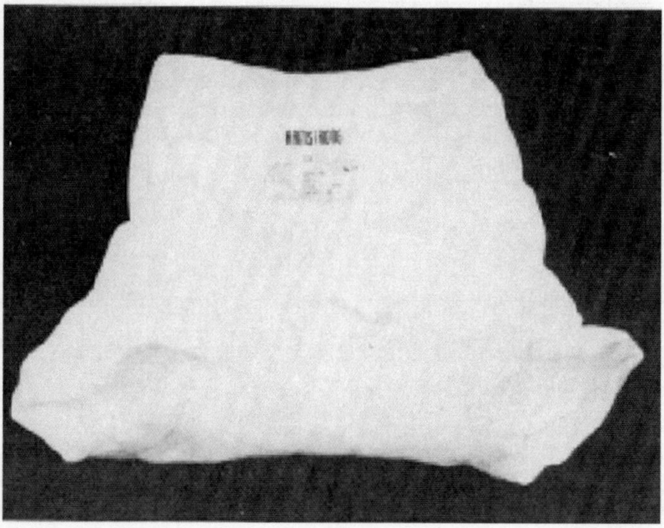

5.1 HYGIENE APOLLO

RÉSUMÉ HOUSEKEEPING

Housekeeping tasks included the charging of batteries, dumping waste water and urine, checking fuel and oxygen reserves, cleaning window and air filters, as well as stowing equipment. (Boeing, 1995-2009; NASA [Debriefing A11], 1969 p. 149)

Apparently the urine collection system did not work properly. Astronauts complained that urine was in the couch areas, on the suits and on the urine dispenser. Astronauts used towels and hot water to clean. (NASA [Debriefing A12], 1969)

Lunar dust became a major problem not only inside the Lunar Module, but also inside the Command and Service Module, because the fine dust floated in microgravity and was too fine for the cleaning system (NASA [Debriefing A12], 1969; NASA [Dust Management], 2006). It seems that the filter system had been upgraded for the Apollo 17 mission. Harrison Schmitt reported that most of the dust was filtered out of the cabin atmosphere (Schmitt, 2009).

Fig. B shows Apollo 17 commander Eugene E. Cernan inside the lunar module. His suit is full of lunar dust.

A:

A: Apollo 16 Astronaut Charles M. Duke Jr. collects lunar samples
B: 'Dirty' Apollo 17 commander Eugene Cernan inside the lunar module on lunar surface after EVA, foto taken by H. H. Schmitt *(credit A: NASA, John W. Young; B: NASA)*

ASTRONAUTS' EXPERIENCES

USABILITY

Gene Cernan (Apollo 17): *"Dust and fatigue were definitely causing problems. Those insistent, fine grains of lunar dirt had worked into the moving parts of our tools, and things were breaking down. Then the makeshift fender gave way and showered us with more dust wherever we drove."* (Cernan, et al., 1999 p. 336)

LIVABILITY

Neil Armstrong (Apollo 11): ***"Our cockpit was so dirty with soot,*** that we thought the suit loop would be a lot cleaner." (NASA [Debriefing A11], 1969 p. 81)

Harrison Schmitt (Apollo 17): "The only dust would have accumulated after we got out of the suit. Once we were in the Command Module and there wasn't much dust transferred from the Lunar Module to the Command Module (...)." (Schmitt, 2009)

FLEXIBILITY

Neil Armstrong (Apollo 11): "I think the idea of having an individual kit where you can place things in individual packages is much better than the large one. And I'd like to see continued effort along this line to come up with better ways of interim stowage." (NASA [Debriefing A11], 1969 p. 149)

B:

5.2 HYGIENE SALYUT

CONCEPT Personal hygiene, toilet, deployable shower, vacuum cleaner
LOCATION Work compartment
SPECIFIC Recycling of water, production of oxygen, shower

RÉSUMÉ PERSONAL HYGIENE

The hygiene unit was located in the large-diameter section of the Work Compartment. It could be separated by a curtain and had its own a ventilation system. The surfaces of the panels were of washable material (Portree, 1995; Vasilyev, et al., 1974).

Personal hygiene equipment was provided for teeth cleaning, shaving, skin cleaning, hair washing, and haircuts. According to Dumitru-Dorin Prunariu, Salyut 6 cosmonauts didn't use toothbrushes on board. They used special tissues rolled around the fingers for teeth cleaning (Prunariu, 2011). Later cosmonauts used an electric toothbrush and 'non-foaming' toothpaste. They shaved with an electric razor with rotating heads that had a vacuum system to suck shaved hair into a chamber. They could also use shaving foam and a razorblade. But cosmonauts didn't use them often, because they had no place to clean them (Prunariu, 2011). Instead, they used daily electric shavers with a built-in pouch for retaining hair, similar to a vacuum cleaner.
Antibacterial cloth with disinfectant solutions was used for washing the skin and hair. To collect hair from haircuts, cosmonauts used the vacuum cleaner (B.J.Bluth, 1987 pp. I-79,80).

A medical kit was provided with medication for motion sickness, fatigue, nasal congestion, bacterial infection, pain, diarrhoea and minor lacerations. It also contained vitamins, medicines for respiratory irregularities, tranquilizers, sleeping pills and drugs to reduce internal stress. The station also had a miniature dentist drill on-board (B.J.Bluth, 1987 p. I-81).

Clothing included heat-protective coveralls, fur boots, sweaters, undershorts, and socks. Underwear was changed once a week (B.J.Bluth, 1987 p. I-156). To simulate the effects of gravity, cosmonauts wore the so-called 'penguin suits' *[Fig. A]* which was worn 16 hours a day, and taken off only when sleeping (B.J.Bluth, 1987 p. IV-10).

A: Salyut 6 cosmonauts Prunariu and Savinykh, 1981 *(credit A: Dumitru-Dorin Prunariu)*

ASTRONAUTS' EXPERIENCES

USABILITY

*"Mechanical or electric shaving of the usual kind is simply impossible, even dangerous in flight – the clipped hairs and hair dust fly immediately into the air and may be inhaled into the lungs. Mechanical shavers developed for the cosmonauts constitute **a "symbiosis" of a razor and a vacuum cleaner**."*
(NASA [Joint Flight], 1974 pp. 10-11)

Dumitru-Dorin Prunariu (Salyut 6): *"Everything [clothes] was specially made for us. My private clothes were only my underwear."*
(Prunariu, 2011)

LIVABILITY

Valentin Lebedev (Salyut 7):
"There is no atmosphere to protect us from the Sun. After two to three minutes under the Sun's rays our skin looks as if we've been on the beach all day."
(Lebedev, 1990 p. 110)

FLEXIBILITY

Valentin Lebedev (Salyut 7):
"(...) we wear our light cotton Penguin jumpsuits. They have elastic cords sewn in to provide additional muscle exercise. They provide a pleasant sensation as they tighten around our bodies during a short space trip. During a long flight, when we live on a station for months at a time, these suits become uncomfortable. Tolia and I decided to remove all the elastic cords."
(Lebedev, 1990 p. 188)

A:

5.2 HYGIENE SALYUT

RÉSUMÉ SHOWER

Starting with Salyut 3, a shower was provided (Portree, 1995). On Salyut 7 the shower was improved. The shower and water supplies were located in the forward Work Compartment next to the exercise treadmill and beneath the sleeping bags.

The shower facility *[Fig. B]* was basically a transparent polyethylene cover with a waterproof zipper stretched between two cylindrical disks. The water container was on the ceiling above the upper disk from which the shower cabin could be folded down.

The disk contained the devices for hot and cold water as well as air supply. The bottom disk of the shower had integrated rubber slippers to keep the cosmonaut in place. (Portree, 1995; Lebedev, 1990). In general the cosmonauts liked to take a shower but had a lot of difficulties with the system.

Shower days were scheduled monthly and while the cleaning of the body itself took only about 15 minutes, the preparation of the shower and the work afterward took hours. (B.J.Bluth, 1987)

Procedure
(B.J.Bluth, 1987 p. I-83)

The shower system was always prepared after breakfast because it took almost the whole day.

1. Cosmonauts switched on the electric heating system and lowered the folded polyethylene cylinder from the ceiling to the floor.
2. The cosmonaut got into the shower cabin and put a clip on his nose, opened a package that contained a soap cloth; then he turned the water on.
3. The water came out from the top and passed through holes in the floor to the waste container.
4. After taking the shower, the cosmonaut rubbed himself and the shower cabin dry.
5. After that the shower cabin was stowed up again in the disk in the ceiling.

A:

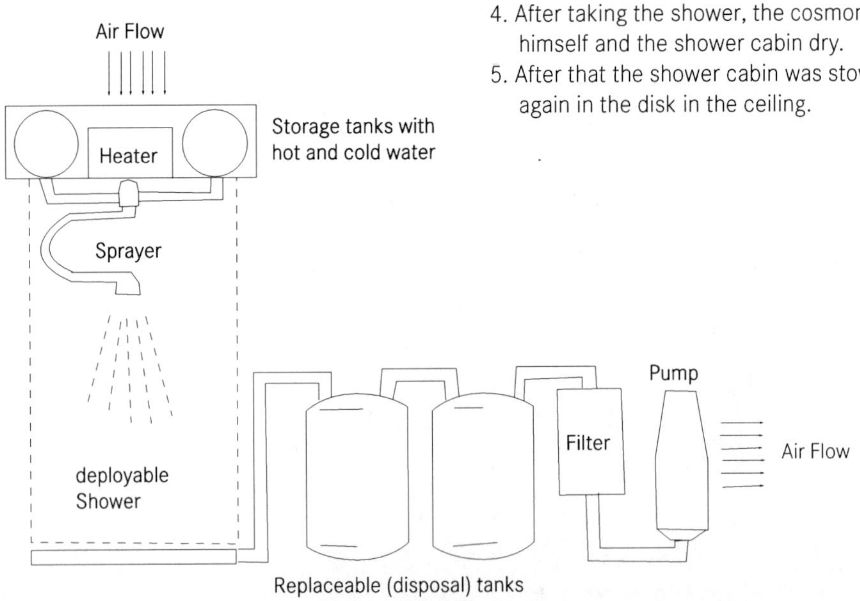

A: Functional diagram of shower system
B: Deployable shower on Salyut

(credit A: Original Image from NASA, adapted by the Author; B: Author)

ASTRONAUTS' EXPERIENCES

USABILITY

Valery Ryumin (Salyut 6): *"When you begin to think of all the preparatory operations you have to do, and then how many post-shower operations you have to perform, the desire to take a shower diminishes. You have to heat the water, in batches, no less. You have to get the shower chamber, set up the water collectors, attach the vacuum cleaner (...)* ***it takes nearly the entire day just for that shower,*** *" he complained."* (Portree, 1995 p. 86)

Lebedev (Salyut 7): *"We went to take a shower, but when I began to zip up the cover, it tore apart."* (Lebedev, 1990 p. 123)

LIVABILITY

Valentin Lebedev (Salyut 7): *"On Earth it is impossible to drown in a shower, but here it is very easy. Therefore while we are taking a shower, we look like scuba divers with snorkels, clips on our noses, and goggles."* (Lebedev, 1990 p. 74)

FLEXIBILITY

Aleksandrov Lyakhov (Salyut 7): ***"It would be a good thing to have*** *(...) a shower on Kosmos (...), a special shower which is always assembled and ready for use."* (B.J.Bluth, 1987 pp. I-83)

B:

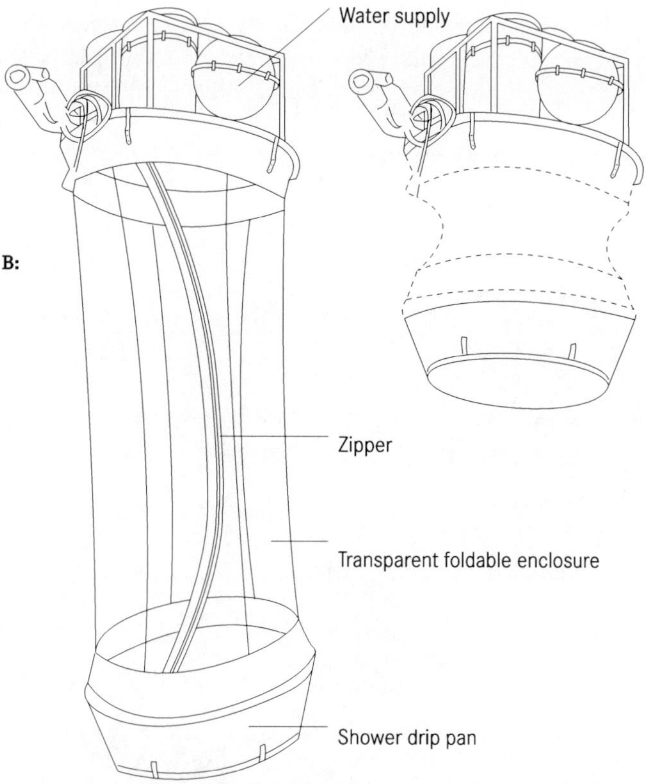

5.2

HYGIENE SALYUT

RÉSUMÉ TOILET

The waste management compartment contained a toilet and washroom. It was located in the rear bottom part of the Work Compartment in the passageway to the Intermediate Compartment. This area could be closed off using a rubber curtain with a zipper (B.J.Bluth, 1987 p. I-78). The toilet bowl was below floor level *[Fig. C]*. This area was equipped with a ventilation system and the walls were of washable material. The principles of the toilet were to collect and process liquid and solid waste separately. The system had a standard toilet bowl and a urine dispenser. The waste management system was designed for use by men and women; according to Bluth it did not present any problems for Savitskaya, the female Salyut 7 astronaut (B.J.Bluth, 1987).

Faecal Procedure

(B.J.Bluth, 1987)

1. To use, cosmonauts put a plastic insert into the bowl.
2. After use it was sealed by a rubber valve.
3. Then the cosmonauts put it into a hermitical sealed rubber bag, which was then stored in a plastic container and
4. finally ejected to space.

Back-up System

The cosmonauts had another system for when the toilet didn't work and for use on board the Soyuz spacecraft. It was a portable device *[Fig. A]* that consisted of a flexible hose with attachable funnels, differently shaped for men and women. The waste was stored in a container until it could be processed with onboard systems. This system was also used on the Mir space station and is still being used as an emergency toilet for the International space station.

A:

Storage Container

Flexible hose with attachable funnels

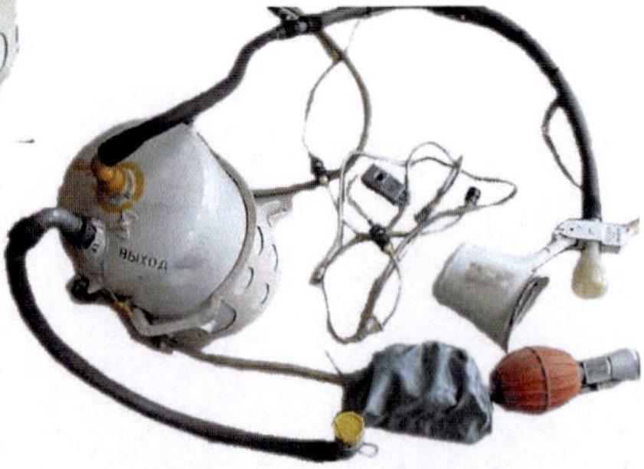

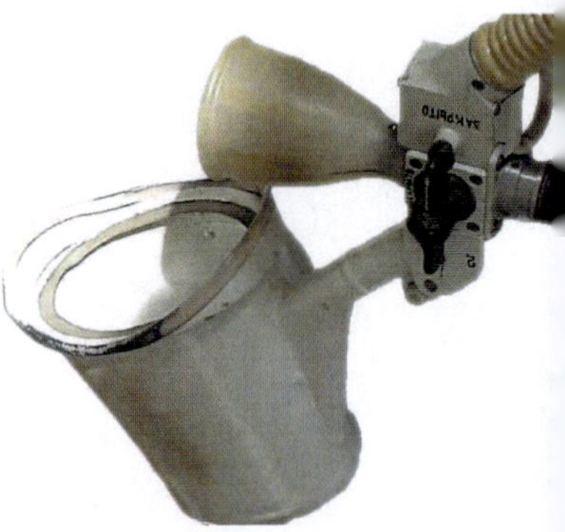

A: Portable toilet system on Soyuz
B: Functional diagram of toilet system
C: Salyut 6 toilet facility

(credit A: Richard Newman, HighTechScience.org;
B: Original Image from NASA, adapted by the Author, 1987;
C: Mark Wade, astronautix.com)

ASTRONAUTS' EXPERIENCES

USABILITY

Valentin Lebedev (Salyut 7):
"The toilet – I literally sit on it like a witch on a broom. Everybody runs into some pretty funny problems with this space bathroom. ***It won't forgive mistakes."***
(Lebedev, 1990 p. 71)

LIVABILITY

Valentin Lebedev (Salyut 7):
"I got up at night to go to the bathroom, but the ASU system was overfilled – the overfill warning signal was on. What a problem! If we were home, we could "go" outside. But that's not a viable option here, so I had to hold it for a whole hour while I pumped the urine out of the ASU."
(Lebedev, 1990 p. 117)

B:

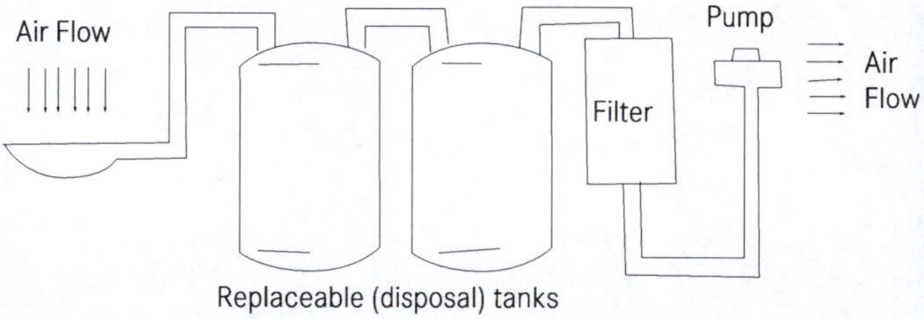

C:

5.2 HYGIENE SALYUT

RÉSUMÉ HOUSEKEEPING

According to Bluth, surface and atmospheric contamination increased with time. Therefore, the cosmonauts had to clean the surfaces regularly with disinfectants (B.J.Bluth, 1987 p. 6). Cosmonaut Valentin Lebedev explains in his diary how they cleaned the station. The cosmonauts used wet napkins with a detergent called katamine for cleaning the table, handrails, bars, door hatches, surfaces of the control panel and other pieces of equipment. They did a wet cleaning once a week.

"Once in a while", they had to do a major cleaning, thus vacuuming behind the panels and other "hard-to-reach places" (Lebedev, 1990 p. 135).

House-keeping activities were usually scheduled on non-working days. (B.J.Bluth, 1987 p. III-27)

Most of the trash was disposed by placing it in the cargo compartment of the Progress freighters. In addition, the space station Salyut had an airlock that could be used for trash disposal. This airlock was also used for scientific experiments that required access to vacuum. (Portree, 1995)

A:

A: Cosmonaut Prunariu during a medical test on board Salyut 6

(credit A: Dumitru-Dorin Prunariu)

ASTRONAUTS' EXPERIENCES

USABILITY

Valentin Lebedev (Salyut 7): ***"The station is a mess*** *- bags with pieces of equipment hang in the scientific-equipment compartment; the intermediate chamber is packed with regenerators, absorbers; and the resupply ship is also choked with regenerators and EDV containers. Only the connection compartment looks clean and spacious with nothing extra scattered about."*
(Lebedev, 1990 p. 133)

Valentin Lebedev (Salyut 7): *"Everybody knows what that [housekeeping] means at home, in the household, trash collects on the floor and dust gathers on the furniture.* ***Here everything floats:*** *dust, pieces of trash, crumbs of food, drops of juice, coffee and tea. It all ends up suspended in the station, with most of it collection on the intake grills of the fans, which we cover with cheesecloth. This way we collect the trash in weightlessness (...)."*
(Lebedev, 1990 p. 135)

LIVABILITY

Valentin Lebedev (Salyut 7): *"We can't smell any odours at all up here, because special filters remove all hazardous contaminants and dust, and we have regenerators with CO_2 absorbents. In general, the air up here is much better than in any city on Earth. It is dry, clean, healthy, ionized by the Sun through the portholes (...) only there is no rain."*
(Lebedev, 1990 p. 168)

Dumitru-Dorin Prunariu (Salyut 6): *"The Ventilators were on all the time, there were a lot of filters, for different smells, but* ***the technical smell of the Space Station always stayed."***
(Prunariu, 2011)

5.3 HYGIENE SKYLAB

CONCEPT Personal hygiene, toilet, deployable shower, vacuum cleaner
LOCATION OWS, floor level, vertical toilet (1.27m x 2.36m x 1.98m)
SPECIFIC No recycling; shower

RÉSUMÉ PERSONAL HYGIENE

The Waste Management Compartment (WMC) of the Skylab station was situated in the Orbital Workshop (OWS), next to the sleep compartments, the wardoom and the experiment work area *[Fig. B]*. The WMC included a faecal / urine collector, a hand washer, a drying station and utility closets for personal hygiene supplies and house cleaning supplies. Stowage containers for various purposes were installed in the Waste Management Compartment, Wardroom, Sleep Compartment, Experiment Compartment, and Forward Compartment (NASA [Skylab LL], 1974). Medical examinations were conducted in the Wardroom (NASA [Bull.3], 1974). The triangular grid of the WMC floor was closed. Two foot restraints that could only be used without shoes for the toilet and the hand washer were provided. Astronauts braced themselves 'between two walls' to fix their location *[Fig. A]*. (NASA [Bull.7], 1974 p. 26; NASA [Bull.9], 1974)

Skylab astronauts used an 'in-orbit hand washer' that consisted of a water dispenser and cloth squeezer to clean themselves (NASA [Skylab LL], 1974 p. SL2-6). Hygiene utilities included tissues, waste collection bags, towels, soap and utility wipes. The towels were made of rayon terrycloth with a stitched colour code for each astronaut. In total 840 reusable washcloths and 420 towels were provided in the stowage dispensers in the waste management compartment (NASA [Skylab], 1977 p. 81). The 'Personal Hygiene Kit' included a razor, shaving cream, hand cream, toothpaste, comb, nail clippers, deodorant, and other personal items (NASA [Skylab], 1977).

Clothing items included 39 jackets, 69 pairs of trousers, 126 shirts, 30 pairs of boots, 18 pairs of gloves, 4 single-piece long thermal underwear called union suits, 201 T-shirts, 45 half union suits, 85 jockey shorts, 112 boxer shorts, 54 knee shorts, 286 pairs of socks and 24 constant-wear (NASA [Skylab], 1977 p. 81).

The crew of Skylab 3 used worn clothes as cleaning rags (NASA [Bull.6], 1974).

A:

B:

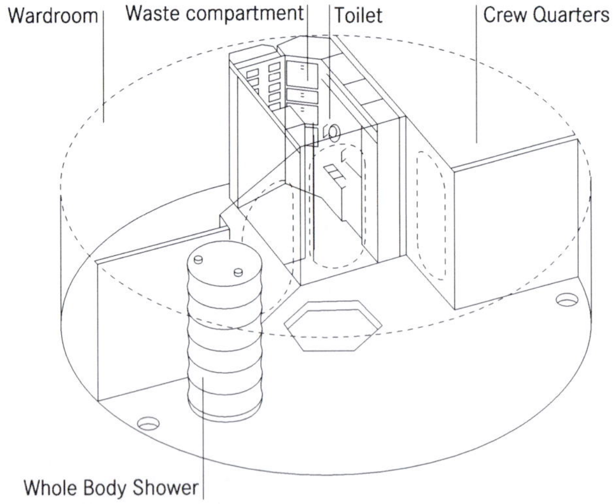

A: Skylab astronaut Jack Lousma shaving in the WMC
B: Hygiene facilities in the OWS
C: Personal hygiene locker, which contains toothbrush, battery operated razor, toothpaste, and hand cream

D: An engineer demonstrates the use of the hand washing fixture in the waste management area of the Skylab Workshop mockup
(credit A, C, D: NASA, B: Original Image from NASA, adapted by the Author)

ASTRONAUTS' EXPERIENCES

USABILITY

Owen Garriott (Skylab 3): *"I begrudged that time, and so I said, I don't need a shower. I just need a washcloth getting wet and give yourself a good body scrubbing. That's all that it takes."* (Garriott, 2000)

Charles Conrad (Skylab 2): ***"Nobody used their trash bag for trash in the sleep compartment.*** *(...) I used it to keep personal gear in."* (NASA [Bull.3], 1974 p.12)

Gerald Carr (Skylab 4): *"Traffic was the thing there [Waste management compartment]."* (NASA [Bull.8], 1974 p. 434)

LIVABILITY

Owen Garriott (Skylab 3): *"I think I'm the only person who went two months and never took a shower, but I was as clean as anybody."* (Garriott, 2000)

Gerald Carr (Skylab 4): *"And we should have soaps and things (...) that got the smells we are used to down here."* (NASA [Bull.8], 1974 p. 419)

FLEXIBILITY

Skylab astronaut: *"You ought to have a place to wash your hands and a place where you defecate and urinate, but you also ought to have a separate place for shaving and taking care of routine primping and - and - and cleaning up, combing hair and washing, brushing your teeth (...)"* (NASA [Bull.8], 1974 p.17)

C:

D:

5.3

HYGIENE SKYLAB

RÉSUMÉ SHOWER

The Skylab space station provided a deployable shower, which was situated in the experiment and work area in the Orbital Workshop (OWS). The shower *[Fig. A, C]* was a cylindrical foldable device that was mounted on the floor. The bottom ring was attached to the floor and contained foot restraints. (NASA [Bull.9], 1974) The upper ring contained the showerhead and hose. When not in use, the shower was collapsed down to its base. The shower combined pressurized water flow with a suction device.

Each Skylab crew seemed to have a different reaction to the shower and these reactions were paradoxically different. The first Skylab crew (Skylab 2) liked the shower a lot. The second crew (Skylab 3) did not like the shower at all. Alan Bean only took two showers; Jack Lousma took one shower and Owen Garriott none on their 59-day-mission (Garriott, 2000). The third crew (Skylab 4) crew all took showers on their rest day (Carr, 2000; NASA [Skylab], 1974).

The shower concept was principally acceptable, but the method of water removal and cleaning the shower was not. "It took too much time and the effort for setting the shower up and putting it down was too inconvenient" (NASA [Skylab LL], 1974 p. SLL 2_47).

Procedure

(NASA [Skylab], 1977; Garriott, 2000)

1. The astronaut filled a pressurized portable bottle with heated water and attached it to the ceiling (A flexible hose connected the water bottle to the hand held shower head)
2. He connected the plumbing and pulled the "shower curtain" up from the floor and attached it to the ceiling.
3. Astronauts got in the spray - water was provided through a push-button shower head, attached to a flexible hose.
4. They washed themselves with liquid soap and sprayed again to wash off the soap
5. Then they had to vacuum everything up and to stow the shower away.

A:

B:

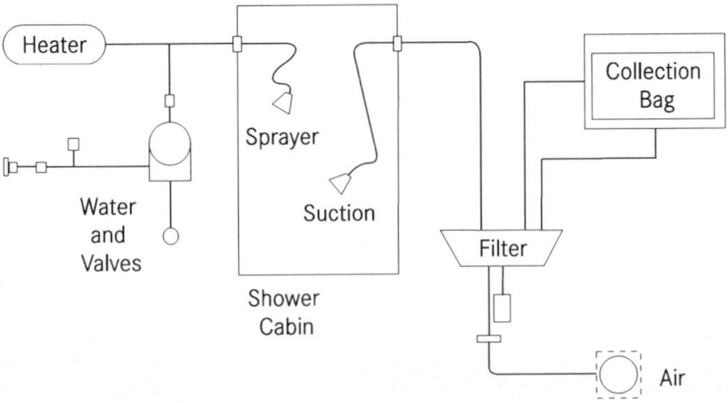

A: Skylab 2 astronaut Charles Conrad after a hot bath in the shower in the OWS
B: Functional diagram of the shower system
C: Shower in the OWS

(credit A: NASA; B: Original Image from NASA, adapted by the Author; C: Author, based on NASA documents)

ASTRONAUTS' EXPERIENCES

USABILITY

Skylab 2: *"The shower really worked well and we felt good after showering."* (NASA [OWS], 1974 p. 424)

Skylab 3: *"I would recommend to have a shower in the future space station, but they have it connected into the plumbing, just like the rest of the water. When you step in to take a shower and then when you get finished, turn off the shower and get out. You just leave the shower with water around on that area: you have revisions made so that the water's automatically sucked off."* (NASA [OWS], 1974 p. 426)

Skylab 3: *"the soap doesn't wash off."* (NASA [OWS], 1974 p. 426)

LIVABILITY

Skylab 2: *"(...) you **froze** your tail off, getting to a towel"* (NASA [OWS], 1974 p. 424)

Skylab 4: ***"I like to enjoy that** [take a shower], but we don't have time for it."* (NASA [OWS], 1974 p. 428)

FLEXIBILITY

Skylab 4: *"The only shortcoming of the shower, I would say, is the suction head. It's just not flexible enough (...) I think it could be very easily redesigned into something quite - quite nice and useful."* (NASA [OWS], 1974 p. 427)

C:

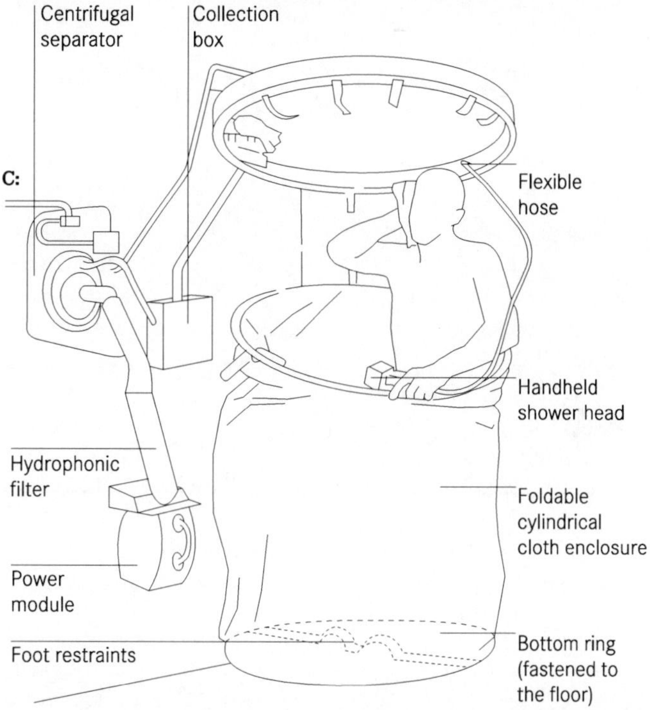

5.3
HYGIENE SKYLAB

RÉSUMÉ TOILET

The Skylab toilet was designed to eliminate body wastes and to support biomedical experiments by sampling and collecting. The system consisted of a faecal-urine collector, collection and sample bags, sampling equipment, and odour-control filters and a fan. The toilet *[Fig.B]* was mounted in the middle of the wall, which obviously posed no problem in microgravity. Between the toilet and the lap belt was the holder for the urine receiver. Urine was collected in three drawers at the bottom. Urinating could be done in a sitting or standing positions. The urine collector was located on the wall below the faecal collector. Foot restraints on the floor were provided, but proved to be of little use. (NASA [Bull.9], 1974; NASA [Bull.10], 1974)

A:

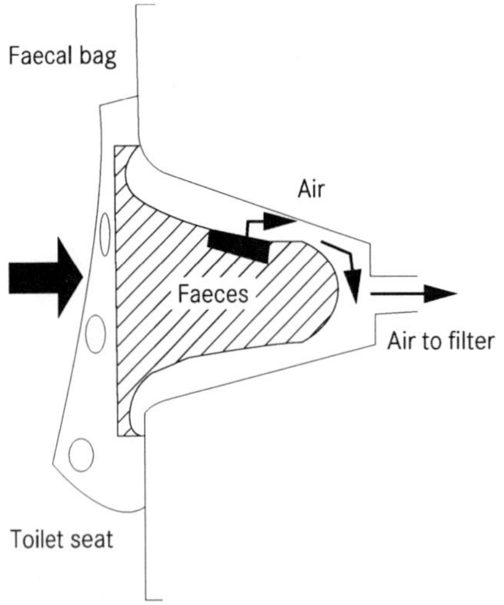

A: Skylab toilet
B: Skylab toilet mounted to the wall

(credit A: Original Image from NASA, adapted by the Author; B: NASA)

Faecal Collection Procedure
(NASA [OWS], 1974 pp. 59-60)

1. The astronaut removed the faecal bags from the stowage container and installed it in the collection module.
2. Then the astronaut sat on a contoured seat and fastened a belt across his lap to hold himself in position.
3. He defecated into the faecal bag.
4. Airflow from the fan separated the faecal matter from his body and deposited it in the faecal collection bag.
5. The used bag was sealed, removed from the collection module and placed in the Specimen Mass Measurement Device (SMMD) where the mass was determined.
6. The sealed bag was then placed in the waste processor for a given time (the time is a function of the mass) during which the solid wastes were processed (dried) by the combined effects of vacuum and heat.
7. The processed materials were then stowed for return to Earth via the command module.

B:

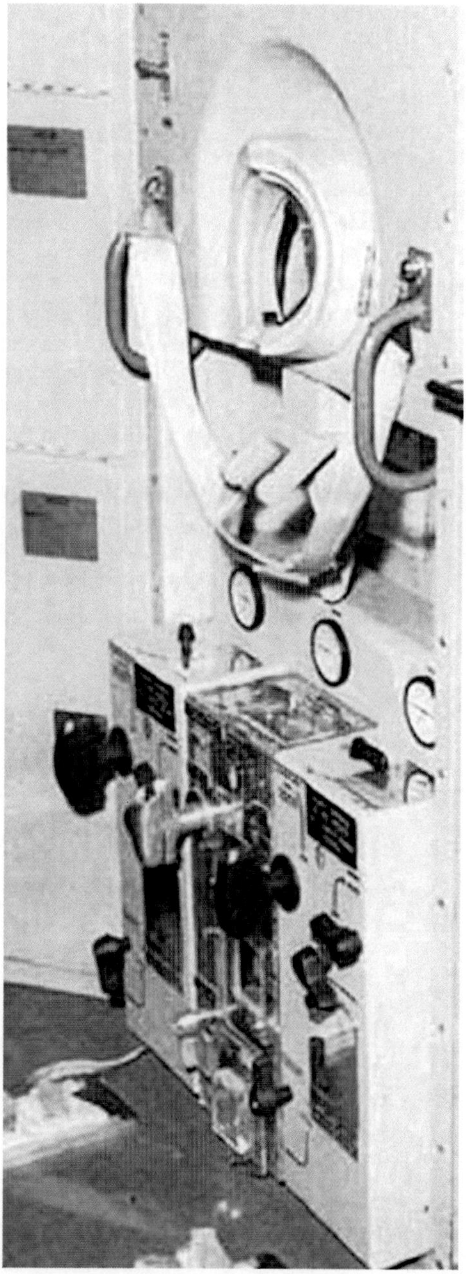

5.3 HYGIENE SKYLAB

Urine Collection Procedure
(NASA [OWS], 1974)

1. The astronaut removed the urine receiver from the drawer and installed it on the collection module.
2. Then he micturated into the inlet-line of the collection module.
3. Airflow carried the urine to the centrifugal separator; where air and urine are separated and the urine is pumped into the urine bag.
4. The urine was collected for a 24-hour period, and the urine bag is replaced at 24-hour intervals. After that period the total volume was determined and a sample extracted in a sample bag.
5. A sample urine (bag) was frozen and stowed in a container for return to Earth in the Command Module (CM).
6. The remaining urine (bag) was removed from the collection module, placed in a trash bag for disposal in the trash airlock.

Contingency Procedures
Contingency procedures were applied when the crew was sick or when the toilet system didn't work. In vomitus collection and in contigengy faecal collection the crewman obtained a contingecy faecal bag from the stowage container and defecated / vomited directly into the bag. The bag was then sealed, mass determined, processed, stowed and returned to Earth for examinations.

In contingency urine collection the crewman urinated directly into the urine bag by way of an elastomer cuff. Again urine was collected for a 24-hour period, volume determined, sampled, and disposed of.

A:

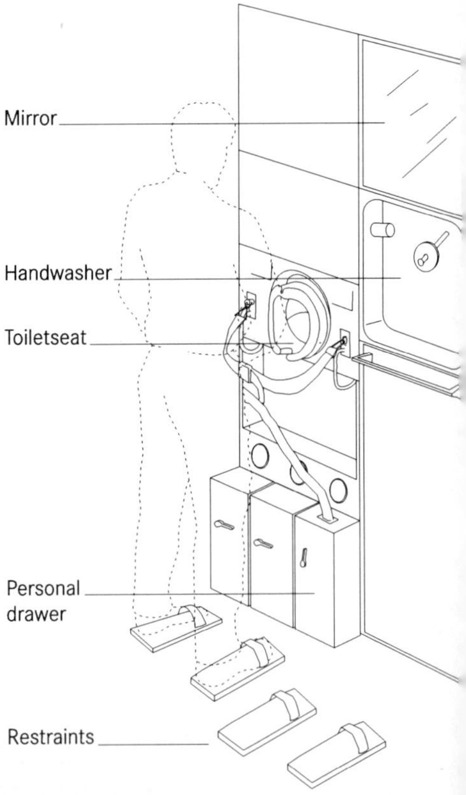

A, B: Skylab astronaut using the toilet system *(credit A: Author, based on NASA documents; B: NASA)*

ASTRONAUTS' EXPERIENCES

USABILITY

Joe Kerwin (Skylab 3): "***We owe our greatest appreciation to the people who designed it,***" *he said. "It has worked much better than anticipated, and it has been essentially trouble free and not terribly time consuming.*" (NASA [Skylab], 1977 p. 80)

Skylab Astronaut: "*The lap strap and handholds are an absolute requirement to the fecal collection equipment working correctly in that you do have to pull the cinch down and hold yourself down very close and firmly on the seat in order for the air flow to work correctly.*" (NASA [Bull.10], 1974 p. 13)

B:

5.3 HYGIENE SKYLAB

RÉSUMÉ HOUSEKEEPING

Considerable time was spent on trash management. The astronauts used a vacuum cleaner to remove things from the walls and clean the workshop surfaces. Skylab astronauts needed temporary stowage places where trash could be collected and disposed "when time allowed". When that time became available, the astronauts moved the trash to the zone below the crew deck where the trash airlock enabled them to put the trash into the "waste tank", which was an unused oxygen tank.

Non-flammable, biologically inert trash was collected in trash bags made of vented fabric. When filled, these bags were stored inside the area adjacent to the waste tank beneath the crew compartment floor. Flammable and biologically active trash was collected in specially designed bags, which were disposed of in the waste tank through the trash airlock.

A:

B:

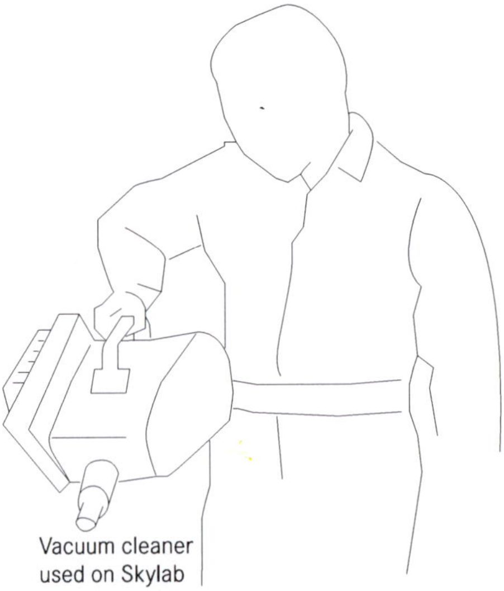

Vacuum cleaner used on Skylab

A: Astronaut Jack R. Lousma vacuum cleaning the filter of the ventilation mixing chamber
B: Astronaut with vacuum cleaner
C: Skylab 4 astronauts William R. Pogue and Gerald P. Carr force a trash bag further down into the OWS disposal tank

(credit A, C: NASA; B: Author)

ASTRONAUTS' EXPERIENCES

USABILITY

Skylab astronaut: ***"There were a lot of housekeeping tasks, that I had never done before,*** *and again, the first time through required more time than we allotted."* (NASA [Bull.7], 1974 p. 56)

"Well, there ain't no place to stow it [replacement filter]; we don't have enough equipment restraints and clops." (NASA [Bull.12], 1975 p. 55)

LIVABILITY

Joe Kerwin (Skylab 3): *"It takes forever to dry both one's self and the wall (...) even using that inadequate little vacuum cleaner that we've got."* (Time, 1973)

FLEXIBILITY

Skylab astronaut: *"And every time we have old shirts and shorts (...) we put them in that bag. Sometimes it gets too full and we shoot it out the trash airlock. But there's always some rags in there; so when it comes time (...) to clean, we usually go over there and (...) use* [the clothes] *instead of trying to do all that cleaning with these tissues."* (NASA, Bull.6, 1974 p. 35)

C:

5.4 HYGIENE SHUTTLE

CONCEPT Sponge baths, toilet compartment, galley rack
LOCATION Middeck
SPECIFIC Short mission duration

RÉSUMÉ PERSONAL HYGIENE

Shuttle astronauts take daily sponge baths using two washcloths, one for washing and one for rinsing, and one towel per day (NASA [Hygiene], 2002). To wash their hair, they use non-rinse shampoo *[Fig. A]*. Facilities for wet towels and trash bags are integrated in the Waste Management Compartment. With the galley, a personal hygiene station is provided. It is located on the left side of the middeck.

The personal hygiene kits are located in the forward Middeck. These kits include personal articles for dental hygiene, hair care, nail care and shaving. Towels and washcloths are restrained using rubber restraints with a Velcro base on the waste management door or middeck walls (NASA [Hygiene], 2002).

A:

A: STS-121 astronaut Lisa M. Nowak washes her hair on the middeck of the Space Shuttle Discovery
B: Astronaut Jean-François Clervoy shaves his face on the middeck of the Space Shuttle *(credit A: NASA; B: NASA, Jean-François Clervoy)*

ASTRONAUTS' EXPERIENCES

USABILITY
Jean- François Clervoy (CNES/ ESA): *"Normally there is no smell, but sometimes when people work hard, or people have been doing a lot of physical activity, there is a perspiration smell, but you get used to it and you don't realize it's there."* (Clervoy, 2009)

LIVABILITY
Julie Payette (NASA): *"You're not allowed to carry anything, actually. For all the hygiene items - the toothbrush, the hairbrush, etc. - you go to the crew equipment provider. You walk into a room and there's clothing, toothbrushes, razors for the boys, things like that, and you choose. "Okay, I'd like this kind of exercise shoes and two pairs of those shorts", and then you don't see them again until you're in space."* (Payette, 2009)

FLEXIBILITY
Gerhard Thiele (ESA): *"After waking we tried to allocate tasks such as washing, cleaning up, getting dressed and so on. Onboard one must grant privacy to the others - on our crew we had two women and with women you are, naturally, even a little bit more respectful."* (Thiele, 2010)

B:

5.4

HYGIENE SHUTTLE

RÉSUMÉ TOILET

The waste collection system collects, stows and processes wastes from male and female astronauts. It is similar to toilets designed for airplanes, but has separate receptacles for the collection of liquid and solid body wastes built into the ergonomic seat. The vacuum toilet works with airstreams compensating for gravity and has an ergonomic seat with a small opening of about 10cm (Clervoy, 2009).

It is located in the middeck of the orbiter within a compartment that can be closed off with two Nomex curtains (NASA [Shuttle Reference], 1981 p. 4-23).

A:

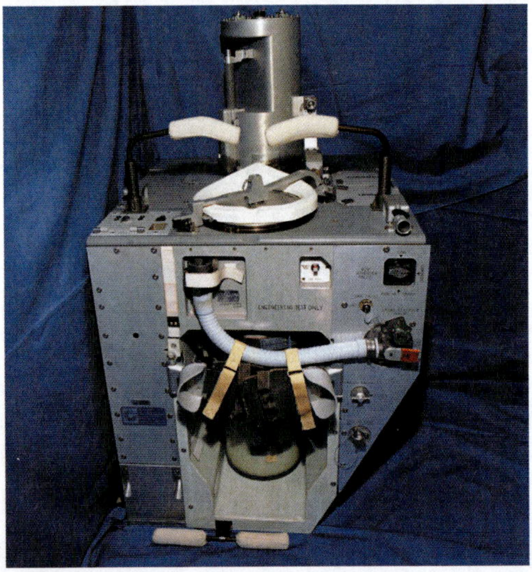

Urine and Faecal Collection Procedure
(NASA [WCS], 2002)

Liquid and solid wastes are collected, processed and stored separately.

1. Before use the commode is pressurized.
2. The Urinal can be used in a standing position or if attached to the commode in a sitting position. Female astronauts use a funnel which is especially designed to fit the female body.
3. If seated, body restraints and handholds are provided.
4. Liquid waste is collected using the urinal, which is a funnel attached to a flexible hose.
5. The fan separator separates the urine from airflow.
6. The liquid is stored in a waste water storage tank.
7. Solid waste is collected using the single multilayer hydrophobic porous bag liner integrated into the commode.
8. Solid waste and sanitary tissues are shredded by high speed tines and with rotating fans.
9. Solid waste is collected in a cylindrical storage tank.
10. Then the commode is depressurized to vacuum-dry the waste.

11. Waste water is vented to space, although future systems may recycle it, such as they do on the ISS. The air is filtered to remove odour and bacteria and then returned to the cabin.

Apollo bags are used as a backup-system in case of a failure of the waste collection system.

A, B: Toilet (WCS) on the Shuttle orbiter *(credit A: NASA; B: Author, based on NASA documents)*

ASTRONAUTS' EXPERIENCES

USABILITY

Jean- François Clervoy (CNES/ ESA): *"(...) but if you don't sit properly, ehm, it's a mess."* (Clervoy, 2009)

LIVABILITY

Jerry Linenger (Mir NASA 3): *"Quiet conversations can be heard coming from the adjacent middeck, and* ***any noise that one might generate can be downright embarrassing.****"* (Linenger, 2000 p. 59)

B:

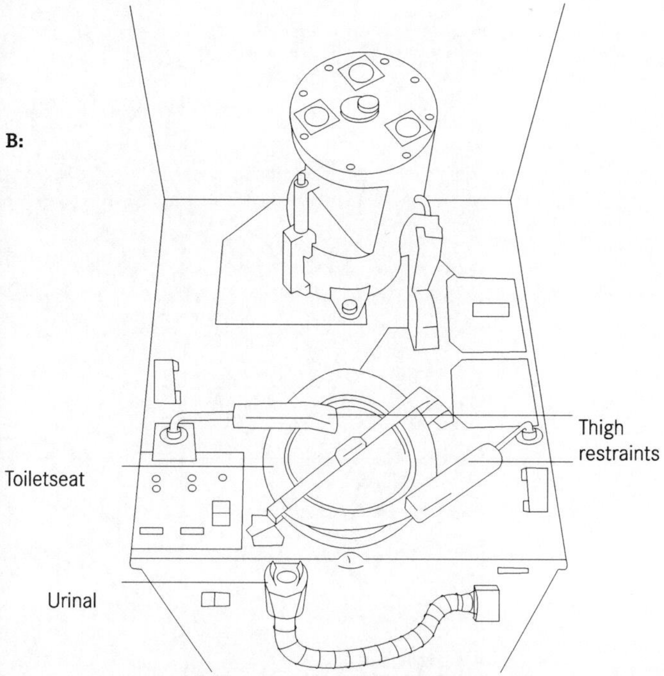

5.4 HYGIENE SHUTTLE

RÉSUMÉ HOUSEKEEPING

The Shuttle does not require extensive on orbit housekeeping. Housekeeping activities include the cleaning of the waste compartment, dumping excess water, replacing the Lithium Hydroxide canisters that scrub carbon dioxide out of the air, purging the fuel cells, dumping of garbage, and cleaning the floors and walls of the dining and hygiene area (NASA [Shuttle Reference], 1981 p. 5-6).

The cleaning equipment includes an Orbiter-powered vacuum cleaner, biocidal cleansers, disposable gloves, and general-purpose wipes (NASA [Shuttle Reference], 1981 p. 5-28). For intermediate storage of waste, trash bags for wet and dry trash are available in the crew compartment.

Daily, these bags are stowed and wet trash is put in a 'wet trash stowage compartment' which is located below the Middeck floor. Trash is brought back to Earth to be disposed of.

A:

A: STS-127 astronaut Christopher Cassidy removes dust particles from the air filter system with a vacuum cleaner
B: STS-100 astronaut Chris A. Hadfield wipes clean one of the overhead windows on the aft flight deck of the Space Shuttle Endeavour *(credit A, B: NASA)*

ASTRONAUTS' EXPERIENCES

USABILITY

Bill Oefelein (NASA, STS-116): *"**The most important part of my job is to keep the orbiter in tip top shape.** As the pilot, I help to maintain all the systems onboard."* (Oefelein, 2006)

LIVABILITY

Tom Jones (NASA): *"The challenge is to put these dozens of steps into just the right sequence so as to minimize wasted effort."* (Jones, 2006 p. 277)

FLEXIBILITY

Gerhard Thiele (ESA): *"Everyone tidies up their own things."* (Thiele, 2010)

B:

5.5 HYGIENE MIR

CONCEPT Toilet, shower, vacuum cleaner
LOCATION Base Block, KVANT 2; Kayutkas
FEATURES Recycling of water, fixed shower

RÉSUMÉ PERSONAL HYGIENE

The lavatory compartment was located in the wall aft of the Work Compartment. It had a spherical hair- and hand-washing unit *[Fig. B]*. The cosmonauts could wash their hair or hands by inserting them through the connected rubber gaskets (Portree, 1995). In addition the two crew quarters had a mirror for shaving and provided space to get dressed. For shaving, an electric razor with rotating heads that had a nozzle to suck in shaved whiskers and epidermal cells was provided (B.J.Bluth, 1987).

For whole body cleansing, cosmonauts used a tin-foil packet that contained soap. They filled it with water, shook it and punctured it a few times in order to put some drops on the body. Then they took a cloth and distributed it over the body (Linenger, 2000 pp. 181-182).

According to Jerry Linenger Mir clothing consisted of a cotton T-shirt, a pair of cotton shorts, sweat socks but no underwear. It was planned for the crew to change clothes every third day, but sometimes there was only enough clothing onboard to change once every two weeks (Linenger, 2000 p. 181). Colin Foale reports that they were provided with new underwear every three days (Foale, 1999 p. 186)

The cosmonauts were also influencing the design of their clothes; they could make choices of colour and pocket locations. Most clothing was disposed after wearing (B.J.Bluth, 1987).

A:

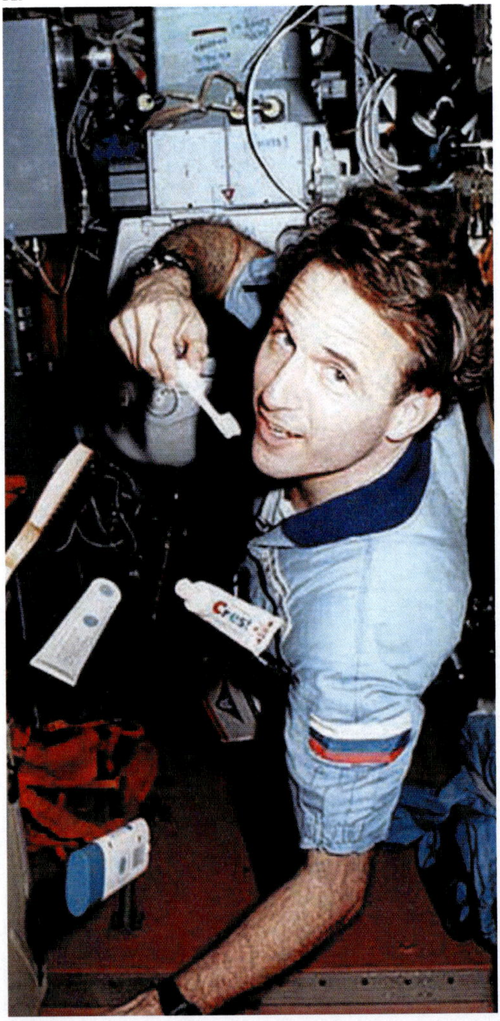

A: NASA/Mir-23 researcher Jerry Linenger brushes his teeth in the Spektr module
B: Waste and lavatory compartment in the Core Module *(credit A: NASA; B: Author)*

ASTRONAUTS' EXPERIENCES

USABILITY

Jerry Linenger (Mir NASA 3): *"Not wanting waste forty-five minutes shaving each day, I would simply squirt some water onto my face, rub it around my hands, and then dry my hands by dabbing them on my uncut hair."* (Linenger, 2000 p. 90)

LIVABILITY

Jerry Linenger (Mir NASA 3): ***"We were too busy to worry about, how we looked or even smelled."*** (Linenger, 2000 p. 183)

Jerry Linenger (Mir NASA 3): *"Five months of no haircut, no shower and only an occasional shave."* (Linenger, 2000 p. 183)

Jean- François Clervoy (CNES/ ESA): *"[When the thermal control system was broken] it was like moisture, very strong moisture smell, like you have in a cave."* (Clervoy, 2009)

B:

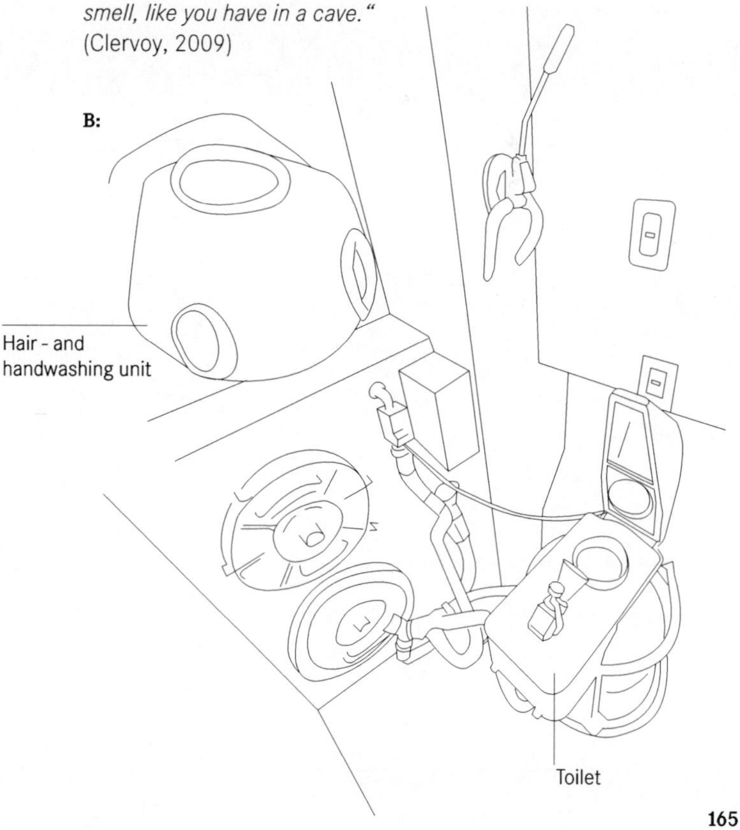

5.5 HYGIENE MIR

RÉSUMÉ SHOWER

In 1989 a shower compartment was launched with the Kvant 2 module. It was a metal compartment with a window facing into the Kvant 2 module *[Fig. A]*. Warm water was provided by a sprayer, used water was processed for reuse. This shower system did not work well, as the water adhered to the user and the inside of the shower cabinet. Cleaning was difficult (Portree, 1995 p. 164).

The shower cabin was later used as a steam-bath. In 1995 the shower cabin was removed and cut to pieces to make room for a gyrodyne (McDonald, 1998 p.19). Valery Polyakov, who stayed on Mir space station 437 days, used wet towels. He reported that "his skin was even better after the flight than before" (ESA [WHC], 2010).

A:

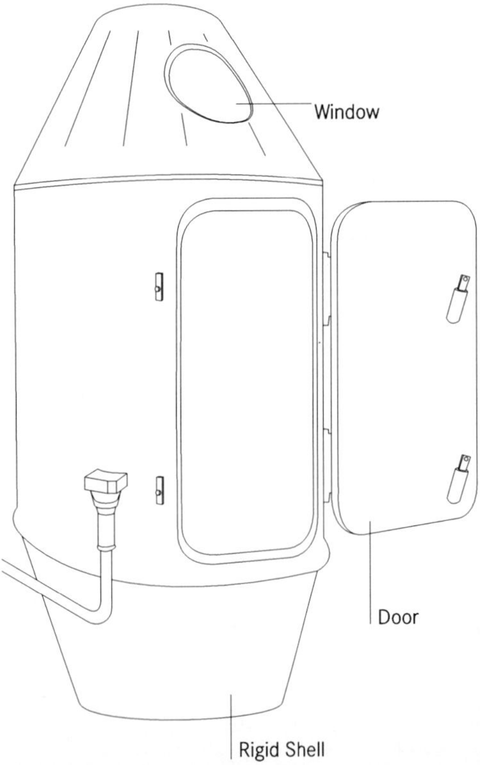

A: Rigid shower in Kvant 2 module on Mir *(credit A: Author)*

ASTRONAUTS' EXPERIENCES

USABILITY

"Water had to be sprayed in the shower cabin and a cosmonaut had to breathe through a special tube put through the wall of the cabin to avoid choking."
(ESA [WHC], 2006)

5.5 HYGIENE MIR

RÉSUMÉ TOILET

The space station Mir had two toilets, one in the base block and one in the Kvant 2 module (Harland, 2005 p. 331).

The toilet in the base block was in the floor, like the Salyut toiltet. It was not preferred "for sanitary and privacy reasons" (Burrough, 1998 p.88).

The toilet compartment in Kvant 2 *[Fig. A]* included the toilet and a 'wash basin' with a transparent cover that had holes for insertion of the head or hands. It could be closed off, using a door.

The Kvant 2-system called ASU collected, isolated, and stored waste. Toilet tissues were vented over-board. Urine was recycled and the purified water was used to generate oxygen. It was safe to drink, but the cosmonauts preferred not to drink it.

A:

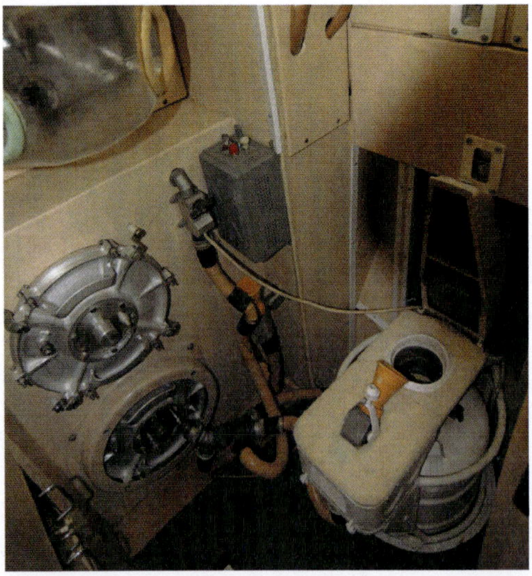

An advanced version of the Mir toilet system is currently in use on-board the ISS in the Zvezda module. The toilet can be used by men and women. Linenger described the procedure in his book "Off the Planet" as follows:

Urine Collection Procedure
(Linenger, 2000 pp. 58-59)

1. The astronaut grabs the urine-collecting hose,
2. Hooks his personal funnel to its free end,
3. Turns the fan on to create some suction
4. Opens a valve and
5. Urinates.

Faecal Collection Procedure
(Linenger, 2000 pp. 58-59)

1. The astronaut enters the toilet area and closes the door.
2. Then he or she sits down on the toilet seat,
3. And restrains the body with spring-loaded thigh-hold-down bars and footholds.
4. A stick is shifted forward to open the sliding cover of the 'hole'.
5. The astronaut defecates and
6. Shifts the stick back to its original position to close the hole.
7. Then the astronaut removed the restraints.

A: Toilet with washbasin in the Core Module of Mir
B: Toilet system on Mir *(credit A: Didier Capdevila, Capcom Espace; B: Author)*

ASTRONAUTS' EXPERIENCES

USABILITY
Reinhold Ewald (DLR/ESA):
"(...) to leave behind a clean toilet is something one is pedantically taking care of, so as not to cause a tense atmosphere amongst the crew members." (Ewald, 2009)

Jerry Linenger (Mir NASA 3):
"Urinating is not particularly difficult (...) but defecating in space is a much more demanding and time-consuming process. ***Not all astronauts are successful on their first try.****"* (Linenger, 2000 p. 58)

LIVABILITY
Jerry Linenger (Mir NASA 3):
"Yes I can drink my own urine if necessary. But I could not stomach the idea of drinking someone else's urine." (Linenger, 2000 p. 190)

B:

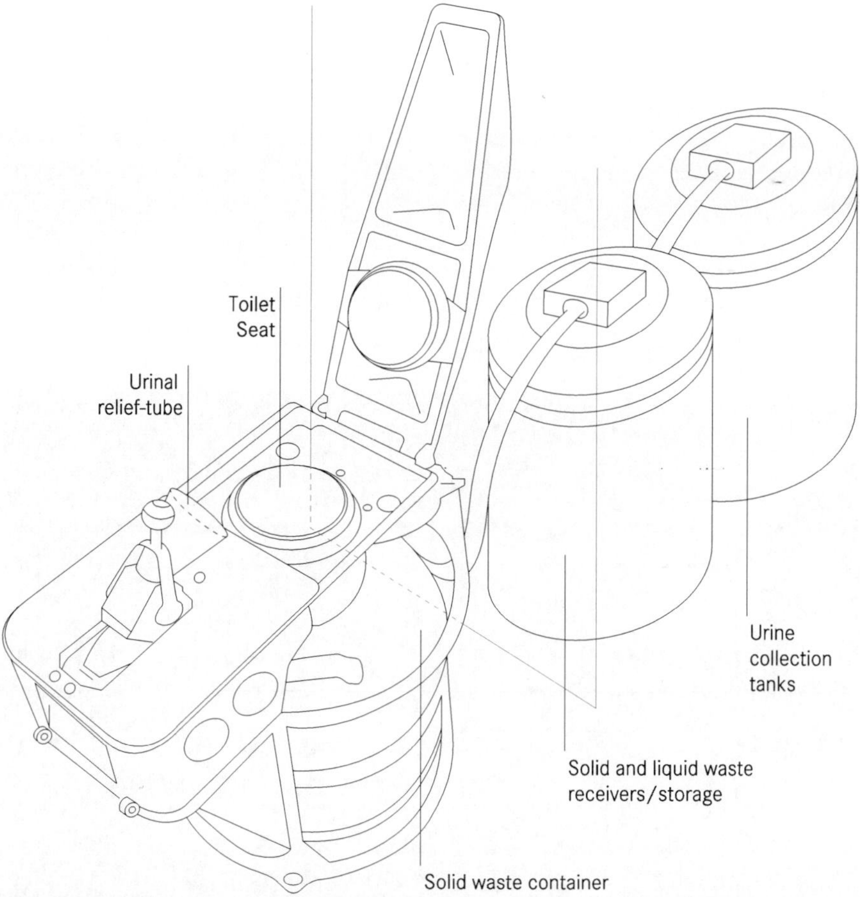

5.5 HYGIENE MIR

RÉSUMÉ HOUSEKEEPING

Housekeeping activities were similar to the Salyut predecessors. During the years storage demands increased. The module Kvant was used to store trash bags and unused equipment. (Burrough, 1998)

Most of the trash was disposed by placing it in the cargo compartment of the Progress freighters.
In addition, the Space station Mir had a science airlock which was used for scientific experiments that required access to vacuum. This airlock was also used for trash disposal (Portree, 1995 p. 105).

A:

A: Stowed items in the Mir Space Station including in-flight maintenance tools strapped down *(credit A: NASA)*

ASTRONAUTS' EXPERIENCES

USABILITY

Jean-Pierre Haignere (CNES/ ESA): *"There was a lot of mess in Mir, a lot of objects (...) all the walls were covered with these objects."* (Haigneré, 2009)

Jean-Pierre Haignere (CNES/ ESA): *"(...) there was a specific technical compactor for cans for instance. It was very important."* (Haigneré, 2009)

Tsibliyev (RSA): *"We worked with a saw til three o'clock but we suceeded. (...) It will take lots of time and effort to clean this area. It is full of slime."* (Burrough, 1998 p. 203)

"[Kvant] has dried up quite a bit (...) but now we have a mushrooming mold. (...) too bad we don't have any oil to fry them up here." (Burrough, 1998 p. 257)

LIVABILITY

Michel Tognini (ESA, Mir Antares): ***"The walls are full of things.*** *You don't see the texture of the wall."* (Tognini, 2009)

FLEXIBILITY

Michael Foale (NASA): *"I use old underwear, T-shirts and pants, and towels to soak up the water, if it is a film on the walls. (...) For great big water balls (...) in very cold corners (...) I use a (...) can with a rubber bladder, out of which I have pumped some air to create lower pressure and some suction inside. (...) It makes a gurgling sound, like bathwater running out."* (Foale, 1999 p. 186)

5.6 HYGIENE ISS

CONCEPT „Sponge baths", two toilets, vacuum cleaner
LOCATION Zvezda Service Module, Node 3 (Tranquility)
FEATURES Recycling of air, water and urine

RÉSUMÉ PERSONAL HYGIENE

The waste collection compartment- a small wash room with a toilet is located in the Service module Zvezda. An additional Waste and Hygiene Compartment is located in Node 3 (Tranquillity). Hygiene supplies, like washcloths, soap, and rinse-less shampoo, razors, toothbrushes and toothpaste are also provided. The astronauts take 'sponge baths' and can use personal hygiene supplies.

When an astronaut gets a haircut, a vacuum cleaner is used to prevent hair from floating free in the module *[Fig. A]*.

Several systems onboard ensure the health and safety of the astronauts during long term missions. The 'Crew Health Care System' (CHeCS) consists of three subsystems (NASA [ISS], 2010):

1. The 'Countermeasures System' provides exercise equipment and monitors the fitness of the astronauts.
2. The 'Environmental Health System' monitors the interior atmosphere, water, acoustics and radiation levels.
3. The 'Health Maintenance System' provides medical equipment and monitoring.

A:

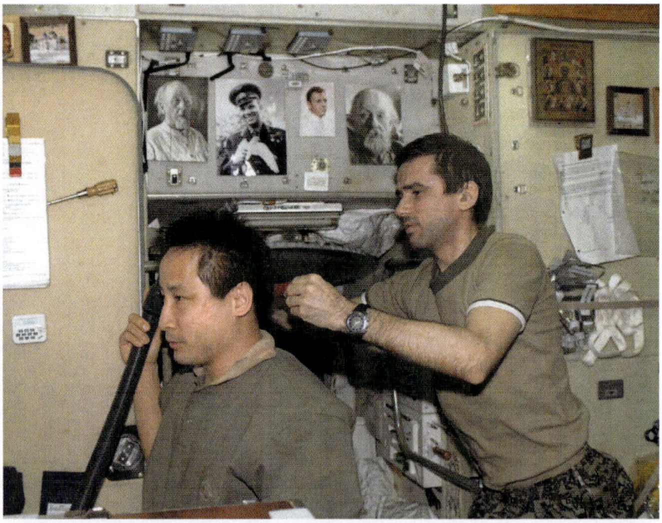

A: Expedition 7 commander Cosmonaut Yuri I. Malenchenko cuts astronaut Edward T. Lu's hair in the Zvezda Service Module, Lu holds a vacuum device *(credit A: NASA)*

ASTRONAUTS' EXPERIENCES

USABILITY

Leroy Chiao (NASA): *"(...) how you might clip your nails in space (...) Here's how I did it.*

First, get a strip of duct tape, and make a loop out of it, with the sticky side out. Find a place to do the clipping, next to an air intake filter. This way, any errant nails should be caught in the air filter, for later removal. Find a good place on the wall to stick your tape loop, and then carefully clip each nail, trying to keep the pieces big, so that you have a chance of holding onto them, instead of having them fly off into the cabin somewhere. Fix each piece of nail onto the sticky tape loop.

When you are finished, remove the tape loop, and fold it onto itself, to contain the nail clippings. Then, use the resulting tape double loop to clean off the air filter, of any nail debris, which got caught there. Wad up the tape ball, and discard it into a dry trash receptacle." (Chiao, 2009)

LIVABILITY

Ed Lu (NASA, Expedition 7): *"We don't have a shower up here, (...) so we wash using no-rinse soap and shampoo and a towel. It is the same stuff they use in hospitals for bedridden patients, and it works really well. That being said I am looking forward to a long hot shower when I get home!"* (Lu, 2003)

5.6 HYGIENE ISS

RÉSUMÉ TOILET

The first toilet was located in the Russian-built Service module Zvezda. The system is an updated version of the design used on the Mir space station *[Fig. A]*.

It is composed of two systems, one for urine and one for solid waste. Liquid and solid wastes are collected separately and processed separately. Each astronaut has a personal urinal funnel and different kinds are used for men and women. The module Zvezda also contains a water recovery system that processes water and could be used. for drinking in emergency, but is used to produce oxygen. (ESA [WHC], 2010)

In 2008, a second Waste and Hygiene Compartment and an advanced water recovery system were installed in the US Destiny Laboratory. This system can recycle urine water, so it can be used in space for drinking, food preparation and washing. It is called a toilet-to-tap system. The Waste and Hygiene Compartment is currently installed in Node 3 (ESA [Node 3], 2009).

Procedure
(NASA [Factsheet], 2006)

1. Before using the toilet, astronauts check alerts or warnings and
2. start the urine / air separator motor.
3. The astronauts position themselves on the toilet seat using restraints and
4. Attach their personal urine funnel to the hose's adapter.
5. A fan creates air flow into the solid-waste container and into the funnel.
6. The urine stream goes to the separator and is mixed with a preservative for storage in a tank
7. A small amount of the urine is distilled to create purified water.
8. The astronaut unhooks the seat liner packet, which is automatically sealed and
9. Stored in the solid waste tank until it is
10. Loaded onto the Progress to be dumped.

As an emergency system a portable toilet system is used. (see chapter Salyut toilet)

A:

A: Toilet compartment in the Zvezda Service Module of the ISS

(credit A: NASA)

ASTRONAUTS' EXPERIENCES

USABILITY

Richard Garriott (Space Tourist): *"Liquid collection is really no problem, it's basically a funnel with a vacuum behind it that you pee into, and the only challenge is how to keep yourself floating away while you're trying to hold the funnel in front of you. Technically a little challenging but it's no big deal (...)*

On the other hand, ***dealing with the solid waste is quite a bit more challenging.*** *The method is still basically the same, but you sit on a miniature toilet which is lined with a plastic bag with perforations on the bottom, and there's a fan below that is there to pull things into the bag. However it doesn't really work that well, it doesn't respond in the same way that gravity assists you in getting rid of solid wastes on the Earth. So that is an area that requires additional strategies!"* (Garriott, 2008)

LIVABILITY

Clayton Anderson (NASA): *"Potty ops are quite interesting. We do it just like you do on the ground, but we use a vacuum hose. Your ability to aim and the level of subsequent clean-up are directly linked in all aspects; and the amount of hose suction is critical! You get the idea, right? And you sure don't want to leave a mess for the next guy! That would not go over very well."* (Anderson, 2007)

Michael Barratt (NASA): *"The taste is great."* [drinking recycled urine] (CBS News, 2009)

FLEXIBILITY

Ed Lu (NASA, Expedition 7): *"The toilet works great, although we did have to do some plumbing repairs last week when the fan unit died."* (Lu, 2003)

5.6 HYGIENE ISS

RÉSUMÉ HOUSEKEEPING

The station is cleaned regularly by the astronauts once a week to remove food waste, collect trash, and to clean surfaces that are frequently touched by astronauts. Housekeeping equipment used on the International Space Station includes two portable vacuum cleaners – one in the U.S. and one in the Russian segment. In addition the crew uses dry wipes, durable wipes, detergent wipes, disinfectant wipes, and utensil detergent wipes, utensil sanitizing wipes, and cleansers. Once used, all the wipes go into the trash collection bags.

Trash in the Russian segment is collected and stowed in empty food containers to be disposed of in the Progress. (NASA [ISS], 1998 p. 297)

A:

A: Expedition 20 astronaut Nicole Stott uses a vacuum cleaner in the Destiny laboratory of the ISS
B: Vacuum cleaner elements *(credit A: NASA; B: Author, based on NASA documents)*

ASTRONAUTS' EXPERIENCES

USABILITY
Greg Chamitoff (NASA): *"We've been doing a lot of work to move equipment, supplies, and trash around for the past few weeks. The Progress [Russian cargo vehicle] and ATV [European Automated Transfer Vehicle - cargo ship] are both departing next week. So we needed to off-load any remaining supplies and put everything there that we need to trash."* (Chamitoff, 2008)

LIVABILITY
Sandra Magnus (NASA): *"We do not have a method of washing clothes and tend to wear them until they are ready to be thrown away. We do not have a way to recycle solid waste or garbage and essentially have the equivalent of "throw it over the side of the ship," since it **all gets burned up in the Earth's atmosphere.**"* (Magnus, 2009)

FLEXIBILITY
Sunita Williams (NASA, Expedition 15): *"You know how it is when you move in, you need to make it "your" home. (...) Part of this rearranging is using the barcode reader. We use a system of barcodes to locate and identify pieces and parts to everything here on the ISS. (...) this way we can monitor when something is moved. It also helps us to put things back correctly when we find something "floating around."* (Williams, 2007)

B:

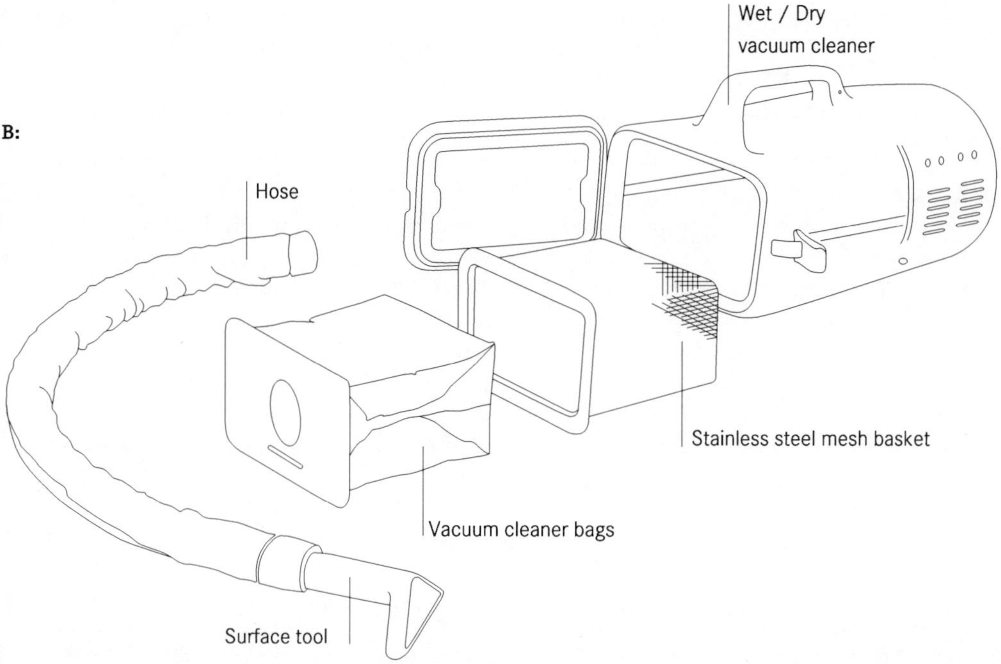

5.7 PERSONAL HYGIENE SUMMARY OBSERVATION

	APOLLO	SALYUT	SKYLAB
Concept	Simple hygiene equipment (shaving)	Full body cleaning (collapsible shower), personal hygiene kit, medical kit	Full body cleaning (collapsible shower), dedicated areas, a hand washer, stowage for personal hygiene items, personal hygiene kit
Review	Apollo astronauts shaved, they liked “getting clean”	Shower was principally appreciated, but the procedure was too long and the material of the foldable shower was not strong enough; Clothing was adjusted by the astronauts	Whole body cleaning was enjoyed, but pre- and after activities were too time-consuming; Conflicting placement of shower in the work area
Potentials	Adequate, sufficient and personal hygiene equipment is necessary; Dust has to be prevented from entering the habitat and delicate equipment	Adjustable and personal clothing is required; Shower system was further developed	Personal hygiene equipment should be close to the hygiene facility (storage); Showers were liked, but the system and procedure was not

Diagram

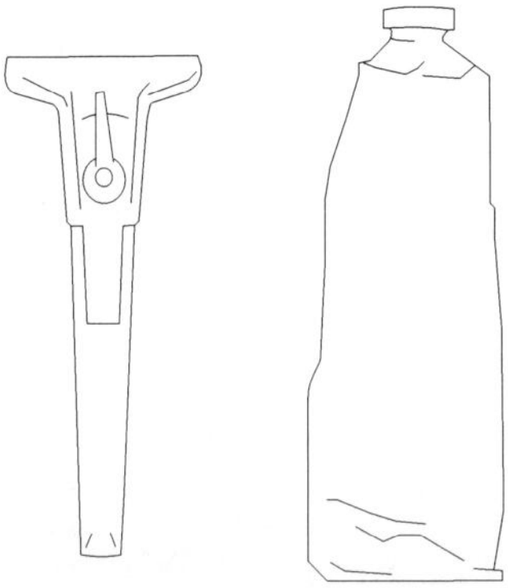

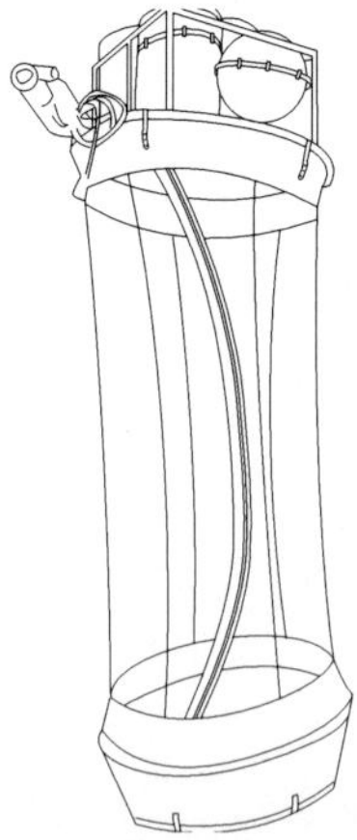

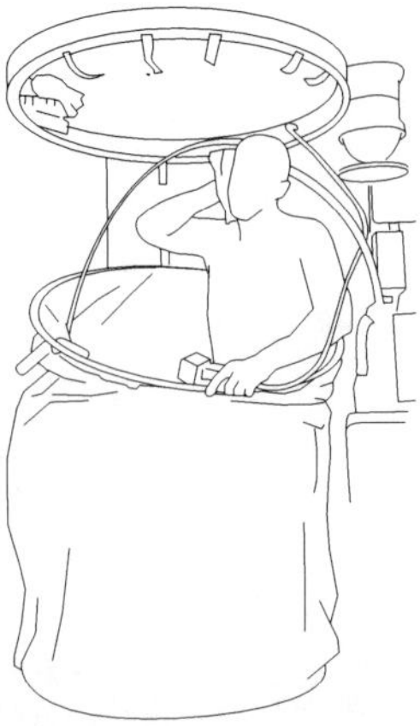

SHUTTLE	MIR	ISS
Daily „sponge baths“, personal hygiene kit	Full body cleaning (shower / sauna), hand wash units, hair-washing unit, electric razor	Daily ‘sponge baths’, personal hygiene kit; Russian segment similar to Mir Core Module
Personal hygiene items can be taken with and used	The shower in Kvant 2 was used as a sauna and later removed;	Astronauts are used to sponge baths - seems adequate for now
Daily sponge baths seem adequate for short time missions	The converted shower to a sauna system seems interesting	Full body cleaning not yet solved for future long term missions

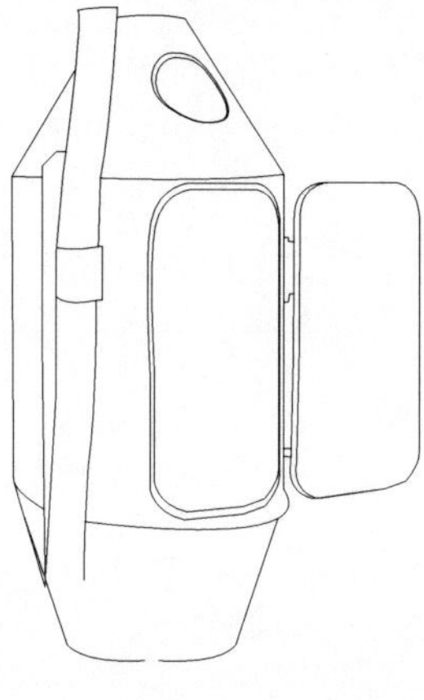

5.7
TOILET SUMMARY OBSERVATION

	APOLLO	SALYUT	SKYLAB
Concept	Apollo bag, UTS, UTA, UCTA, diaper	Toilet compartment with curtain	Toilet compartment with toilet mounted in the middle of the wall; Faeces collected in bags, then vacuum dried and stored; Urine collected with individual receivers
Review	The use of the Apollo bags was not appreciated; Urine on the clothes, couches and the urine dispenser posed a problem	Faeces were collected in sealed containers and vented to space once a week; Toilet system on Salyut 7 was used by a female cosmonaut	Toilet should be far from sleep compartment to minimize noise disturbance to sleeping crewmembers
Potentials	For use in emergency situations	Mobile toilet system is used as an emergency toilet	Toilet should be separated from other areas; Individual thermal control for the waste management compartment was requested

Diagram

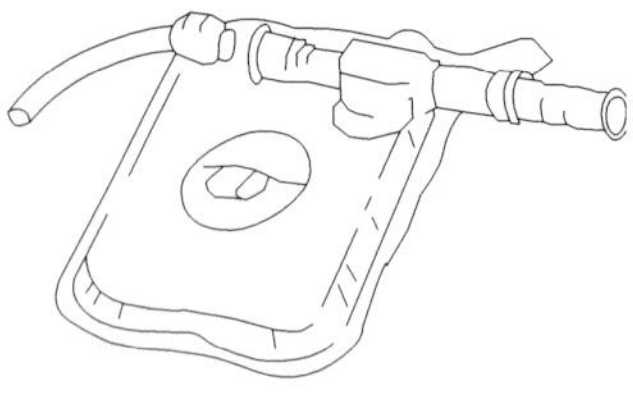

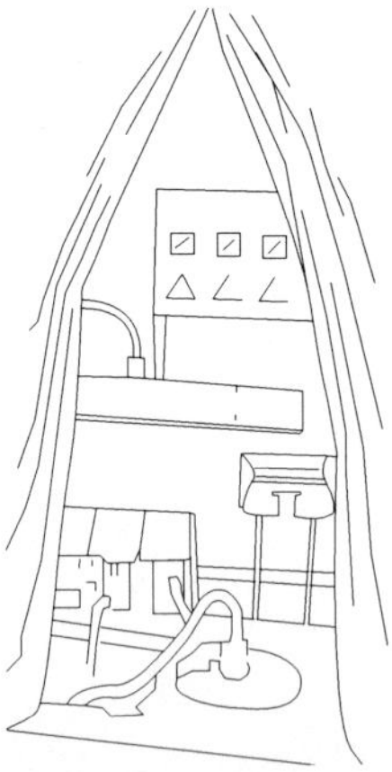

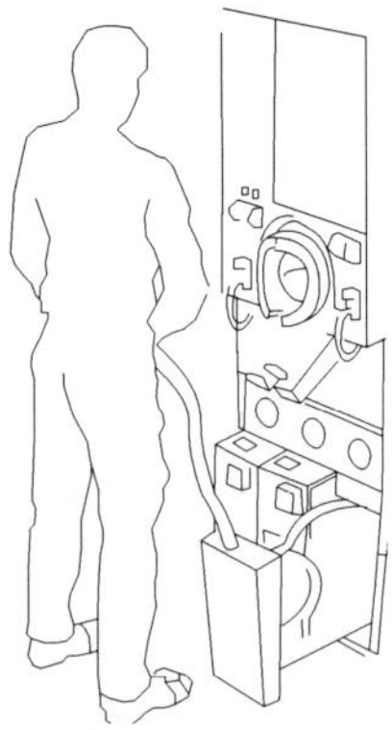

SHUTTLE	MIR	ISS
Toilet compartment	Toilet compartment; process similar to Salyut missions	Since 2009 two toilet compartments
Faeces collected in container, vacuum dried and stored; Urine is stored in the waste water tank and is vented when full	Two toilets in two compartments; Recycling of urine for non-drinking water	One Russian style, one American style toilet; Urine recycling for the US toilet
Toilet acoustically close to other activities; Some astronauts complained about the toilet being adjacent to the galley - they had sanitation / contamination concerns	Two toilet compartments in different modules can provide increased privacy	Two toilets efficient in case of emergency or broken toilet; Also, these toilets provide dissimilar redundancy because they have different designs and failure modes

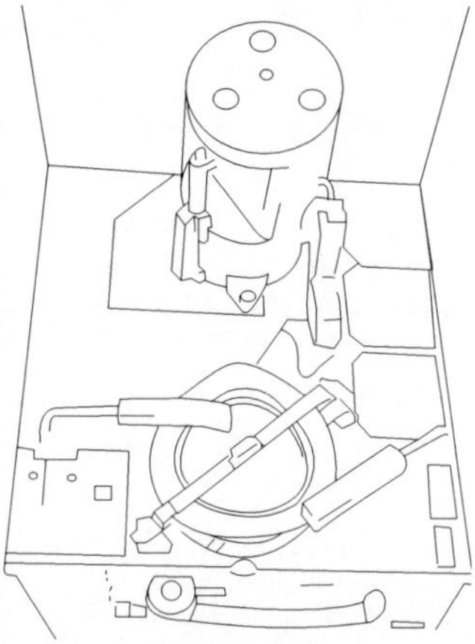

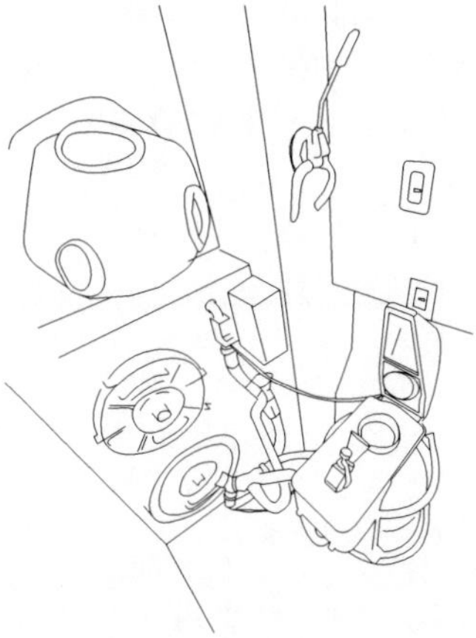

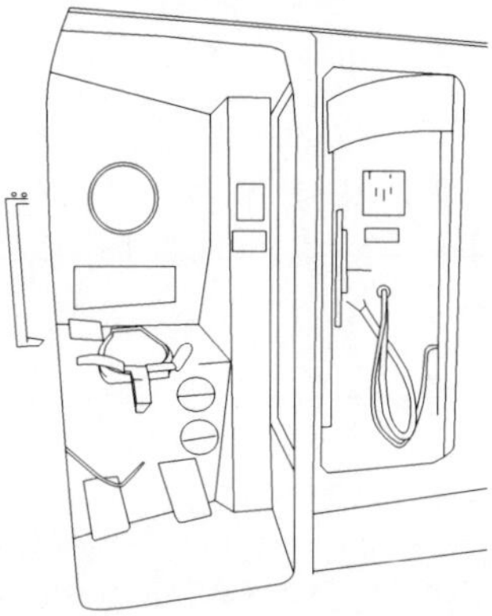

5.7 HOUSEKEEPING SUMMARY OBSERVATION

	APOLLO	SALYUT	SKYLAB
Concept	Charging of batteries, checking reserves, dumping waste, etc.	Vacuum cleaner, wet cloth, disinfectants	Vacuum cleaner, tissues; Waste was disposed through the trash airlock
Review	Lunar dust caused problems; Little tools, like scissors got lost	Waste was collected and ejected from the station; Water was recycled through the condensate regeneration system to produce oxygen	Vacuum cleaner was inefficient; Considerable time was spent on trash management, need of temporary storage; Conflicting placement of shower and work
Potentials	Lunar dust has to be prevented from entering the habitation module and delicate equipment	Efficient housekeeping strategies are required; "The station is a mess"; the place behind the panels was discovered to "hide" stuff	Some hardware was of multi-use; Different storage for different kind of wastes

Diagram

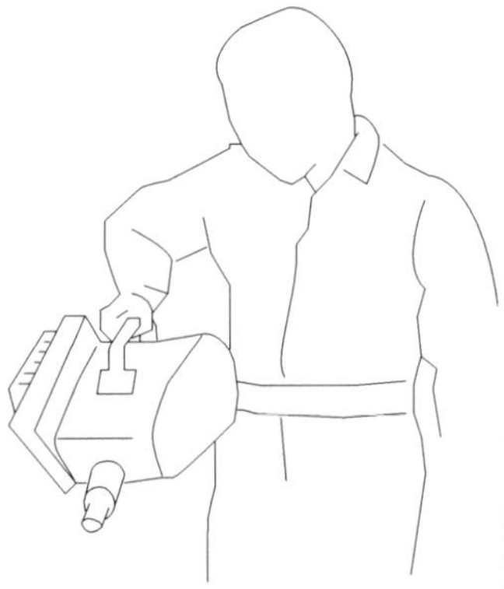

SHUTTLE	MIR	ISS
Limited on- orbit housekeeping; trash is dumped overboard	Vacuum cleaner, wet cloth, disinfectants, trash airlock	Vacuum cleaner, wipes, detergents, trash collection bags
Astronauts maintain systems onboard; Trash is dumped	"A lot of mess", walls were covered with objects after a few years	Regular cleaning of the station, stowage and disposal is still a challenge
Limited management as it returns to Earth	Storage system to be considered for long usage	Storage is still a challenge; Temporary storage places for resupplies needed; Recycling of waste for future missions

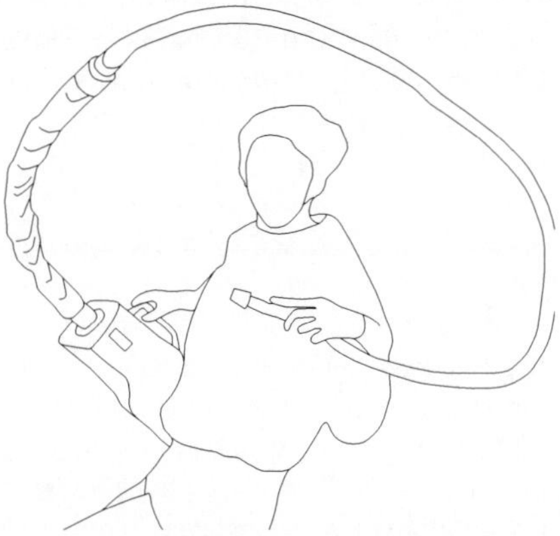

5.7 HYGIENE SUMMARY OBSERVATION

Limitations on hygiene facilities and waste management problems have been high on the list of discomforts reported by individuals in confinement experiments. According to a report by Connors et al (1999 p. 66), problems associated with hygiene and waste management were well known and will receive continued attention.

A Shower in Micro-Gravity

According to NASA-Guidelines (Allen, et al., 2003 p. 44), the ideal personal hygiene facility should be as "easy to use as a ground-based facility." Full body cleaning facilities for use in microgravity were developed for the Salyut, Skylab and Mir space stations. During the Salyut (3 -7) missions and on Skylab an inflatable shower system was used. Bluth reported, that the Salyut 7 cosmonaut Lyakhov proposed to have a shower on Kosmos "which is always assembled and ready for use" (B.J.Bluth, 1987 pp. I-83). This idea was implemented for the Mir space station, but the shower was later converted to a steam room, due to technical problems, and in the end removed in order to provide space for other equipment.

Most of the cosmonauts and astronauts considered taking a shower as a pleasant experience, but the preparation and the work after was too much effort compared to the result (NASA [Skylab CS], 1974 p. 430; Zimmermann, 2003 p. 156; B.J.Bluth, 1987 pp. I-84). Presently, no shower is in use on the ISS, astronauts take 'sponge-baths'.

Showers used in space so far mimicked gravity with forced airflow to move the water in the cabin. Skylab astronauts requested a design "permitting the crewman to actually wash with the water rather than having to soak up everything in a washcloth" (NASA [Skylab LL], 1974 p. SL2-6). But the attempt to transport the shower concept from Earth to space has failed. In terms of future design developments the 'sauna concept' of the Mir space stations was imaginative.

In 1985 Rafael Garcia and others from JSC's Man-Systems Division have developed a shower concept that is based on anthropometric measurements and the requirement of recycling water. "A spry nozzle with a squeeze lever on the side directs water onto a wash cloth or directly onto the body" (Deason, 1987). According to Garcia, all design worked well and were considered for the space station whole body shower. Other micro gravity shower systems were developed at the ISU (Mohanty, et al., 2001) and the TU Munich (Vogler, 2002), but none of them installed.

A Toilet in Microgravity

A number of waste management facilities have been developed and used. The Skylab toilet system worked well for the astronauts, although some found it strange that the seat was mounted on the wall rather than the floor. Their recommendations for future concepts included a higher airflow for faecal collection, the seat to be fabricated in a softer material and a less noisy urine separator device. The Space Shuttle uses a vacuum toilet with an ergonomic toilet seat. This kind of toilet can be used in any gravity environment. The International Space Station provides two toilet facilities to be used by men and women. Although there are different reports on the toilets' usability, the process seems to be technically a little challenging at the beginning, but the crew become accustomed to the process and adapt to the systems.

Hygiene and toilet are personal activities
The Apollo 11 astronauts requested individual hygiene kits, instead of having only one to share (NASA [Debriefing A11], 1969). Personal hygiene becomes a high priority within a small and confined volume. Cosmonaut Lebedev (1990, p. 194) wrote about Savitskaya in his diary: "Sveta spent a long time making herself beautiful in the transport vehicle."

During the Skylab missions a foldable and transparent shower that was located in the middle of the Orbital Workshop was used by the all-male crew. The men considered it a pleasant experience, but the activity was too long and too time-consuming (NASA [TM], 1974). In addition the shower in the middle of the crew deck concept would probably not have worked that well for a mixed gender crew.

Astronauts also had suggestions on the placement of the toilets and hygiene facilities. On Skylab missions the power module of the faecal and urine collector disturbed sleeping crewmembers. "To minimize noise disturbance to sleeping crewmembers" astronauts suggested locating the hygiene area as far away from the sleeping area as possible" (NASA [Skylab CS], 1974 p. 117; NASA [Skylab LL], 1974 p. SLL2-6). On Mir, "for sanitary and privacy reasons, no one was ever enthusiastic about using a toilet two feet from the dinner table" (Burrough, 1999 p. 88).

Noise is often an issue. This is confirmed by Jerry Linenger's (2000, p. 59) experiences with the toilet on the Space Shuttle. He noted, that "noise that one might generate" while on the toilet could be heard outside and were thus be embarrassing for the one inside the toilet compartment. In addition to noise, smell is an important issue to be dealt with. The Waste Management Compartment in Skylab was enclosed with aluminium sheet to isolate any odours (NASA [Bull.18], 1975 p. 10) "We care about smell, but we care in advance, so we don't have strong smells in space" (Clervoy, 2009). Good smells he remembers include the soft smells from personal hygiene products.

Interesting is, that Skylab astronauts complained about inadequate lighting and wrong placement in the WMC. They could not read while using the faecal collector. (NASA [Bull.18], 1975 p. 19]

Clothing
Different kinds of clothing are available for astronauts and have been specially developed to meet specific requirements. Astronauts don't wear their personal clothes taken from Earth, but can choose from a range of clothes supplies in advance of their flight. After wearing, clothes are disposed via a re-entering Progress spacecraft: "When a piece of clothing has been worn as many times as possible, it's placed in a bag for disposal" (NASA [Spacewear], 2003).

Sometimes astronauts adjust their clothes. During the Salyut mission, Salyut cosmonauts wore a special designed 'Penguin Suit' that had elastic bands sewed in to simulate the force of gravity during the day. According to Lebedev (1990 p.188), at least one crew adjusted this suit, because Tolia Berezovoy and he felt more comfortable without the elastic bands. ISS astronaut Donald Pettit found an interesting way of re-using his underwear. He used it as substitute for soil to grow tomato and basil seeds. The 'hand-made sprouter' made of worn underwear, toilet paper and needles worked well and grew seeds within 2 days. (NASA [Laundry], 2003)

Skylab astronauts requested to have place for clothing during the sleep periode close to the place where they slept (NASA [Bull.3], 1974 p. 13). They also mentioned the importance of garments

5.7 HYGIENE SUMMARY OBSERVATION

as tools in order to support the astronauts activities during the day (NASA [Bull.6], 1974 p. 54). Skylab astronauts utilized dirty clothes as cleaning rags for housekeeping (NASA [Bull.6], 1974)

Several studies on clothing and laundry systems in microgravity have been conducted. One example is the VEST Experiment in 2002 (Energia, 2000-2011) and the GOAL Experiment in 2005 (ESA [GOAL], 2005) that evaluated different garments for use in space flight. Recently, in 2010 the General Services Administration in the US launched a competition on a 'Simple Microgravity Laundry System'. But so far none of these concepts was implemented because of costs, volume and mass restraints, and no urgent needs.

Housekeeping
Lunar dust presented a problem during the Apollo missions. The fine, magnetic and electrostatically charged and abrasive dust posed a hazard for the human as well as technical equipment.

During the Soviet space station era it became obvious that with operational time, surface and atmospheric contamination increased. To keep the contamination level down, the interior of the Salyut stations and the housekeeping strategies evolved from the first station to Salyut 7 (B.J.Bluth, 1987).

On Skylab, astronauts used a vacuum cleaner for cleaning the space station. It was designed in such a way that the blower design could also be used for the faecal collector and the shower (NASA [Skylab LL], 1974 p. SLL2-6).

Vacuum cleaners were used also on the Mir space station and are currently used on the Space Shuttle and the ISS.

Concerning trash management, Skylab astronauts suggested that food containers be designed to consume minimum volume and trash to be separated into biological active and inactive material, because active material needs to be disposed of daily, whereas inactive material can be stored in a "temporary collection site" to be disposed of "when time allowed" (NASA [Skylab LL], 1974 p. SLL1-6).

Different Users and Standards
The Skylab station was considered very clean and "you just don't get as dirty up her", because everything was new and the station was inhabited by only 9 men in total. (NASA [Bull.4], 1974; NASA [Bull.6, 1974 p. 43)

Especially for long-term missions housekeeping is very important. The sensation of astronauts on-board Mir is described by Burrough as follows: "Close your eyes and the place smelled faintly of sweat, which was not entirely surprising." (Burrough, 1999, p. 86) Mir Cosmonaut Sergei Krikalev made the observation that "levels of cleanliness and odours varied according to the standards of the resident crew". Krikalev stated that the missions he was part of had a relatively high standard of housekeeping. They kept unused equipment behind the walls and avoided nets of full equipment in the station. (Portree, 1995 p. 106)

This statement matches a review on cultural factors for spaceflight by Kring (2001). Personal hygiene and clothing are among the issues that are influenced by and vary with culture. According to Shuttle astronaut Gerhard Thiele (2010), today astronauts tidy up their own things and act responsibly toward the other crewmembers.

Efficient Use of Resources

Salyut and Mir space stations were the first stations to incorporate water recycling systems. They used a system for collecting condensed water from perspiration and respiration out of the cabin atmosphere. On Salyut 6, fresh water was brought from Earth with the Progress (B.J.Bluth, 1987 p. I-94) but already 50% of the used water was recycled (Davis, et al., 2008 p. 273). According to Bluth the life support system was improved for the space station Mir. Oxygen was recycled and moisture condensate, wash water, and also urine were recycled for use as potable water. Urine water was kept in special coloured containers in case of emergencies, but "for psychological reasons" it was "not preferred" (B.J.Bluth, 1987 p. I-95; Linenger, 2000; Foale, 1999 p. 76). It seems that the Russian have also been working on urine recycling technology to recycle urine, but it never flew in space (Clervoy, 2009).

The International Space Station has an 'Environmental Control and Life Support System (ECLSS) that provides a pressurized and habitable environment. Oxygen is generated by electrolysis of water. Solar generated electricity splits water into hydrogen gas and oxygen gas and hydrogen is vented into space. The ECLSS also removes carbon dioxide and trace contaminants from the air, produced by the astronauts and the experiments from the air (NASA [ISS], 1998).

In May 2009, astronauts on board the International Space Station celebrated a milestone towards self-sustainability. For the first time Astronauts had been drinking water that was recycled from their urine, sweat and water condensed from exhaled air. The recycling system was installed earlier in November 2009 (CBS News, 2009).

A high degree of self-sustainability is a prerequisite for future long-term missions. Firstly, it will not be possible to bring enough oxygen, water, or other vital gases to the Moon or Mars to satisfy the need for a long-term human stay. Secondly a closed-life cycle on board minimizes costs by reducing the need to ship resupplies. In space and planetary habitats self-sufficiency refers to a concept of a closed system or autonomy.

Jean- François Clervoy considers the future design of a spacecraft with the requirement "to treat everything (...) just by itself" (Clervoy, 2009).

5.8 COMPARISON USABILITY MATRIX

APOLLO	SALYUT	SKYLAB
Availability and Equipment		
Personal hygiene equipment (shaving); Apollo bag, UTS, UTA, UCTA, diaper	Shower; Toilet; Housekeeping facilities, washable walls, vacuum cleaner	Shower; Toilet; Housekeeping facilities, in-orbit hand washer, vacuum cleaner
Spatial Arrangement		
All activities in same module	Deployable shower in the Large Diameter Work Compartment; Hygiene and toilet area in the rear	Deployable shower in Experiment Area (OWS); Hygiene and toilet in the Waste Management Compartment
Object Management		
No individual storage	Set of selected hygiene equipment	Personal storage locker, personal hygiene kit
Ergonomic Safety		
Astronauts did a wet shave; Lunar dust was a problem	Snorkels, goggles and restraints for showering	UV light covers on windows; Restraints for the shower and the toilet; Lack of restraints in the WMC

SHUTTLE	MIR	ISS
Hygiene station (Galley); Toilet; Housekeeping facilities, vacuum cleaner	Shower (Kvant 2), hair- and hand-washing unit Two toilets; Housekeeping facilities, vacuum cleaner	Personal hygiene equipment Two toilets; House-keeping facilities, vacuum cleaner
Middeck, near the galley; No dedicated area for personal hygiene	Fixed shower in Kvant 2; Two toilets in two different modules (Base block, Kvant 2)	Personal Hygiene (Node 3, Zvezda, FGB) Waste and Hygiene Compartment (Node 3) Personal dressing area in Crew Quarters Two toilets in two different modules (Zvezda module, Node 3)
Personal hygiene kit	Panelized compartment system; Personal hygiene kit	Rack system; Personal hygiene kit
Ergonomic seat and restraints for the toilet	Handles as restraint for the toilet	Ergonomic seat and restraints for the toilet

5.9
COMPARISON LIVABILITY MATRIX

APOLLO	SALYUT	SKYLAB
Territoriality and Privacy		
'Apollo bags'; No privateness in the module	Toilet and hygiene compartment could be closed off with a rubber curtain	Transparent shower in the Orbital Workshop (man-only-crew) Toilet compartment could be closed off
Sensory perception		
Urine, faeces, and gas odours	Seemed not to be a major problem	Faecal / urine collector and ATM cooling pumps disturbed sleep; WMC was enclosed with aluminium sheet to isolate odors
External relations		
Command and Lunar Module had windows to the outside	None, except space station windows	None in the dedicated hygiene areas;
Internal relations		
"Next to each other", all activities in same module	Shower cabin was transparent; Toilet had a curtain	Shower cabin was transparent; Toilet compartment was spatially enclosed but adjacent to sleep area

SHUTTLE	MIR	ISS
Toilet compartment can be closed off	shower had a rigid opaque enclosure; Toilet could be closed off with a door	Hygiene and Toilet compartment can be closed off
Closure of toilet is not sound-proof, no specific smells reported	Toilet compartment is not sound-proof (embarrassing), moisture smell reported	Soft smells of hygiene products; No specific smells reported
None in the dedicated hygiene areas	None in the dedicated hygiene areas	None in the dedicated hygiene areas
Toilet can be closed off, but noises are sometimes embarrassing	Rigid shower had a window	Can be closed off from group domain with a door

5.10 COMPARISON FLEXIBILITY MATRIX

	APOLLO	SALYUT	SKYLAB
Spatial Flexibility	All functions in the same volume	Dedicated area for toilet and temporary for the shower (deployable)	Dedicated area for toilet and temporary for the shower (deployable)
Object Flexibility	Different systems used Some tools could be used inside the module and during EVA	Deployable shower	Deployable shower
Individual Flexibility	Different bags for the Command and Lunar Module, some shaved some did not, but equipment was available	Cosmonauts adjusted the gravity simulating 'penguin suit'	Personal hygiene could be done anywhere

SHUTTLE	MIR	ISS
Dedicated area for toilet	Permanent shower in Kvant 2; Two toilets in two different modules	Various places for personal hygiene; Two toilet compartments, with different toilet designs in two different modules
Galley can be removed	Shower was later used as steam-bath	Various equipment for personal hygiene
Individual hygiene equipment is allowed	Personal hygiene could be done anywhere	Astronauts use their personal items

5.11 DESIGN DIRECTIONS

The most important findings related to the human activity 'HYGIENE' are summarized and structured in the form of design directions. These directions summarize the main points in a short sentence. The paragraph following refers to the empirical background of experiences made in extraterrestrial habitats.

USABILITY

Integrate easy-to-use full body cleansing

For long term missions a full-body cleansing facility needs to be provided that works, is pleasant and takes the microgravity conditions into account.

Use resources efficiently

All materials and equipment need to be designed in such a way, that they are reusable, transformable (Cascadian cycle) or recyclable (closed-loop system). In view of planetary settlements infrastructure can be shared and in-situ-resource-utilization and technology can be collectively used.

Integrate redundancy

At least two toilet systems that are connected to the on-board Waste management and Life Support System prevent critical situations in case one is broken.

Integrate 'empty' storage space

Sufficient storage space is needed. Empty, available storage space needs to be provided in times of increased demands. When resupplies are delivered, a lot of objects have to be transferred from one place to another and temporary storage demands increase. For long term missions, temporary 'empty'storage can support reconfiguration of the station.

Integrate personal storage

An exclusive private stowage area for personal hygiene equipment and items is necessary for each crewmember - preferably neyt to the places where astronauts sleep.

Design for easy housekeeping

To limit the amount of housekeeping activities for astronauts, surface finishes have to be chosen that are easy to clean. Equipment and facilities should be easy to maintain or repair.

LIVABILITY

Design for maximum level of privacy

Private full-body and partial-body cleansing activities have to be integrated into the design for future space exploration missions. The waste management area should be spatially, acoustically and visually separated from other activity zones.

Take care of odours and sound

Discomforting odours and sound are to be avoided, while 'nice smells' and sound support habitability.

FLEXIBILITY

Provide temporary enclosures

In a very small volume with overlapping functions, temporary increased need for privacy (personal hygiene activities) can be achieved with mobile partitions in various locations.

Provide individual restraints

Easy-to-use restraints to hold the body in a certain position and to fasten equipment are required for hygiene activities.

Allow individual clothing

In view of long-term missions, a laundry system has to be developed and tested, as not all wearables can be taken along and people cannot wear the same clothes all the time. The possibility of cleaning dirty clothes on-board would also support astronauts in bringing their personal and individual clothes, which can be an additional support for their psychological well-being.

Take varying user standards into account

Standards and traditions for personal hygiene and housekeeping vary across the planet. The design of an internationally operated habitat should allow adjustments according to the crew on duty in terms of spatial and time allocation of hygiene activities.

5.12 REFERENCES

Allen, Christopher S., et al. 2003. Guidelines and Capabilities for Designing Human Missions : NASA / Johnson Space Center, 2003. NASA / TM -2003-210785.

Anderson, Clayton Conrad. 2007. Clayton Anderson's ISS In-Flight Journal. [Online] NASA, 2007. [Cited: 5 10 2009.] http://www.nasa.gov/mission_pages/station/expeditions/expedition15/journal_anderson_02.html.

B.J.Bluth, M. Helppie. 1986. Soviet Space Stations as Analogs. Washington D.C., US : NASA Headquarters, 1986. NASA Grant NAGW-659.

Boeing. 1995-2009. Apollo 11 - Mission Log Book. [Online] Boeing, 1995-2009. http://www.boeing.com/defense-space/space/apollo11/logbook.html.

Burrough, Bryan. 1999. Dragonfly - NASA and the crisis aboard Mir. London, GB : Fourth Estate Limited, 1999. ISBN 1 84115 087 8.

Carr, Gerald P. 2000. Oral History Transcript - Gerald P. Carr. [interv.] Kevin M. Rusnak. Huntsville, Arkansas, 25 October 2000.

CBS News. 2009. Space Station Recycling Urine To Water. [Online] © MMIX, The Associated Press, 20 05 2009. [Cited: 5 06 2009.]http://www.cbsnews.com/stories/2009/05/20/tech/main5029346.shtml?source=RSSattr=SciTech_5029346.

Cernan, Eugene A. 2007. Oral History Transcript - Eugene A. Cernan. [interv.] Rebecca Wright. Houston, Texas, 11 December 2007.

Chamitoff, Greg. 2008. Greg Chamitoff's Journal. [Online] 8 February 2008. [Cited: 20 April 2008.] http://www.nasa.gov/mission_pages/station/expeditions/expedition17/journal_greg_chamitoff_3.html.

Chiao, Leroy. 2009. Astronaut Leroy Chiao's Blog. [Online] 25 May 2009. [Cited: 25 July 2009.] http://leroychiao.blogspot.com/2009/05/life-in-space-rest-of-story.html.

Clervoy, Jean- François. 2009. Transcript not published [interv.] Sandra Häuplik-Meusburger. Paris, France, June 2009.

Connors, Mary M., Harrison, Albert A. and Akins, Faren R. 1999. Living Aloft, Human Requirements for Extended Spaceflight. Washington, D.C. : NASA, Ames Research Center, 1999. SP-483 / ISBN: 1-4102-1983-6.

Davis, Jeffrey R., Johnson, Robert and Stepanek, Jan. 2008. Fundamentals of Aerospace Medicine. US : Lippincott Williams & Wilkins, 2008. 0781774667.

Deason, Billy. 1987. High Altitute Hygiene - Space Station shower gets most from every drop. Nov 6. , 1987, p. 3.

Energia. 2000-2011. Human Life Research on ISS Russian Segment: VEST Experiment. [Online] S. P. Korolev Rocket and Space Corporation „Energia", 2000-2011. http://www.energia.ru/en/iss/researches/medic/17.html.

ESA [GOAL]. 2005. Garments for Orbital Activities in weightlessness (GOAL). [Online] 2005. A. Dominoni, Spin Design, Politecnico di Milano, Italy.

ESA [Node 3]. 2009. Node 3 and Cupola Information Kit. European Space Agency, Directorate of Human Spaceflight. 2009. http://esamultimedia.esa.int/multimedia/ISS/ESA_Node_3_cupola_Infokit.pdf.

ESA [WHC]. 2010. Waste and Hygiene Compartment. [Online] 3 February 2010. [Cited: 5 October 2010.] http://asimov.esrin.esa.it/SPECIALS/node3/SEMEHHSJR4G_0.html.

Ewald, Reinhold. 2009. Transcript not published. [interv.] Sandra Häuplik-Meusburger. Cologne, Germany, July 2009.

Foale, Colin. 1999. Waystation to the Stars - The Story of Mir, Michael and Me. London, GB : Headline Book Publishing, 1999. ISBN 0 7472 6331 0.

Garriott, Richard. 2008. Richard Garriott: Space tourist. [interv.] Emily Baldwin. Astronomy Now Online. 13 November 2008. http://www.astronomynow.com/081113RichardGarriottInterview.html.

Garriott, Owen K. 2000. Oral History Transcript - Owen K. Garriott. [interv.] Kevin M. Rusnak. Houston, TX, USA, 6 November 2000.

Haigneré, Jean-Pierre. 2009. Transcript not published. [interv.] Sandra Häuplik-Meusburger. Paris, France, June 2009.

Hall, Edward T. 1966. The Hidden Dimension. N.Y. : Anchor Books, 1966. ISBN 0-385-08476-5.

Harland, David Michael. 2005. The story of the spacestation MIR. Berlin : Springer Praxis Books in Space Exploration, 2005. ISBN: 0387230114 .

Jones, Tom. 2006. Skywalking, an Astronaut's memoir. U.S.A. : HarperCollins Publisher, 2006. ISBN 0-06-088436-3.

Lebedev, Valentin. 1990. Diary of a Cosmonaut - 211 Days in Space. United States and Canada : Bantam Air & Space Series, 1990. Vol. Vol.4. ISBN: 0-553-28778-8. http://www.spaceflight.esa.int/users/index.cfm?act=default.page&level=11&page=iss01-sme7h.

Kanas, Nick M.D. und Manzey, Dietrich Ph.D. 2003. Space Psychology and Psychiatry. Dordrecht, The Netherlands : Springer/ Microcosm Press, 2003. ISBN 978-1-4020-6769-3.

Kring, Jason. 2001. Multicultural factors for international spaceflight. Journal of Human Performance in Extreme Environment. 2001, pp. 11-32.

Linenger, Jerry M. 2000. Off The Planet - Surviving five periolous months aboard the Space Station MIR. U.S.A. : McGraw-Hill, 2000. ISBN 0-07-136112-X.

Lu, Ed. 2003. Greetings, Earthlings - Ed's Musings from space. [Online] NASA, Expedition 7, 09 08 2003. [Cited: 20 5 2009.] http://spaceflight.nasa.gov/station/crew/exp7/luletters/lu_letter9.html.

Magnus, Sandra. 2009. Sandra Magnus' Journal . [Online] NASA, 6 January 2009. [Cited: 10 Februar 2010.] http://www.nasa.gov/mission_pages/station/expeditions/expedition18/journal_sandra_magnus_6.html.

McDonald, Sue. 1998. Mir Mission Chronicle, November 1994 - August 1996. Houston Texas, US : NASA, Lyndon B. Johnson Space Center, 1998. NASA TP-1998-208920.

Mohanty, Susmita and Rycroft, Michael J. 2001. Design Concepts for Zero-G Whole Body Cleansing on ISS Alpha - Part II: Individual Design Project. Houston, Texas 77058-3696 : Host Institution: Lyndon B. Johnson Space Center, International Space University, 2001. NASA/CR-2001-208931.

NASA [Bull.3]. 1974. Skylab Experience Bulletin No. 3: Architectural Evaluation for Sleeping Quarters. Houston, Texas, US : NASA, Lyndon B. Johnson Space Center, 1974. JSC-09537.

NASA [Bull.6]. 1974. Skylab Experience Bulletin No. 6: Space Garments for IVA Wear. Houston, Texas, US : NASA, Lyndon B. Johnson Space Center, 1974. JSC-09540.

NASA [Bull.7]. 1974. Skylab Experience Bulletin No. 7: An Overview of IVA Personal Restraint Systems. Houston, Texas, US : NASA, Lyndon B. Johnson Space Center, 1974. JSC-09541.

NASA [Bull.8]. 1974. Skylab Experience Bulletin No. 8: Cleansing Provisions within the Waste Management Compartment. Houston, Texas, US : NASA, Lyndon B. Johnson Space Center, 1974. JSC-09542.

NASA [Bull.9]. 1974. Skylab Experience Bulletin No. 9: Foot Restraint Systems. Houston, Texas, US : NASA, Lyndon B. Johnson Space Center, 1974. JSC-09543.

NASA [Bull.10]. 1974. Skylab Experience Bulletin No. 10: Body Restraint Systems. Houston, Texas : NASA, Lyndon B. Johnson Space Center, 1974. JSC-09544.

NASA [Bull.12]. 1975. Skylab Experience Bulletin No. 12: Temporary Equipment Restraints. Houston, Texas, US : NASA, Lyndon B. Johnson Space Center, 1975. JSC-09546.

NASA [Bull.14]. 1975. Skylab Experience Bulletin No. 14: Personal Hygiene Equipment. Houston, Texas, US : NASA, Lyndon B. Johnson Space Center, 1975. JSC-09548.

NASA [Bull.18]. 1975. Skylab Experience Bulletin No. 18: Evaluation of Skylab IVA Architecture. Texas, Houston, US : NASA, Lyndon B. Johnson Space Center, 1975. JSC-09552.

NASA [Debriefing A11]. 1969. Apollo 11 - Technical Crew Debriefing. Houston, Texas : NASA, Manned Spacecraft Center, 1969. 276.

NASA [Debriefing A12]. 1969. Apollo 12 - Technical Crew Debriefing. Mission Operations Branch - Flight Crew Support Division. Houston, Texas : Manned Spacecraft Center, 1969. p. 308.

NASA [Debriefing A17]. 1973. Apollo 17 Technical Crew Debriefing (U). prepared by: Mission Operations Branch Flight Crew Support Division. Houston, Texas : Manned Spacecraft Center, 1973. MSC-07631.

NASA [Dust Management]. 2006. The Apollo Experience Lessons Learned for Constellation Lunar Dust Management. Houston, Texas, US : Sandra A. Wagner, Johnson Space Center, 2006. NASA/TP-2006-213726.

NASA [Hygiene]. 2002. NASA Human Spaceflight - Personal Hygiene Provisions. [Online] 2002. http://spaceflight.nasa.gov/shuttle/reference/shutref/crew/hygiene.html.

5.12 REFERENCES

NASA [Factsheet]. 2006. Living and Working in Space - Factsheet. Houston, Texas, USA : NASA Lyndon B. Johnson Space Center, 2006. FS-2006-05-019-JSC.

NASA [ISS]. 1998. International Space Station Familarization. Houston Texas, USA : Lyndon B. Johnson Space Center, 1998. TD9702A.

NASA [Joint Flight]. 1974. On the Treshold of the Joint Flight - Translation of: „V Preddverii Sovmestnogo Poleta, Nauka i Zhizn", No. 10, October 1973, pp. 68-73. Washington, DC, USA : NASA, 1974. NASA TT F-15, 269.

NASA [Laundry]. 2003. Astronaut's Dirty Laundry. [Online] 10 April 2003. http://www.nasa.gov/vision/space/livinginspace/Astronaut_Laundry.html.

NASA [Shuttle Reference]. 1981. Space Shuttle News Reference.Houston, Texas, USA : U.S. Government Printing Office, 1981. 723-000/562.

NASA [Skylab LL]. 1974. Lessons Learned on the Skylab Program. Houston, Texas, USA : NASA, Lyndon B. Johnson Space Center, 1974. JSC-09096.

NASA [OWS]. 1974. MSFC Skylab Orbital Workshop Vol. III. Alabama : Skylab Program Office, George C. Marshall Space Flight Center, 1974. Technical Memorandum. NASA TM X-64813.

NASA [Skylab]. 1977. Skylab, our first space station. Washington D.C., US : NASA, Scientific and Technical Information Office, 1977. Prepared by George C. Marshall Space Flight Center, Edited by Leland Belew. NASA-SP-400.

NASA [Spacewear]. 2003. Living in Space / Space Wear. [Online] 02 04, 2003. [Cited: 01 08, 2009.] http://spaceflight.nasa.gov/living/spacewear/index.html.

NASA [Skylab CS]. 1974. MSFC Skylab Crew Systems Mission Evaluation - Technical Memorandum. Alabama : George C. Marshall Space Flight Center, 1974. Technical Memorandum. NASA TM X-64825.

NASA [WCS]. 2002. Human Space Flight - Waste Collection System. [Online] July 2002. http://spaceflight.nasa.gov/shuttle/reference/shutref/orbiter/eclss/wcs.html.

NASA. 1972. Apollo Experiences Report - Crew Provisions and Equipment Subsystem. Houston, Texas, USA : Manned Spacecraft Center, 1972. NASA TN D-6737

Oefelein, Bill. 2006. Bill Oefelein's STS-116 Mission Blog. NASA. [Online] 19 December 2006. [Cited: 5 December 2008.] http://www.nasa.gov/mission_pages/shuttle/shuttlemissions/sts116/oefelein_blog.html.

Payette, Julie. 2009. Astronaut Julie Payette talks to Kate Fillion about her next trip to space - the luggage, the food, the office politics, the calls home. [interv.] Kate Fillion. Astronomy Now Online. 4 February 2009. http://www2.macleans.ca/2009/02/04/maclean's-interview-julie-payette

Portree, David S.F. 1995. MIR Hardware Heritage. Houston, Texas : NASA, Lyndon B. Johnson Space Center, Information Services Division, 1995. NASA RP 1357.

Prunariu, Dumitru-Dorin. 2011. Transcript not published. [interv.] Sandra Häuplik-Meusburger. Vienna, Austria, 2011.

Sauer, Richard L. and Jorgenson, George K. 1975. Biomedical Results of Apollo - Waste Management System. Houston, Texas, US : Lyndon B. Johnson Space Center, The Boeing Company, 1975. pp. Section VI, Chapter 2. NASA SP-368.

Schmitt, Harrison. 2009. Transcript not published. [interv.] Sandra Häuplik-Meusburger. Vienna, Austria, 2009.

Thiele, Gerhard. 2010. Transcipt not published. [interv.] Sandra Häuplik-Meusburger. Vienna, Austria, December 2010.

Time. 1973. Science:Living it up in Space. [Online] 25 Juni 1973. http://www.time.com/time/magazine/article/0,9171,907462,00.html.

Tognini, Michel-Ange Charles. 2009. Transcript not published. [interv.] Sandra Häuplik-Meusburger. Cologne, Germany, July 2009.

Vasilyev, M. P., et al. 1974. Salyut Space Station in Orbit, Translation of ‚Salyut na Orbite`, Mashinostroyeniye Press, Moscow, 1973, 160 pages. Washington, D.C. : NASA, National Aeronautics and Space Administration, 1974. NASA TT F-15,450.

Vogler, Andreas. 2002. The Munich Model: Creating an Environment for Space Architecture Development. Houston, Texas, US : AIAA Space Architecture Symposium, 2002. AIAA 2002-6122.

Williams, Sunita Lyn. 2007. Flight Engineer Sunita Williams' Space Station Journals. [Online] NASA, 2007. [Cited: 20 12 2009.]

Zimmermann, Robert. 2003. Leaving Earth, Space Stations, Rival Superpowers, and the Quest for Interplanetary Travel. Washington, D.C., USA : Joseph Henry Press, 2003. ISBN 0-309-08548-9.

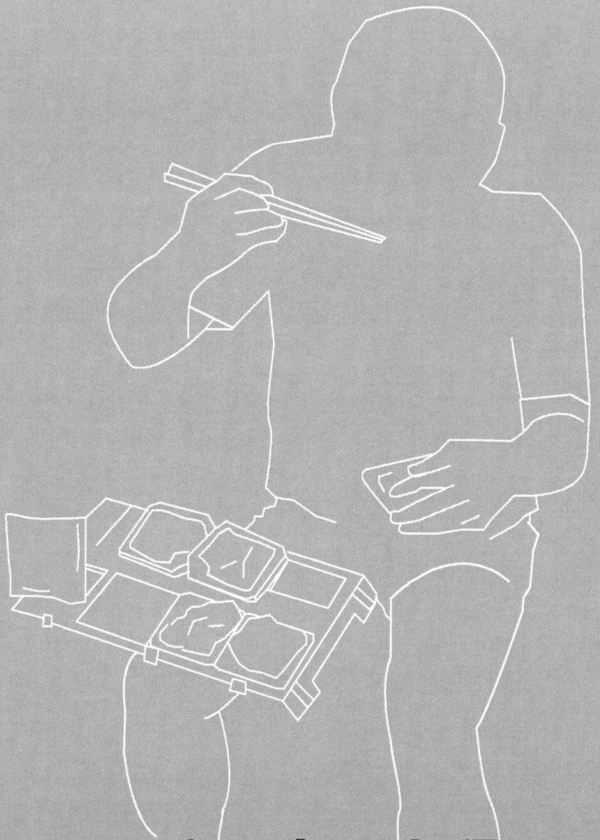

PART 6: HUMAN ACTIVITY

FOOD

Sub-activities associated with 'Food' include preparing, growing, and consuming food and drinks, collecting, storing and processing waste; as well as associated translation through the space habitat and stowage. Food systems for outer space habitats have changed to a high degree over the last 40 years. The food eaten today is mainly prepared on Earth.

6.1 FOOD APOLLO

CONCEPT Rehydratables, bite-sized, ready-to-eat, thermostabilized food
LOCATION Command Module / Lunar Module
SPECIFIC Improvement of food and packaging

RÉSUMÉ

Design Criteria for the food system of the early Apollo missions resulted from the experiences of the Mercury and Gemini missions. A summary of the design drivers and the resulting design requirements (in brackets) is listed below (Smith, et al., 1974 p. 5):

- Light Weight (Use of dehydrated food to reduce weight)
- Low Volume (Use of dehydrated and compressed food to limit stowage volume)
- Individual Crew Acceptance (Use a wide variety of food flavours to preclude monotony)
- Least Preparation time (Use of dehydrated food, that can be quickly rehydrated; Use of ready-to-eat bite-size food that require no preparation time)
- High Food-residue stabilization (Use of nontoxic germicidal agents to prevent micro-organism growths in used food packages; Use of re-sealable packages for food)
- Wholesomeness (Use certified, top-quality food raw materials)
- High Gastrointestinal compatibility (Use food that is easy to digest)
- High Nutrient content (Use food that provides high energy levels compared to its weight and volume)
- Long-term stability (Use food that is physically, chemically and microbiologically stable throughout the mission; Use packaging that protects the food against physical damage)
- Constrained Volume (Use food for minimal metabolic maintenance when space-suited)
- Shape and Size (Use food that fits stowage-volume configurations)

A:

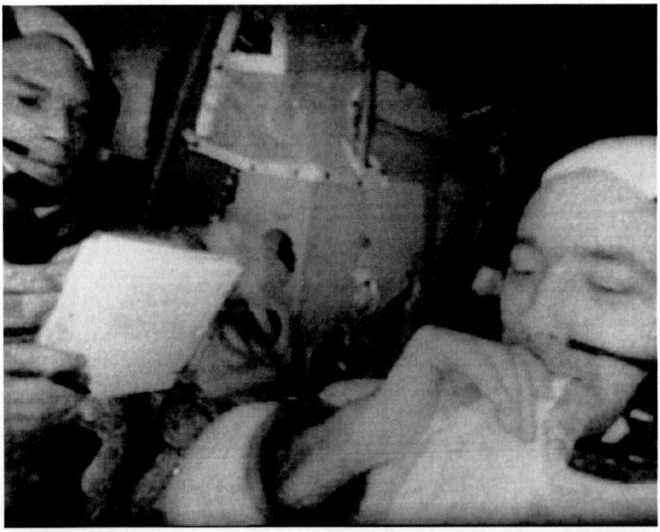

A: Apollo 9 commander James A. McDivitt is drinking from a space food pouch
B: Apollo Food (1968-1972)

(credit A, B: NASA)

The **menu** had to provide balanced nutrition for the three astronauts for a mission length of 14 days. For the stay on the lunar surface a menu for two days was prepared. The menu consisted of bite-sized, rehydratables and semisolid thermostabilized food and beverage powder (Smith, et al., 1974 p. 11).

The dehydrated *bite-sized food items* were ready-to-eat cubes with a protective and edible coating. They were rehydrated with saliva in the mouth and came in various tastes, such as meat, cheese and fruits.

The *rehydratables* were precooked and dehydrated. They could be rehydrated using water in less than 15 minutes before consumption. Tuna, salmon or chicken salad as well as shrimp cocktail was served this way.

The *semisolid thermostabilized food* was packed in flexible metal tubes and was a special high-nutrient-defined formula.

Hot water was available only in the Command Module; in the Lunar module astronauts consumed cold food.

B:

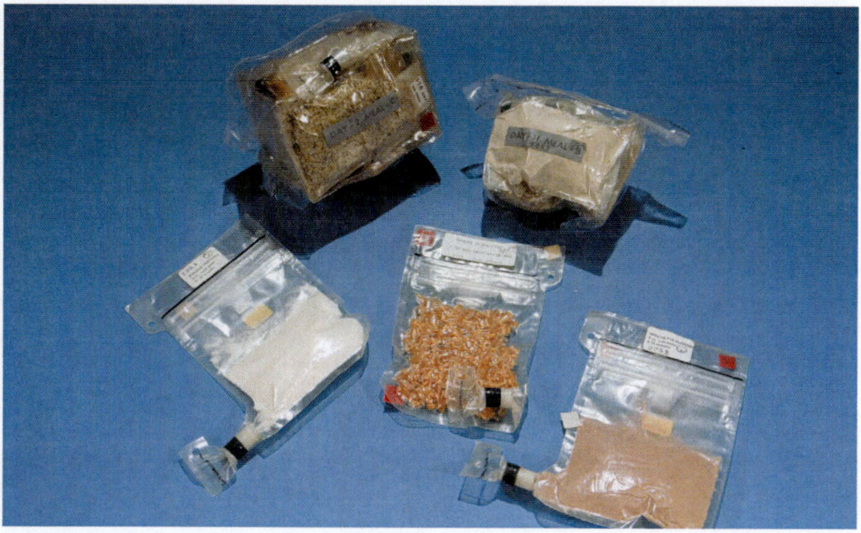

6.1 FOOD APOLLO

The sealed food packages could only be opened by cutting with scissors, which were lost sometimes (Cernan, et al., 1999 p. 308). Rehydratable foods were packed in plastic bags with a valve for water insertion (NASA [Bull.19], 1976 p. 3).

Food systems as well as packaging were improved throughout the Apollo program. Later a "rehydratables-food spoon-bowl package" *[Fig. A, B]* for food that astronauts could eat with a spoon and the "wetpack" a ready-to-eat meal was introduced. Also sandwiches were provided for the first part of the flight and new kinds of flavours were introduced (Smith, et al., July 1974 pp. 5, 41).

The **food-stowage** and a water dispenser were located in the lower equipment bay of the Command module and in one compartment in the Lunar module (Smith, et al., July 1974 p. 23). The food containers were colour-coded for each crewmember. There was no space dedicated to preparing food or to eating in either the Command Module or the Lunar Module.

A:

B:

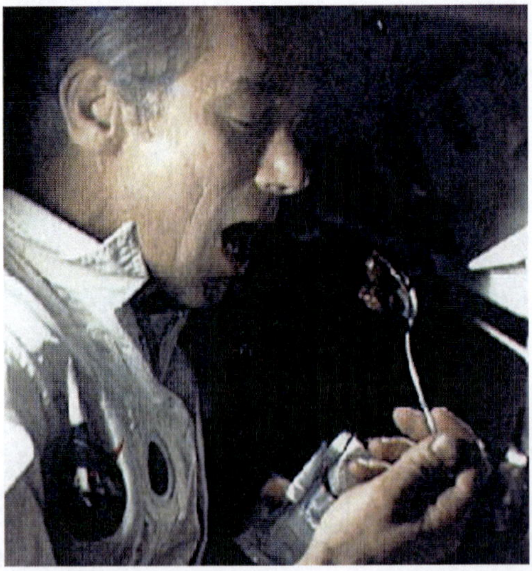

A: Apollo rehydratable food spoon-bowl package
B: Astronaut Eugene Cernan eating a meal (chocolate pudding) aboard Apollo 17 spacecraft *(credit A, B: NASA)*

ASTRONAUTS' EXPERIENCES

USABILITY

Buzz Aldrin (Apollo 11) enjoyed the shrimp cocktail, explaining later that, *"The shrimp were chosen one by one to be sure they would be tiny enough to squeeze out of the food packet, and they were delicious!" (Eat Me Daily, 2009)*

LIVABILITY

CapCom Gordon Fullerton (Apollo 17): *"Scissors, scissors, who's got the scissors?" [teased Fullerton, when Ron Evans lost his scissors) (Cernan, et al., 1999 p. 308)*

Harrison Schmitt (Apollo 17): *"The food was good (...) It didn't have much taste (...) but also NASA had decided, they didn't want to have a lot of spice in the food. They didn't want to stimulate your intestinal system. You were always looking for something that had a little more taste, and the bacon squares probably had the most taste, so they disappeared quickly." (Schmitt, 2009)*

FLEXIBILITY

Charles Duke (Apollo 16): *"And so we had a gallon of water in there that we could drink, you know. A little high-energy foodbar that was Velcroed, that came up inside the helmet that you could snack on. And that– so that kept you nourished." (Duke, 1999 p. 71)*

6.2 FOOD SALYUT

CONCEPT Canned, dehydrated, freeze-dried, fresh food
LOCATION Large-diameter Work Compartment
SPECIFIC Greenhouses: Biogravistat, Malakhit, Magnetogravistat, Fiton Hothouse, Oasis, Oasis 1M, Oasis 1AM, Vazon, Svetoblock

RÉSUMÉ

Food and packaging guidelines included low weight and volume. Food was prepared to last up to 18 months and mostly packed to bite-sized pieces, which were made edible by adding water and kneading the food to a pasty consistency. Some food was coated to prevent formation of crumbs. The menu consisted of canned, dehydrated, or food in aluminium tubes and changed every six days. In addition, cosmonauts took vitamins and sometimes artificial stimulants, tranquilizers and sleeping pills (B.J.Bluth, 1987 p. I-118). Visiting crews brought fresh food. Vodka was allowed on board (Zimmermann, 2003 p. 174). The menu developed over time, become more varied, tastier, and of higher nutritive value. Also requests of crew members were incorporated (B.J.Bluth, 1987).

A:

The development of a facility to grow plants *[Fig. A]* as a source for food, for oxygen production, and for enjoyment during long-term missions was of great importance. On Salyut 1, a greenhouse was introduced. It was a compact greenhouse, called Oazis that used artificial soil and light. It also allowed ventilation of the plants' roots and metered the moisture, light, nutrients, and temperature. On Salyut 4, cosmonauts planted peas and onions. Vazon was a flowerpot unit in which small plants such as onions could be grown. Other greenhouse facilities included the Biogravistat, Malakhit, Magnetogravistat, Fiton, Hothouse, and Svetoblock. (Porterfield, et al., 2002; B.J.Bluth, 1987; Casado, May 2006; Portree, 1995).

On Salyut 7 the pantry system replaced completely pre-assembled and package units. Food was delivered as separate units and cosmonauts could select the food within a calculated caloric ratio (B.J.Bluth, 1987 p. I-125).

The large-diameter Work Compartment contained a folding table *[Fig. B]* that had special openings for heating food tubes. It also had restraints for keeping the food in place. Two electrical ovens, containers for disposal, can openers, spoons, forks and devices to restore dried food were provided (B.J.Bluth, 1987).

A: Expedition 5 cosmonaut Victor Savinykh with plants onboard Salyut 6
B: Cosmonauts Prunariu, Kovalionok, Savinykh onboard Salyut 6

(credit A: Spacefacts, J. Becker; B: Dumitru-Dorin Prunariu)

ASTRONAUTS' EXPERIENCES

USABILITY

Valentin Lebedev (Salyut 7): *"I'm losing weight every day. I don't know what's happening to me! Today I weighed myself: 70 kg, even less than before. I try to eat more, and my appetite seems fine." (Lebedev, 1990 p. 148)*

Valentin Lebedev (Salyut 7): *"The food here is usually excellent; offering delicious taste and a wide variety." (Lebedev, 1990 p. 212)*

LIVABILITY

Valentin Lebedev (Salyut 7): *"It is amazingly pleasant onboard to look after plants and to observe them. (...)* ***They are simply essential to men in space****." (B.J.Bluth, 1987 p. I-43)*

Valery Ryumin (Salyut 6): *"(...) the* ***'green friends'*** *of man grown by the cosmonauts (...) are not only the subject of (...) study but also - and I stress this - a not unimportant psychological factor exerting a positive effect on (...) emotions." (B.J.Bluth, 1987 p. I-43)*

Valentin Lebedev (Salyut 7): *"We admitted that we feel sad and uncomfortable without our garden and without our dear plants." (Lebedev, 1990 p. 201)*

FLEXIBILITY

Valentin Lebedev (Salyut 7): *"So we live a wonderful life up here - we can design any menu we want to. All we have to do is float to a panel, open it, and pick out whatever we want. We eat whenever we want to, or just stuff some goodies into our pockets and chew them while we work." August 31, 1982 (Lebedev, 1990)*

Valentin Lebedev (Salyut 7): *"Today we planted tomatoes, coriander, radishes, and cucumbers in our space garden inside the Malakhit chamber, which we named Orbita." (Lebedev, 1990 p. 229)*

B:

6.3
FOOD SKYLAB

CONCEPT Table for three to prepare frozen and dehydrated food, freezer
LOCATION OWS, lower level
SPECIFIC Dedicated area with window, special designed table

RÉSUMÉ

The Skylab station had a dedicated area for food preparation and dining, called the wardroom, after the comparable officers' dining area on naval ships. It was situated in the lower level of the Orbital Workshop (OWS) and had a big porthole to look outside. The wardroom was about 14.5m^3 with a ceiling height of 1.98m (NASA [Bull.18], 1975 p. 10) It wardroom was equipped with a food freezer and refrigerator. Astronauts could select their individual menu from a wide variety of frozen and dehydrated foods.

The menu included the following food types (NASA [Bull.19], 1976 p. 4):
- Rehydratable
- Thermostabilized
- Frozen
- Beverages
- Natural Form

In the centre of the wardroom was a table that could accommodate all three crewmembers at the same time. The table provided three different restraints *[Fig. C]*, thigh restraints, restraints that could be used with the triangular shoes, and Velcro restraints that could be used bare footed. (NASA [Bull. 9], 1974)

Each astronaut could heat his food individually in a special designed food tray *[Fig. A, B]* on a table with a triangular layout. Four large and four small openings held the food packages in place and three of them were used for heating the food. (NASA [Bull.19], 1976 p. 4)

The table was designed by the industrial designer Raymond Loewy in order to avoid creating hierarchical positions for crew members during long missions. Food was eaten using forks, knives and spoons, which were held in place on the table by magnets. Liquids were drunk from squeezable plastic containers. (Loewy, 1980) The Skylab astronauts ate most meals together in the wardroom. According to astronaut Weitz (2000, p. 12-35), the menu was repeated every six days.

A:

B:

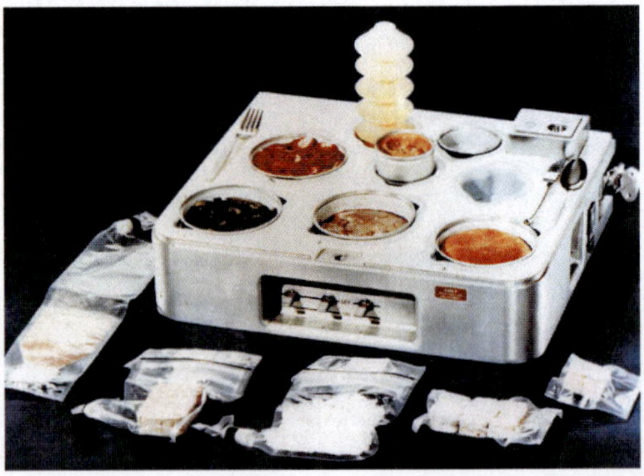

A: Skylab food heating and serving tray in stowed position
B: Tray with food, drink, and utensils
C: Astronaut restrained to the table
D: Skylab 2 astronauts Joseph P. Kerwin, Paul J. Weitz, and Charles Conrad eat space food in the wardroom of the OWS Trainer

(credit A, B: NASA; C: Original Image from NASA, adapted by the Author; D: NASA Johnson Space Center)

ASTRONAUTS' EXPERIENCES

USABILITY

Paul Weitz (Skylab 2): *"So I thought the food was quite good. Plus, we had frozen strawberries,* ***we had ice cream****."*
(Weitz, 2000)

Paul Weitz (Skylab 2): *"We also find the wardroom very convenient for posting items for general crew reference."*
(NASA [Bull.4], 1974 p. 36)

LIVABILITY

Jack Lousma (Skylab 3):
"Sometimes we had breakfast together, sometimes we didn't, because sometimes the sun was up and we had to get somebody on the solar telescope right now, right at breakfast time. So it was a little haphazard in eating breakfast."
(Lousma, 2001)

FLEXIBILITY

Alan Bean (Skylab 3):
"The problem came up if you used them [magnets] on the little food table, which wasn't very big. You had more things than the area would allow (...)

What you need in future design is a table that's a combination table for making changes and a work table, and with good lighting."
(NASA [Bull.12], 1975 pp. 38, 39)

C:

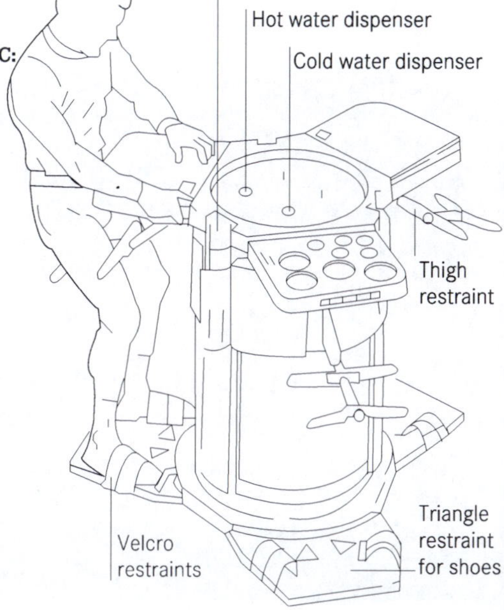

D:

6.4 FOOD SHUTTLE

CONCEPT Rehydratable, thermostabilized, irradiated, fresh food
LOCATION Middeck, locker trays, galley
SPECIFIC Fresh food locker; Greenhouses: Plant Growth Unit (PGU), Plant Generic Bioprocessing Apparatus (PGPA), Kit C Plant Growth Chambers (EPO-Kit C)

RÉSUMÉ

The types of food available include rehydratables, thermostabilized, irradiated, and natural form and fresh food items.

Astronauts select their menu about five months before flight. They can eat the standard Shuttle 7 day menu, but also substitute some items or design their own from over 200 food items. The personalized menu is analyzed for nutritional content and corrected for any nutrient deficiencies (NASA [Food], 2006 p. 2). The food that astronauts eat in space is similar to what they eat on Earth. Only items that need refrigeration are not taken to space. (NASA [Shuttle Food], 2009)

Shuttle astronauts are also provided with fresh food, such as bread, fruits and vegetables that are stored in the 'fresh food locker.' Individual prepared meals are stored in locker trays, each marked with a specific coloured dot for identification.

The Galley rack is located on the Shuttle's Middeck *[Fig. D]*. It is a modular unit that can be removed. It includes an oven to heat food, a rehydration station, a cold and hot water dispenser and the provision of hygiene water. Heating the food takes about 20 to 30 minutes (NASA [Shuttle Reference], 1981 p. 5-21). While eating, astronauts use the food service tray that can be attached to the leg or lap with Velcro *[Fig. B, C]*.

A:

B:

A: Astronaut Ellen Ochoa, STS-110 mission specialist, prepares a meal on the middeck of the Space Shuttle Atlantis
B: Astronaut Jerry L. Ross, STS-110 mission specialist, with a tray of food on the middeck
C: Space Shuttle Tray with food and utensils taken in the food tasting lab
D: Galley Configuration (concept)

(credit A, B, C: NASA; D: Original Image from NASA, source: Space Shuttle News Reference; adapted by the Author)

ASTRONAUTS' EXPERIENCES

USABILITY

Jean- François Clervoy (CNES/ ESA): *"We don't have strong smells in space, because we on purpose don't select food that has strong smell."* (Clervoy, 2009)

LIVABILITY

Gerhard Thiele (ESA): *"Midday one eats while working. The designated one hour break, intended so that one could eat in peace, in reality didn't exist on our flight."* (Thiele, 2010)

D:

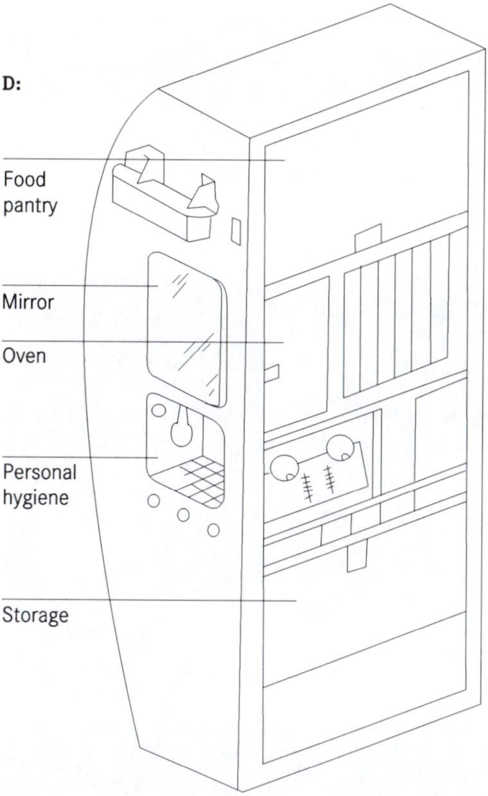

C:

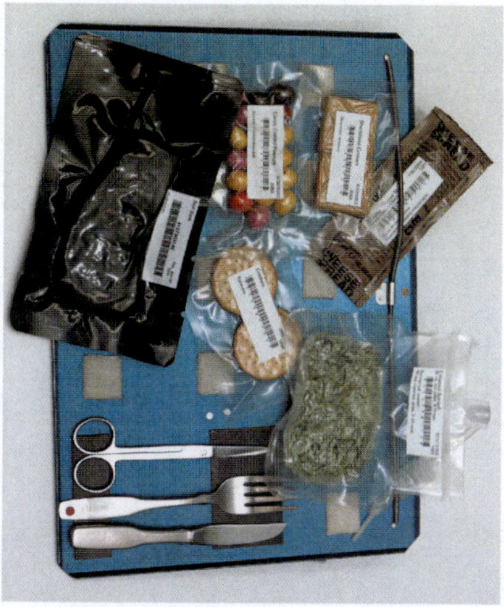

6.5 FOOD MIR

CONCEPT Packaged bite-sized pieces, dehydrated, alcohol
LOCATION Base block, Kristall
SPECIFIC Greenhouses: SVET Space Greenhouse (SG), SVET-2 SG, SVET-GEMS, SVET Light, Svetoblok-G, Svetoblok-S, Svetoblok-M, Greenhouse Light, Chlorella-A

RÉSUMÉ

Most of the food was prepared by dehydration, removing 90 % of the water to last for up to 18 months. The food items were packed in bite-sized pieces.

The Progress resupply ship regularly delivered new and fresh items. According to Bluth (1987, p. 6), cosmonauts could select their food individually for each day, so long as it was in the range of expected calorie input. They also took a vitamin daily and used artificial stimulants, tranquilizers and sleeping pills sometimes. Although alcohol was theoretically not allowed, it is reported, that sometimes they received vodka or wine on the Progress freighters.

The Mir base block was equipped with the food cabinet and a refrigerator, where the food was stored. A table was available for food preparation and eating. In addition to straps to restrain food items *[Fig. B]*, the table had an integrated fan that sucked air down to prevent food from floating away (Foale, 1999 p. 73).

Similar to Salyut stations, a number of greenhouse technologies were tested and continuously updated. The so-called 'SVET greenhouse' was located in the Kristall module.

A:

B:

A: Mir Austromir-91 cosmonauts Franz Viehböck and Tokhtar Aubakiroc drinking
B: A. Volkov and F. Viehböck in the base block
C: Mir 97 crew having lunch in the base block

(credit A, B: Franz Viehböck; C: DLR, Reinhold Ewald)

ASTRONAUTS' EXPERIENCES

USABILITY

Jean-Pierre Haigneré (CNES/ESA): *"**Food was just taking medicine**, it was uninteresting (…) sometimes a little bit of wine or whiskey (…) when we had something to celebrate we did (…) we did have fresh food or a special meal, wine for three people. It was nice to make the life easier, more fun." (Haigneré, 2009)*

LIVABILITY

Jerry Linenger (Mir NASA 3): *"It was not only, that the fresh fruit tasted delicious, but it was seen as a gift, as "**the aroma of the good earth**." (Linenger, 2000 p. 97)*

Mike Foale (NASA): *" (...) and rape seeds, which I planted three days ago, after a lot of checkout in the Greenhouse, are showing some little sprouts with leaves. Very encouraging." (Foale, 1999 p. 82)*

"Before retiring, Sasha [Alexander Kaleri] liked a social drink or tea or other beverage." (Foale, 1999, p. 100)

FLEXIBILITY

Michel Tognini (ESA, Mir Antares): *"We talk a lot. Its work talking, not leisure talking. At dinner at night we have a time, even if you are busy, you set this time to make jokes and to have fun but rest of the time you work a lot." (Tognini, 2009)*

C:

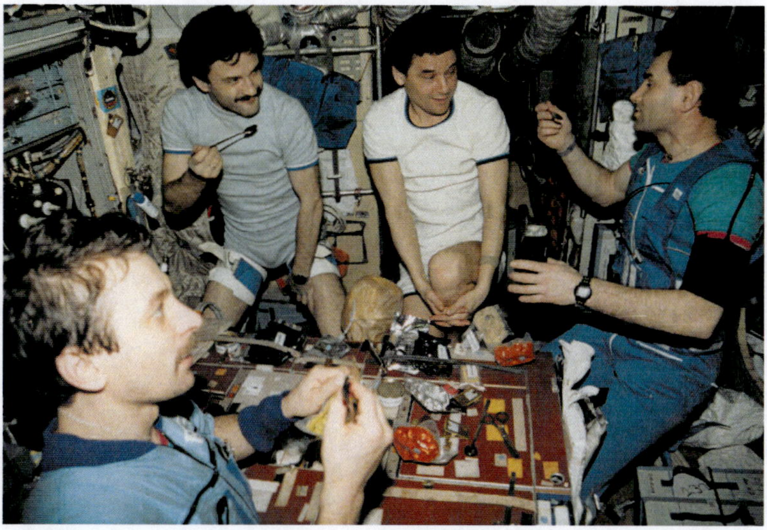

6.6 FOOD ISS

CONCEPT Thermostabilized, rehydratables, intermediate moisture, natural form, fresh, irradiated food; beverages

LOCATION Service Module Zvezda, US Lab

SPECIFIC Wide variety of food; Advanced Astroculture System, LADA-12 Greenhouse, EMCS Advanced Astroculture System, LADA-12 Greenhouse, EMCS

RÉSUMÉ

Food facilities include the dining area and food preparation hardware. A galley with a table, a hot water dispenser and food storage are located in the Service Module Zvezda. There was planned to be a Wardroom in the US Habitation module, but that module was never built.

Thermo-stabilized rehydratables, intermediate moisture, pre-cooked, fresh and irradiated food and beverages are provided (NASA [ISS], 1998 p. 300). In the early years of the space station food items could be selected from the U.S. menu and the Russian menu. Now, astronauts can also choose items from the Japanese menu and a European menu. Today the menu is based on a 16-day-plan that consists principally of Russian and U.S. items. According to Kitmacher (NASA [ISS], 2006 p. 90) the most popular in-orbit foods are shrimp cocktail, tortillas, barbecue beef brisket, breakfast sausage links, chicken fajitas, vegetable quiche, macaroni and cheese, candy-coated chocolates, and cherry blueberry cobbler. The favourite beverage is lemonade.

In addition astronauts are provided with one 'bonus container' in which any food items to increase variety can be placed, that pass the NASA microbiological tests and need not to be refrigerated (Magnus, 2009). Another option of adding variety to the menu is using condiments, such as sauces, pesto and mustard brought by former crewmembers.

The table in the Zvezda module has restraints to hold food items in place. Duct tape and Velcro is used a lot *[Fig. C]*.

A:

B:

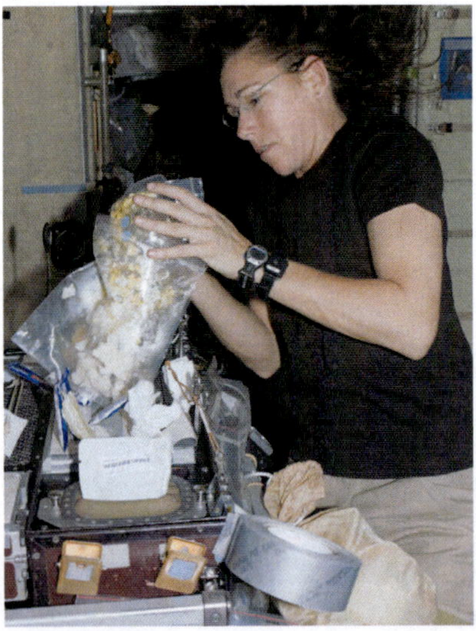

A: The Expedition Four and STS-110 crewmembers share a meal in the Zvezda Service Module on the ISS
B: Astronaut Sandra Magnus, Expedition 18 flight engineer, prepares to eat a Christmas meal at the galley in the Zvezda module
C: Table with Food restraints

(credit A: NASA; B: NASA and its International Space Station Partners; C: Author)

ASTRONAUTS' EXPERIENCES

USABILITY

Sandra Magnus (NASA, Expedition 18): *"If you open too much, and there is not enough liquid in the package, out flies the food and you spend the rest of your mealtime chasing your meal around the cabin and making a mess in the process.* ***This falls under the category of bad space etiquette****."*
(Magnus, 2009)

LIVABILITY

Sunita Williams (NASA, Expedition 15): *"The smell of space…We had a long discussion about this over the dinner table at the end of our EVA on Thursday. Why does the space suit, tools and airlock, for that matter, all smell like something metallic when we come in from an EVA?"*
(Williams, 2007)

FLEXIBILITY

Sandra Magnus (NASA, Expedition 18): *"So the variety of food is pretty good and you can increase the variety by mixing and matching things, and in my case doing some space 'cooking' (...)* ***So it is possible to cook in space*** *with a few hours, lots of dry and wet wipes and the basic tools of duct tape, plastic bags, foil pouches, and a small knife. It is fun and certainly an adventure!"*
(Magnus, 2009)

Peggy Whitson (NASA, Expedition 16): *"I think I got bored with the food much sooner than I did during my first flight. Yuri has indicated the same boredom to me. Obviously, we have to stay healthy in order to accomplish our mission…so we eat anyway…something I refer to as "sport eating. (...) Our crew motto is,* ***"It's all about the sauce."***
(Whitson, 2008)

C:

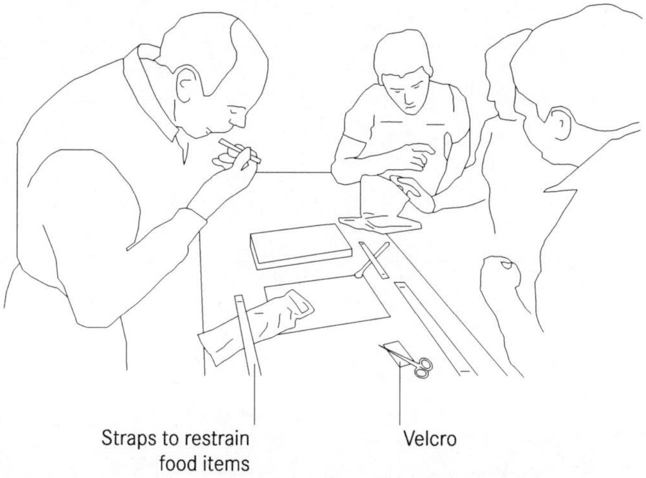

6.7 SUMMARY OBSERVATION

	APOLLO	SALYUT	SKYLAB
Concept	No dedicated area; Dehydrated and bite-sized food	Wardroom with table, electric heaters, two ovens; canned dehydrated or bite-sized food; Greenhouses	Wardroom with table for three, individual heating devices; Window
Review	Development of food systems	The requirement for fresh and tastier food was incorporated; "Planting" was liked for recreation, a table for having meals together was introduced; Cosmonauts liked, they could choose their menu and time to eat	Astronauts requested to prepare food to their own taste; Astronauts prepare and consume food simultaneously; Food and drink spills occured frequently
Potentials	Food system development with regard to long term flights of for emergencies	Requirement for fresh food; Development of greenhouses is important for long-term missions	Individual food stowage items; A large simple window was appreciated during meal time; Several restraints for eating

Diagram

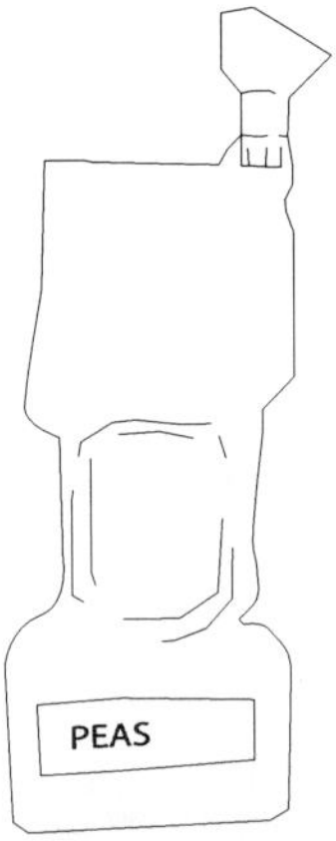

SHUTTLE	MIR	ISS
Tray, fresh food locker; Rehydratables thermostabilized, irrigated and natural form foods	Food cabinet, refrigerator, foldable table; Greenhouses	Preparation and eating area for all in the Zvezda module dining area
Wide range of food items; On short missions astronauts don't necessarily eat together	Not all meals are taken together; Alcohol, but not officially	Wide range of food is available (Russian, U.S., Japanese, European); Astronauts like to eat together
Rotating meals for short term missions; Variety of food	Greenhouse development and integration into the design	Monotony of limited food rotation could make meals seem boring; Encourage space cooking; Greenhouse Development

6.7 SUMMARY OBSERVATION

Food Systems

Astronauts require high nutritional, well-balanced and tasty food to maintain health and be active. Food must be stored, whilst maintaining its quality. It must be easy to prepare, but still appealing and finally consumed. Trash has to be recycled or thrown away.

Food systems for outer space habitats have changed to a high degree over the last 40 years. The improvement of food packaging was vital for the quality and variety of the food. Mercury, Gemini and Apollo astronauts ate bite-sized, de-hydratables and liquids in aluminium tubes. The later ones had hot water for the first time and with the already improved food packaging such as the 'spoon bowl' quality of food menus constantly improved. On Salyut missions cosmonauts were even allowed some vodka. The food eaten today by the Shuttle astronauts and at the ISS is mainly prepared on Earth. On-orbit astronauts have the possibility to heat food in an oven.

Food Preparation and Space Cooking

Salyut stations were the first ones to include a table for having dinner together. Skylab had a dedicated Wardroom area and a specially designed table for food preparation while sitting together. An exclusive area for food preparation seemed also important to Jean-Pierre Haigneré, as food items can easily contaminate adjacent areas, when free floating (Haigneré, 2009).

Today a variety of food is available for astronauts, but still available food can get boring if you are on a long-term mission. According to Shayler et al. (2005 p. 309), already Salyut cosmonauts were "cooking" on-board, leading to a "renewed pleasure".

To increase the variety of tastes astronauts are inventive in creating new meals by mixing food ingredients. They are doing space "cooking". Sandra Magnus (2009), astronaut on Expedition 18 has two logs in her online-journal about her experiences with cooking in space. Her favourite food item is the tortilla, because for her it allows a lot of variation. She uses utensils already on-board. "So it is possible to cook in space with a few hours, lots of dry and wet wipes and the basic tools of duct tape, plastic bags, foil pouches, and a small knife. It is fun and certainly an adventure!"

Eating Together

On Skylab, astronauts had, for the first time, a large dedicated area for food preparation and dining. They were eating together on a specially designed table, eating with knives, forks and spoons. From then on, a table for having meals together has been considered of importance by the crew and became a requirement. Although it is still difficult because of mission task scheduling to have dinner together, there is a wide agreement of its importance for the crew.

On Skylab missions the crewemembers refused to 'float' over the table, as it was seen as inappropriate behaviour. They used the leg clamp-sitting in conjunction with the triangular foot restraints to keep them "in their seats" (NASA [Bull.10],1974 p. 7]. The wardroom on Skylab provided ample space for the three astronauts. But pictures from the Mir space station and also from the ISS show that astronauts often have to arrange themselves 'around' the table in order to fit.

To have meals together as a crew is a very important social activity. The importance of having meals together for the crew is illustrated in a letter, that Mike Foale wrote from the Mir space station to his parents. He wrote that they [the crew] have three meals a day together. They enjoy social conversation and discuss for example "the differences of our languages". (Foale, 1999 p. 82)

Greenhouses

Since the pioneering Oasis greenhouse on Salyut 1, plant growth facilities have been included on space stations for science experiments in almost every space programme (Salyut, Mir, ISS and STS). The first plants to have been exposed to the space environment were Chlorella and the seeds of onions, peas, wheat and corn onboard the satellite Sputnik 2 in 1957. The first vegetables in space have been grown on the Salyut station in the SVET greenhouse. Since 2002, the LADA Greenhouse is installed on the International Space Station to study how plants grow in microgravity.

It has been demonstrated that plants can be grown in microgravity. Almost every space program has included experimental greenhouses to investigate not only technical and biological feasibility, but also habitability related benefits of plant growth activities in space. The 'human' aspect of plant growth facilities assumes increasing importance. Space agencies are investigating and developing relevant technologies and operational concepts related to aspects of nutrition and psychology, with the explicit aim of "provid[ing] reliable, low-cost, stimulating products for crewmember well being (...) to allow crewmembers to eat space-grown vegetables [and] to maximize crew mental health" (NASA [VPU], 2009).

Aside from nutritional and life support system applications, these benefits include sensory and spatial enhancement of the spacecraft environment both through plants as such and the design of their growth chambers, and meaningful occupation through individual interaction.

Plants as Green Friends

In addition to scientific and life-support value, there is anecdotal evidence that cosmonauts and astronauts enjoy handling plants and observing them grow. For the Salyut cosmonaut Valentin Lebedev plants were "like pets". On his mission his sleeping bag was attached to the ceiling next to the Oasis greenhouse, in order to look at the plants before going to sleep (Zimmermann, 2003 p. 170).

Also Shuttle-Mir astronaut Mike Foale accordingly "loved these experiments" [with the Greenhouse], it "reduced his irritability" (Foale, 1999 p. 78. The tending of plants provides regular occupation and interaction of the crew with living material in a technologically mediated habitat. Even though the appreciation of 'gardening' may differ from individual to individual, the presence of growing plants may benefit the mental health of the whole crew. The extent to which this applies is still to be examined (Häuplik-Meusburger, et.al [Greenhouse], 2010).

Alcohol, Cigarettes and Drugs

Although alcohol was not officially allowed onboard space stations, there is evidence, that astronauts on Salyut and Mir space station did drink alcohol (Burrough, 1999). Also visiting cosmonauts brought alcohol to the station as a gift (Haigneré, 2010; Prunariu, 2011; Viehböck, 2010). According to Salyut 6 cosmonaut Dumitru-Dorin Prunariu, the ethyl alcohol that was on board for cleaning rubber gaskets was also used mixed with water for drinking (Prunariu, 2011). According to Burrough "vodka was stored inside half-litre "drink bags" that were sent up on Progress supply ships under the guise of 'psychological support' materials (Burrough, 1999, p. 87). Cigarettes were completely forbidden, but at least one cosmonaut did smoke a cigarette on board the Salyut 6 space station, and "the bad smell had persisted for about three hours" (Prunariu, 2011). Although astronauts may not be supplied with alcohol and cigarettes, the use of drugs (medical or substances on-board) will be an issue, especially on long-term missions.

6.8 COMPARISON USABILITY MATRIX

	APOLLO	SALYUT	SKYLAB
Availability and Equipment	Ready-to-eat food; Improvement of food systems	Wardroom with table, electric heaters, two ovens; canned dehydrated or bite-sized food; Progressive development of Greenhouses	Wardroom / Galley, special designed table for three crewmembers, individual heating devices
Spatial Arrangement	No dedicated area (CM, LM, Space suit)	Table for eating together in the Large Diameter Work Compartment	Table for preparing food and eating together in the Galley of the OWS; 'Climbing over' other crewmembers
Object Management	Colour coded food containers; Scissors to open the packages got lost	Food containers, integrated heating device for tubes and cans in the table	Coloured storage containers, squeezable food packages (storage); Table was too low
Ergonomic Safety	Scissors to open the packages	Restraints on the table	Restraints and magnets to hold everything in place

SHUTTLE	MIR	ISS
Galley (Rack), fresh food locker; Food trays	Food cabinet, refrigerator, table; Greenhouses	Galley with table
Middeck, Galley; In the work area	Table in the Mir Base Block, SVET Greenhouse in Kristall; Next to exercise devices;	Table in the Service module Zvezda; Galley in the US Lab
Locker trays, food trays, colour system	Food rations (tons of cookies, that nobody eats)	Different objects used for food preparation
Velcro and duct tape are used to restrain objects, food tray is restraint to upper leg	Restraints, contamination of food items possible	Velcro and duct tape "every-where"

6.9 COMPARISON LIVABILITY MATRIX

	APOLLO	SALYUT	SKYLAB
Territoriality and Privacy			
	No personal privacy in the module (but not requested)	Area for preparation, eating and planting	Dedicated social area for food preparation and eating (Wardroom)
Sensory perception			
	Spices were not used in order to prevent diarrhoea; Not a lot of taste	Planted orchids, tomatoes, cucumbers, onions, etc.; Dealing with plants was a pleasure for cosmonauts	Appealing and functional design of food system and galley; Frozen food was preferred
External relations			
	Space craft windows	Many space station windows	Large window with view from the dining table to the outside
Internal relations			
	All activities in same module	Greenhouse provided surrogate view	Spatially divided from work and other areas, but visually connected

SHUTTLE	MIR	ISS
Galley (rack) in Middeck	Dedicated area for food preparation and eating	Dedicated area for food preparation and eating (Zvezda)
Wide range of food available	Plants in greenhouse, artificial stimulants; Dinner together	Sauces and mixing food flavours; Dinner together; "Space cooking"
None from Middeck	Space station windows	Space station windows
No special relations	No separation and close to other functions; Greenhouses	In the Core Module of the "Russian part"

6.10 COMPARISON FLEXIBILITY MATRIX

	APOLLO	SALYUT	SKYLAB
Spatial Flexibility	No dedicated area	Cosmonauts could eat where they wanted	Dedicated area for preparation and eating
Object Flexibility	Food items could be chosen	Foldable table; Variety of Greenhouses – but only one or two at one time	Foldable containers, multiple and different restraints
Individual Flexibility	Best items "vanished quickly"	Cosmonauts could select food within caloric ratio	Each astronaut could select his own menu

SHUTTLE	MIR	ISS
Galley is fixed, no dedicated place to eat	Dedicated area for food preparation and eating together in the Core Module	Dedicated area to prepare food and for eating together
Eating tray has Velcro	Restraints, foldable table	Astronauts are inventive in using diverse objects for cooking
Astronauts eat anywhere and can select their menu	Cosmonauts could select food within caloric ratio	Wide range of food

6.11 DESIGN DIRECTIONS

The most important findings related to the human activity 'FOOD' are summarized and structured in the form of design directions. These directions summarize the main points in a short sentence. The paragraph following refers to the empirical background of experiences made in extraterrestrial habitats.

USABILITY

Integrate food growing systems

Mass and volume considerations have been very important for designing the food system. For future missions bioregenerative life support systems in which food is produced and waste is recycled will play a major role. Considering the high cost of resupplying space stations and the planetary habitats in the future, the ongoing improvement of greenhouse technologies is critical (Horneck, et al., 2000; Porterfield, et al., 2002 p. 183).

Develop cooking devices

So far, no Zero-G or Partial-G cooking devices that reflect astronauts needs have been developed. In addition to a dedicated area for food preparation and a place that accommodates all crewmembers to eat together, in the future especially designed facilities will improve the astronaut's habitability by supporting them in food experimentation.

Integrate an area for food preparation

The area for food preparation has to include the necessary equipment for the preparation, serving and eating of food includes eating surfaces, multiple restraints and utensils (NASA [HIDH], 2010) as well as cleaning and waste management (dish washer). A design that considers easy cleaning of the area as well as the equipment is also important.

Separate food preparation from other activities

If available volume allows, the area where food is prepared should be separated from other activities, such as hygiene, work, sleep or exercise. In any case ample space shall be provided to allow other crewmembers passing by.

LIVABILITY

The kitchen is a place for socializing

Having dinner together is a social activity shared by every culture. Designs of extra-terrestrial habitats are often derived from familiar social rules, because people carry their habit and customs with them. The area where astronauts eat should allow all crewmembers to have a meal together and provide adequate translation paths. One has to keep in mind, that astronauts reported, they dislike talking to a colleague next to him upside-down while having dinner together. Social roles are taken to space. Astronauts bring their culture and social norms with them, which includes the rituals of communal dining.

Use the greenhouse to increase habitability

In view of long duration missions, plant growth facilities should not be regarded as costs and penalties. Their future appears to extend beyond providing desirable add-ons, but as an essential habitability component that may even substitute for a lack of outwards-facing windows and compensate in some small degree for separation from the home ecosystem on the Earth.

Integrate surrogate windows

As a 'window to something living,' an integrated greenhouse could provide this surrogate view in all mission scenarios (short and long-term). In addition plant displays and the virtual display of natural scenes or materials can provide 'surrogate

A: Windows and surrogate views (a: real window, b: local view to group domain, c: virtual window, screen, media, d: greenhouse
B: Trabant Greenhouse

(credit A: Author; B: Häuplik-Meusburger, Holzgethan)

views' to achieve restorative effects (Heerwagen, et al., 1986 p. 626; Levi, et al., 1999 p. 212). Where the inclusion of windows or the view outside is limited, for instance during deep space transfers, or where lingering in front of windows designed for operational purposes is not advised due to higher radiation exposure, the greenhouse can offer a multipurpose and therapeutic alternative. (Häuplik-Meusburger, et al. [Greenhouse], 2010)

FLEXIBILITY

Support Individual interaction with plants

Monitoring greenhouse facilities and interacting with plants may become an important pastime aboard spacecraft, especially during long-term missions, where some of the current favourite pastimes of astronauts become limited (e.g. communication with ground and window gazing).

Integrate personal greenhouses

The development of small modular growth units could be of advantage for long-term missions. In addition to a large greenhouse facility, small plants could be extracted at liberty and transported in small portable growth units. They can function as 'trabant' greenhouses for easy access in the galley, act as decor for personal quarters and common spaces or provide live food supplement in pressurised rovers. (Häuplik-Meusburger, et al. [Greenhouse], 2010).

A:

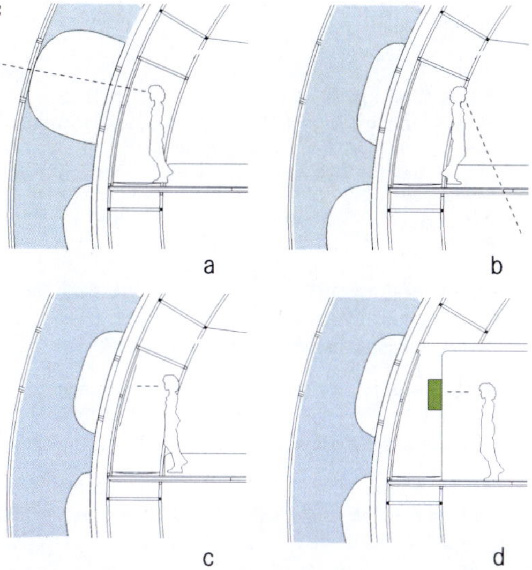

B:

6.12 REFERENCES

B.J.Bluth, M. Helppie. 1986. Soviet Space Stations as Analogs. Washington D.C., US : NASA Headquarters, 1986. NASA Grant NAGW-659.

Burrough, Bryan. 1999. Dragonfly - NASA and the crisis aboard Mir. London, GB : Fourth Estate Limited, 1999. ISBN 1 84115 087 8.

Casado, Javier. May 2006. Horticulture in space. Spaceflight - The Magazine of Astronautics and Outer Space. Cultivating the future - growing food in space, May 2006, Vol. Vol 48, No. 5.

Cernan, Eugene A. 2007. Oral History Transcript - Eugene A. Cernan. [interv.] Rebecca Wright. Houston, Texas, US, 11 December 2007.

Clervoy, Jean- François. 2009. Transcript not published [interv.] Sandra Häuplik-Meusburger. Paris, France, June 2009.

Duke, Charles M. Jr. 1999. Oral History Transcript: Charles M. Duke. [interv.] Doug Ward. Houston, Texas, US, 12 March 1999.

Eat Me Daily. 2009. Eat Me Daily- Food of the Apollo 11 Lunar Landing. [Online] 16 July 2009. http://www.eatmedaily.com/2009/07/food-of-the-apollo-11-lunar-landing/.

Foale, Colin. 1999. Waystation to the Stars - The Story of Mir, Michael and Me. London, GB : Headline Book Publishing, 1999. ISBN 0 7472 6331 0.

Haigneré, Jean-Pierre. 2009. Transcript not published. [interv.] Sandra Häuplik-Meusburger. Paris, France, June 2009.

Häuplik-Meusburger, Sandra; Peldszus, Regina; Holzgethan, Verena [Greenhouse]. 2010. Greenhouse design integration benefits for extended spaceflight. Acta Astronautica. Volume 68, 2010, Vols. Issues 1-2, January-February 2011, pp. 85-90.

Heerwagen, Judith H. and Gordon, H. September 1986. Adaptations to Windowlessness: A Study of the Use of Visual Décor in Windowed and Windowless Offices. Environment and Behaviour. September 1986, Vol. 18, No. 5, pp. 623-639.

Horneck, G., et al. 2000. HUMEX - Study on the Survivability and Adaptation of Humans to Long-Duration Interplanetary and Planetary Environments. 2000. ESTEC / Contract No. 14056/99/NL/PA.

Lebedev, Valentin. 1990. Diary of a Cosmonaut - 211 Days in Space. United States and Canada : Bantam Air & Space Series, 1990. Vol. Vol.4. ISBN: 0-553-28778-8.

Levi, Daniel and Kocher, Sara. March 1999. The Future Effects of Information Technolgoy on Our Relationship to Nature. Environment and Behaviour. March 1999, Vol. Vol. 31, No. 2, pp. 203-206.

Linenger, Jerry M. 2000. Off The Planet - Surviving five periolous months aboard the Space Station MIR. U.S.A. : McGraw-Hill, 2000. ISBN 0-07-136112-X.

Loewy, Raymond. 1980. Industrial Design. London, GB : British Library Cataloguing in Publication Data, 1980. ISBN 0-947795-38-3.

Lousma, Jack R. 2001. Oral History Transcript - Jack Lousma. [interv.] Carol L. Butler. Houston, Texas, 7 March 2001.

Magnus, Sandra. 2009. Sandra Magnus' Journal . [Online] NASA, 6 January 2009. [Cited: 10 Februar 2010.] http://www.nasa.gov/mission_pages/station/expeditions/expedition18/journal_sandra_magnus_6.html.

NASA [Bull.4]. 1974. Skylab Experience Bulletin No. 4: Design Characteristics of the Sleep Restraint. Houston, Texas, US : NASA, Lyndon B. Johnson Space Center, 1974. JSC-09538.

NASA [Bull.9]. 1974. Skylab Experience Bulletin No. 9: Foot Restraint Systems. Houston, Texas, US : NASA, Lyndon B. Johnson Space Center, 1974. JSC-09543.

NASA [Bull.10]. 1974. Skylab Experience Bulletin No. 10: Body Restraint Systems. Houston, Texas : NASA, Lyndon B. Johnson Space Center, 1974. JSC-09544.

NASA [Bull.12]. 1975. Skylab Experience Bulletin No. 12: Temporary Equipment Restraints. Houston, Texas, US : NASA, Lyndon B. Johnson Space Center, 1975. JSC-09546.

NASA [Bull.18]. 1975. Skylab Experience Bulletin No. 18: Evaluation of Skylab IVA Architecture. Texas, Houston, US : NASA, Lyndon B. Johnson Space Center, 1975. JSC-09552.

NASA [Bull.19]. 1976. Skylab Experience Bulletin No. 19: Food System. Houston, Texas, US : NASA, Lyndon B. Johnson Space Center, 1976. JSC-09553.

NASA [Food]. 2006. NASA-Facts: Space Food. JSC : Lyndon B. Johnson Space Center, USA, 2006. FS-2006-04-015-JSC.

NASA [HIDH]. 2010. HIDH - NASA Human Integration Design Handbook, Initial Release. Washington, DC, USA : NASA JSC, 2010. NASA/SP-2010-3407.

NASA [ISS]. 1998. International Space Station Familarization. Houston Texas, USA : Lyndon B. Johnson Space Center, 1998. TD9702A.

NASA [ISS]. 2006. Reference Guide to the International Space Station. [ed.] Gary, editor Kitmacher. Washington, DC, US : Library of Congress Cataloging-in-Publication Data, 2006. ISBN 0-9710327-2-6.

NASA [Shuttle Food]. 2009. STS-118 Education Resources: Michele perchonok, Shuttle food System Manager. [Online] 20 January 2009.

NASA [Shuttle Reference]. 1981. Space Shuttle News Reference. Houston, Texas. : U.S. Government Printing Office, 1981. 723-000/562.

NASA [VPU]. 2009. Validating Vegetable Production Unit (VPU) Plants, Protocols, Procedures and Requirements (P3R) Using Currently Existing Flight Resources (Lada-VPU-P3R). [Online] May 2009. http://www.nasa.gov/mission_pages/station/science/experiments/Lada-VPU-P3R.html.

Porterfield, D. M., Neichitailo, G. S., Mashinski, A. L., & Musgrave, M. E. 2002. Spaceflight Hardware for Conducting Plant Growth Experiments in Space: The Early Years 1960-2000. Advances in Space Research (Volume 31, Issue 1, 2003, Pages 183-193), pp. 183-193.

Portree, David S.F. 1995. MIR Hardware Heritage. Houston, Texas : NASA, Lyndon B. Johnson Space Center, Information Services Division, 1995. NASA RP 1357.

Prunariu, Dumitru-Dorin. 2011. Transcript not published. [interv.] Sandra Häuplik-Meusburger. Vienna, Austria, 2011.

Schmitt, Harrison. 2009. Transcript not published. [interv.] Sandra Häuplik-Meusburger. Vienna, Austria, 2009.

Shayler, David J. and Moule, Ian. 2005. Women in space - Following Valentina. Berlin, Heidelberg, New York : Springer-Praxis Books in Space Exploration, 2005. ISBN 1-85233-744-3.

Smith, Malcolm C. Jr., et al. July 1974. Apollo Experience Report - Food Systems. Houston, Texas, USA : NASA, Lyndon B. Johnson Space Center, July 1974. TN D-7720.

Thiele, Gerhard. 2010. Transcript not published. [interv.] Sandra Häuplik-Meusburger. Vienna, Austria, December 2010.

Tognini, Michel-Ange Charles. 2009. Transcript not published. [interv.] Sandra Häuplik-Meusburger. Cologne, Germany, July 2009.

Weitz, Paul J. 2000. Oral History Transcript - Interview with Paul J. Weitz. [interv.] Rebecca Wright. 26 March 2000.

Whitson, Peggy. 2008. Peggy Whitson's Journal - Exercise in Space. NASA. [Online] 18 January 2008. [Cited: 5 February 2009.]

Williams, Sunita Lyn. 2007. Flight Engineer Sunita Williams' Space Station Journals. [Online] NASA, 2007. [Cited: 20 12 2009.] http://www.nasa.gov/mission_pages/station/expeditions/expedition15/journal_sunita_williams_11.html.

Zimmermann, Robert. 2003. Leaving Earth, Space Stations, Rival Superpowers, and the Quest for Interplanetary Travel. Washington, D.C. : Joseph Henry Press, 2003. ISBN 0-309-08548-9.

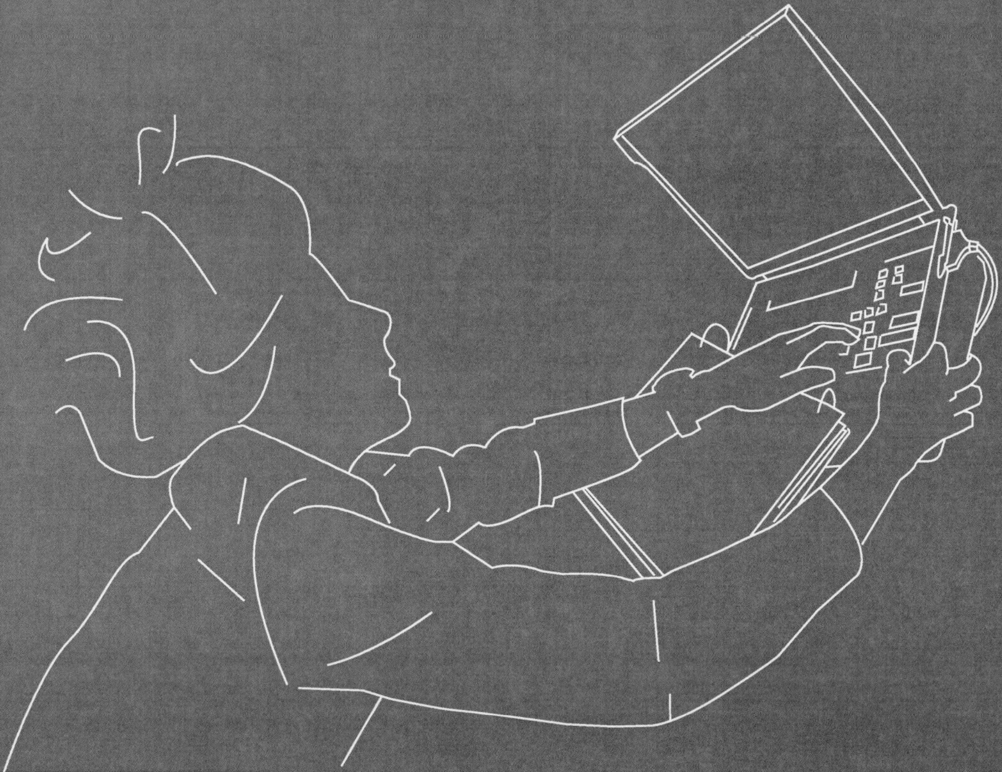

PART 7: HUMAN ACTIVITY

WORK

The category 'Work' includes the sub-activities operation, work task, conducting experiments and communication; education and training.

7.1 WORK APOLLO

CONCEPT Experiments on the Lunar surface
LOCATION Lunar Surface; Lunar Orbit
SPECIFIC Microgravity and 1/6g

RÉSUMÉ

The prime objective was to land a man on the Moon and to return safely to Earth, but in addition Apollo astronauts conducted various experiments on the lunar surface.

Lunar Surface Experiments

So-called ALSEPs (Apollo Lunar Surface Experiment Packages) were placed on the lunar surface by Apollo 12, 14, 15, 16 and 17. Each mission landed at a different site and the composition of the ALSEPS varied, including experiments on lunar seismology, soil mechanics, atmosphere and gravity amongst others (Beatie, 2001). Apollo astronauts installed the experiments and left them on the lunar surface, as some were designed to operate over a long period. From Apollo 15 onward, a lunar roving vehicle (LRV) that was deployed by the astronauts supported their mobility.
Another task was to collect samples of lunar rocks. In total 382 kg were collected by the astronauts and returned to the Earth with them. All Apollo astronauts were pilots, only one scientist, Harrison Schmitt has been to the Moon.

Extra-Vehicular Activities

On every Apollo lunar mission, the two astronauts on the surface went EVA on the moon in lunar gravity. Following trans-Earth Injection (TEI) on the return trans-Earth Coast on every mission except Apollo 13, the Command Module Pilot (who had stayed in the CSM during the landing) had the job of doing an EVA to recover the externally mounted camera *[Fig. B]*. All three crew suited up, the cabin was depressurized, the hatch opened, and the CM pilot went out EVA on an umbilical to retrieve the film magazines (Woods, et al., 2008).

Maintenance

So called Inflight Maintenance was not an 'integral element', but the Apollo astronauts had a small standard set of mechanical tools (NASA [Bull.5], 1974.

A:

A: Apollo 12 astronaut Richard F. Gordon attaches a telephoto lens to a camera
B: Astronaut Ronald Evans photographed during transearth coast EVA
C: Apollo 17 Scientist-Astronaut Harrison H. Schmitt collects lunar rake samples

(credit A, B, C: NASA)

ASTRONAUTS' EXPERIENCES

USABILITY

Gene Cernan (Apollo 17):
*"Learning how to walk was **like balancing on a bowl of Jell-O**, until I figured out how to shift my weight while doing a sort of bunny hop."*
(Wise, 1988 p. 322)

Gene Cernan (Apollo 17):
"Stripping off my gloves was a painful process, and I wasn't surprised to discover knuckles and backs of my hands were blistered with a fiery red rawness. My fingers felt almost broken and I had to flex them to see if they still worked."
(Cernan, et al., 1999 p. 327)

LIVABILITY

Al Worden (Apollo 15):
*"The EVA itself was kind of unique. It's sort of a unique perspective. I did have a chance to stand up on the outside [of the Service Module] and look [around, and] (...) **I could see (...) the Moon and the Earth at the same time**. And if you're on Earth, you can't do that. If you're on the Moon, you can't do that. **It's a very unique place to be**."*
(Worden, 2000)

FLEXIBILITY

Harrison Schmitt (Apollo 17):
"We were busy as explorers (...) You try to take advantage of the time you have." (Schmitt, 2009)

B:

C:

7.2 WORK SALYUT

CONCEPT Experiments, maintenance
LOCATION Large Diameter Compartment
SPECIFIC Permanent evaluation of hardware

RÉSUMÉ

The work compartment on Salyut 7 had two areas: the Instrument area and the Living area.

The familiar Earth circadian rhythm of work and rest with an extended rest period was used. Experience showed that even when adapted to space, work still took longer than on Earth. According to Bluth (1987 p. 6), music was used to relax and also as a work stimulant. It helped to cover back ground noise from the ventilators and life support system of the station (Prunariu, 2011).

Maintenance and Repair

Maintenance and Repair was very important for the Soviets. In her report 'Soviet Space Station Analogs' B.J. Bluth (1987 p. I-50) cites Kidger Neville: " Repairs are costly but nowhere near as costly as launching a new station." Much of the onboard equipment had to be repaired, maintained or replaced by the cosmonauts.

Beginning with Salyut 1 the table *[Fig. B]* for dining could be used as a workbench for minor maintenance and repair jobs (Portree, 1995). It was located in the middle of the module. Rubber straps on the foldable table allowed the cosmonauts to restrain equipment and tools (Connors, et al., 1999; Lebedev, 1990; Prunariu, 2011).

Communication

The Salyut 6 had two external cameras for docking manoeuvres and a television camera near the EVA hatch to monitor EVA activities. Inside the station they had two television cameras, which were used for reports, for inspections of the station and for psychological support (B.J.Bluth, 1987 p. II-5). Communication with mission control (FCC) was limited by frequent loss of signal. According to Bluth (1987 p. II-6) a "walkie-talkie system" was used for internal communication.

Onboard Training

The cosmonauts had refresher courses for safety reasons, training for complicated work by the ground personnel and training for skill maintenance. The onboard crew was also trained by visiting crews. To do so, videotapes, photographs and also training models were used (B.J.Bluth, 1987 p. III-119).

A:

A: Inside Salyut 7
B: Salyut 7 cosmonauts Vladimir Aleksandrovich Dzhanibekov and Viktor Petrovich Savinykh freezing

(credit A: Didier Capdevila, Capcom Espace; B: Spacefacts, J. Becker)

ASTRONAUTS' EXPERIENCES

USABILITY

Valentin Lebedev (Salyut 7): *"Today I didn't have a chance to even glance at the Earth." (Lebedev, 1990 p. 31)*

"Our flight is so long that ***a lot of our training is forgotten****, our skills get lost, and a man comfortable in his surrounding, begins to relax." (Lebedev, 1990 p. 228)*

Dumitru-Dorin Prunariu (Salyut 6): *"If you lose something you might find it on the ventilator net. All objects float and get stuck in the ventilator net." (Prunariu, 2011)*

LIVABILITY

Valentin Lebedev (Salyut 7): "Many little details, such as photographs on the panels, children's drawings, flowers, and green plants in the garden, turn this high-tech complex into our warm and comfortable, if a little bit unusual, home." (Lebedev, 1990 p. 217)

"I sing songs as I fly through the station doing my work." (Lebedev, 1990 p. 226)

"I think our fatigue grows because our interest in work is fading. I don't even want to look out a porthole anymore." (Lebedev, 1990 p. 251)

B:

FLEXIBILITY

Valentin Lebedev (Salyut 7): *"However, we never found the focusing ring (from the camera), so we had to make one ourselves. We held a competition for the best ring:* ***The best one would be installed.*** *Tolia won." (Lebedev, 1990 p. 144)*

"We tried to hide as much stuff as possible behind the panels. As we did so, we noticed a lot of impractical features that were hard to foresee on Earth. A lot of empty space behind the panels is not used at all. No one realized we would need a place to store various pieces of equipment such as brackets, poles, and holders when we finished using them. Besides that, bags, belts, locks and other items have been permanently fastened here, preventing our using the existing space." (...)

"When a crew arrives, the cosmonauts may rearrange things to make themselves feel at home. This kind of work usually requires moving a lot of material and equipment. ***Sometimes this requires sawing of metal,*** *which not only litters the station but also takes lots of time and effort." (Lebedev, 1990 p. 137)*

7.3 WORK SKYLAB

CONCEPT Experiments
LOCATION OWS, Experiment area, MDA
SPECIFIC Horizontal, two-level

RÉSUMÉ

Skylab was an orbital laboratory and a lot of work included manually controlled operations and monitoring scientific observations, such as solar flares (NASA [Skylab], 1977). Another scientific objective was the study of long-term effects on humans, including biomedical and behavioural experiments. The equipment also included science devices for physics, psychophysics, biomedical research, photography, and astronomy (NASA [Skylab], 1977).

The experiment work area of the Skylab station was situated in the Orbital Workshop (OWS) upper deck in the Dome Area. It was separated from the Crew Deck by the triangle grid served as the ceiling of the Crew Deck and the floor of the Dome Area or upper deck.

The Crew Deck included the waste compartment, the wardroom, the shower, exercise equipment, and the sleep compartments.

Maintenance and Repair
The initial planning foresaw only a limited degree of in-flight maintenance. In reality each of the three crews completed unscheduled repairs effectively that were critical to the Skylab mission (NASA [Skylab], 1977). A table or workstation for maintenance items was not foreseen. The crew improvised and used the Wardroom table when necessary.

Communication
In 1973 the Skylab-4 crew "went on strike" and stopped communication with mission control in Houston to take the day off, which eventually lead to a modification of the work schedule. The crew was now given a list of tasks, instead of a detailed time plan (Linenger, 2000 p. 132; NASA [Skylab LL], 1977.

A:

A: Skylab 3 astronaut Alan Bean operating the Ultraviolet Stellar Astronomy experiment in the Skylab Airlock Module
B: Astronaut Edward G. Gibson during EVA on Skylab 4 *(credit A, B: NASA)*

ASTRONAUTS' EXPERIENCES

USABILITY

Joe Kerwin (Skylab 2):
"We hardly used the portable handholds. Almost anything serves as a handhold when you're arriving at a location." (NASA[Bull.1, 1974 p.12)

Charles Conrad (Skylab 2):
"(...) and we inadvertently turned off a fair amount of switches in the beginning, mostly with our feet (...) After we got a little better about managing our feet, we didn't turn too many of those off." (NASA[Bull.1, 1974 p. 34)

LIVABILITY

Jack R. Lousma (Skylab 3):
"The things you needed to do the most to prolong your life on Skylab were the things that got the least priority: eating on time, sleeping on time, exercising on time. All of those things we got done. We made sure that we didn't go to bed, we would never quit, until it was all done. We [seldom] left anything until tomorrow. ***The three things you should do the most, we (...) worked."*** (Lousma, 2001)

FLEXIBILITY

"The performance of a given experiment should be as independent of the spacecraft as possible. The day-to-day activities in the spacecraft should not be severely restricted because of various operation constraints of the experiment." (NASA [Skylab LL], 1974 p. SLL 2-27)

B:

7.4 WORK SHUTTLE

CONCEPT Experiments, assembling, maintenance and repair of ISS
LOCATION Middeck, Bay, in outer space
SPECIFIC 1g and microgravity

RÉSUMÉ

The astronauts on a shuttle missions are mainly busy with conducting experiments, maintenance activities in space and assembling the station. The crew also conducts missions related to the repair, release or capture of satellites. The Flight Deck is used for piloting the orbiter and to monitor orbiter systems *[Fig. B]*. Two seats are installed for the pilots. From the Aft flight deck windows, the astronauts stand on the flight deck floor and can overlook the payload bay.

Airlock
The shuttle has one external airlock, which allows exit to the payload bay and provides access for extravehicular activities. It can host two crewmembers and includes mobility aids for performing a variety of tasks. The airlock can also be configured according to different mission operations. It can for example be substituted with a docking module when direct docking of two vehicles is required or with a transfer tunnel to an additional module, such as the Spacelab in the payload bay.

Spacelab was a scientific laboratory developed by the European Space Agency to conduct experiments in the fields of medicine, manufacturing, astronomy and pharmaceutics. It was transported in the Shuttle's cargo bay and returned to Earth after each mission. It could be reused about 50 times. (NASA [Shuttle Reference], 1981 p. 1-6 and 3-12)

Storage
Stowage facilities include rigid and flexible containers. Among the rigid containers are four permanently installed lockers on the Flight Deck and interchangeable modular lockers of different sizes in the Middeck. Flexible containers include soft stowage bags and retention nets. If the Middeck Accommodations Rack (MAR) is onboard it is located in the area where the galley can be installed. It provides additional storage volume for payloads and experiments and has electrical power and thermal control. (NASA [SS PressKit]) Colour codes are used to distinguish specific items for each crewmember, such as food and clothing.

A:

A: Randy Bresnik, STS-129 mission specialist, during an EVA to remove micrometeoroid and orbital debris shields from the Quest Airlock
B: Barbara R. Morgan, STS-118 mission specialist, uses a computer on the aft flight deck of Space Shuttle Endeavour *(credit A, B: NASA)*

ASTRONAUTS' EXPERIENCES

USABILITY

Jean- François Clervoy (CNES/ESA): *"(...) In space we manipulate hundreds of objects every day.* ***And keeping continuously awareness of where they are is a challenge.****"* (Clervoy, 2009)

Hans-Wilhelm Schlegel (ESA): *"In reality storage [reqirements] change over time (...) and that is a critical issue."* (Schlegel, 2009)

LIVABILITY

Bill Oefelein (NASA, STS-116): *"We have been very busy re-wiring the space station, adding a new piece of hardware to the ISS truss, and transferring a lot of gear. The days are long, but we have a great crew and we work well together."* (Oefelein, 2006)

FLEXIBILITY

Michel Tognini (ESA): *"[the laptop] it is a very good machine. You can open it and you can remove all the parts yourself (...) you can repair it."* (Tognini, 2009)

B:

7.5
WORK MIR

CONCEPT Experiments, maintenance
LOCATION Core module and science modules
SPECIFIC Horizontal and vertical

RÉSUMÉ

The work schedule was preset from Mission control in Moscow. In the beginning, the timelines of the Cosmonauts were very strict with every minute pre-planned, as Jerry Linenger reports (2000 p. 132). In following years, the cosmonauts were allowed some autonomy in their work schedule and could decide themselves upon the most efficient sequence.

Scientific work was carried out in the dedicated modules Kvant, Kvant 2 and Kristall. The Mir core module was used for computing, communication, maintenance and repair. Small maintenance and repairs were conducted using a workbench in the form of a steel table with a vice that could be mounted on it and an extension for soldering (B.J.Bluth, 1987 p. I-4). The module Kvant was later used as a 'cellar for junk' (Foale, 1999 p.74).

Like on the Salyut stations, music was not only used for relaxation, but also as a work stimulant (B.J.Bluth, 1987 p. 6).

Education and Training

Mir cosmonauts stayed for a long-term in orbit. The crews could not train and prepare in advance for everything that might happen. The Russians implemented an advanced 'on-the-job-training' instead of trying to train for every eventuality (Portree, 1995).

A:

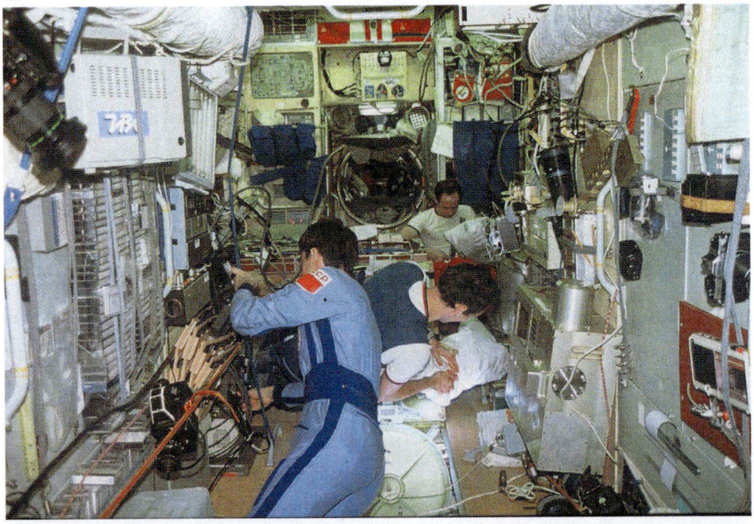

A: Austromir-91 crew at work
B: Thomas Reiter with the Munich Space Chair in the Spektr Module
C: Scientific experiment during the Austromir-91 mission *(credit A, C: Franz Viehböck; B: ESA /RSA/TUM 1995)*

ASTRONAUTS' EXPERIENCES

USABILITY

Reinhold Ewald (DLR/ESA): *"We get recommendations from ground control. Set it up in so and so module, over there somewhere, and fix yourself here or there (...) But in a way, where and how one works is still a matter of experience." (Ewald, 2009)*

LIVABILITY

Jerry Linenger (Mir NASA 3): *"I wanted to complete all mission goals - no exceptions - and to go beyond the stated objectives whenever possible. People were depending on me, vast sums of money were invested on getting me up here, and* ***if I had to become something of a robot, so be it****."* (Linenger, 2000 p. 90)

Michel Tognini (ESA): *"My crew went to bed usually at three and four o'clock in the morning because they had so much work to do (...) and we have to wake up at 7."* (Tognini, 2009)

FLEXIBILITY

Jerry Linenger (Mir NASA 3): *"Mir was too crowded with six people onboard, its life support system pushed too hard. It would be nice to complete the twenty-day crew overlap and get back to routine."* (Linenger, 2000 p. 100)

Franz Viehböck(Austromir): *"(...) Well, you are constantly confronted with situations that challenge the human intellect and where you have to improvise, just like in daily life."* (Viehböck, 2009)

B:

C:

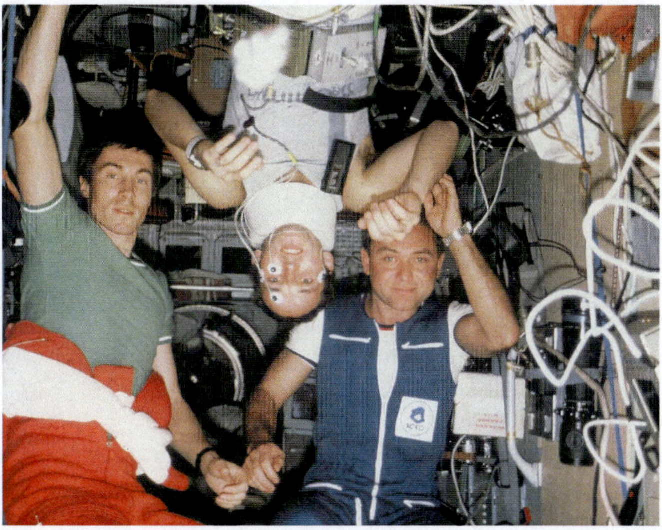

7.6 WORK ISS

CONCEPT Experiments, maintenance
LOCATION All modules
SPECIFIC Biggest space station complex ever

RÉSUMÉ

The crew's main purposes are activities related to research, medical experiments and maintenance of the station. Scientific work is primarily done in the U.S. Lab Module, the ESA Columbus lab module, the Zvezda Service Module and the Japan Experiment Module. Each lab is equipped with specific equipment related to the research of biology, physical and Earth sciences amongst others. Computer work stations, a microgravity science glove box and the 'Nadir window' for Earth observation are located in the U.S. Lab.

A:

EVA Activities

The procedure before an Extra-vehicular Activity *[Fig. A]* starts the day before to activate and configure the airlock (NASA [ISS], 1998 pp. 250-251):

1. The two astronauts scheduled for an EVA have to sleep in the airlock overnight to perform the pre-breathing to prevent decompression sickness. This activity is called 'Campout'.
2. On the day of the EVA the astronauts don their suit and have an in-suit pre-breathe.
3. The astronauts depress the chamber and
4. conduct their EVA.
5. After ingress they doff their suits, dry them and store the EVA equipment
6. On the day after the EVA, the astronauts recharge the EMU with Oxygen and water.
7. After the last EVA, the airlock equipment is powered down and the racks are secured.

EVA equipment includes a mini-workstation (MWS) that can be attached to the EMU. It is used to carry working tools.

A: Astronaut Bresnik installs an adaptor near the Columbus module
B: Astronaut Sunita L. Williams, Expedition 15 flight engineer, enters data in a computer in the Destiny laboratory

(credit A, B: NASA)

ASTRONAUTS' EXPERIENCES

USABILITY

Carl Walz (NASA, Expedition 4):
"We worked both night and day (remember we get sunrise 16 times a day), using our helmet lights to illuminate our paths when necessary."
(Walz, 2002)

LIVABILITY

Greg Chamitoff (NASA):
"The next 10 days will undoubtedly be extremely busy, fun, and exhausting. ***We'll basically skip next weekend*** *and rest again after my current roommates depart." (Chamitoff, 2008)*

Dan Bursch (NASA, Exp. 4):
"Working closely with someone is a big jump from an acquaintance. Living with someone is a big jump from working with them. And ***living and working together with only two other people*** *for several months is yet another big jump." (Bursch, 2002)*

FLEXIBILITY

Hans Schlegel (ESA):
"(...) But we are still developing the minimal requirements (...) We are in the middle of developing it, because, just like on Earth; a bedroom for one family might be a study for the next, and the former study might become the bedroom or vice-versa." (Schlegel, 2009)

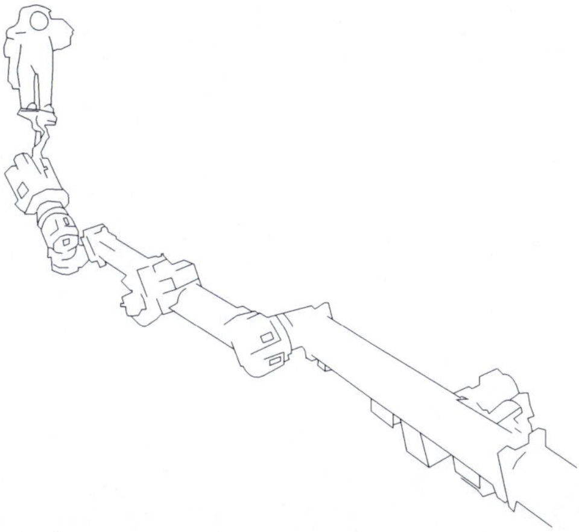

B:

7.7 SUMMARY OBSERVATION

	APOLLO	SALYUT	SKYLAB
Concept	Work was the main issue	Small diameter and large diameter work compartment	Experiment work area in the Orbital Workshop, MDA
Review	Working in 1/6 g still needs further research	On-board education and training is requested for long term missions; Colours were used for orientation, the cosmonauts decorated their "workplace" with personal items	Astronauts requested standardizations of design of all displays and controls; A table was requested for maintenance of items, the crew wanted a working table in addition to the wardroom table
Potentials	Explorers need flexible schedule for research (to be able to explore); Space suit gloves development (Stiff and sore fingers after work in the suit)	On-board-training was effective; Many of the on-board systems could be repaired by cosmonauts	Displays and controls should be standardized; Additional (flexible) work surfaces for maintenance and repair

Diagram

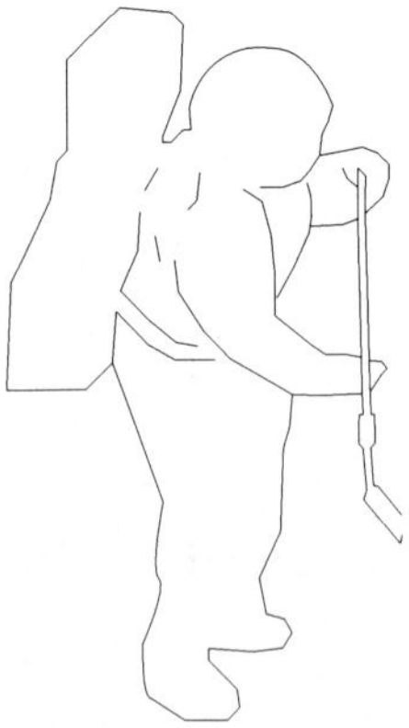

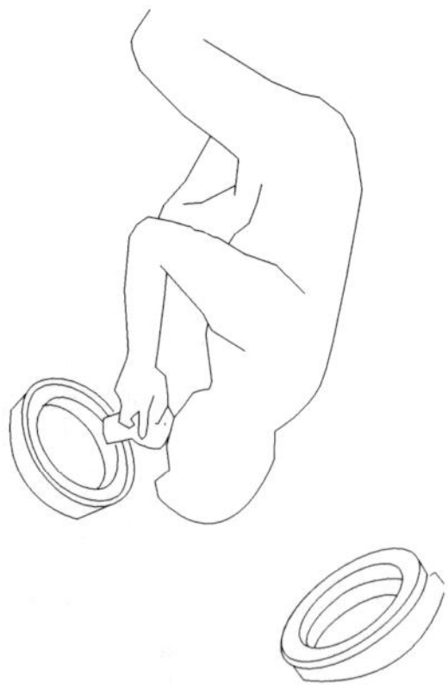

SHUTTLE	MIR	ISS
Orbiter for assembling the ISS, maintenance and transportation	Main module with adjacent science modules	International science activities in multiple scientific modules
For short term missions; Interior configuration can be adjusted according to the mission (payload, airlock)	Long life time, was planned for two crewmembers	Many different experiments, high quality research; Regular update of equipment; Still not finished
Shuttle stops working; Re-usable launcher that can be maintained on Earth	"Worked" 14 years; Many experiences for long-term missions could be used	International cooperation; Research for long-term human missions ongoing; Storage is still a topic to be solved!

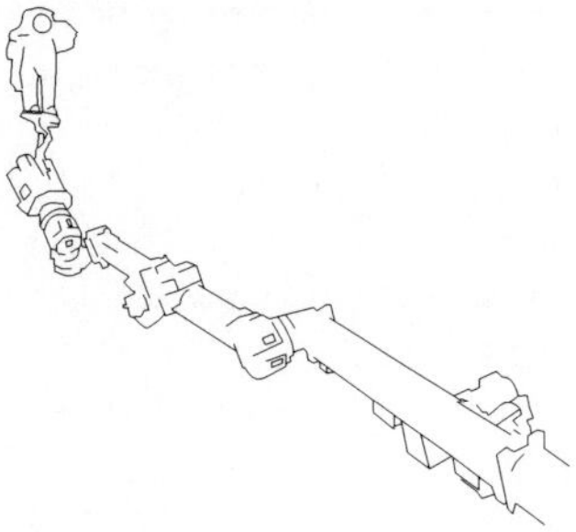

7.7 SUMMARY OBSERVATION

Work is Mission Priority

Activities on past space missions were highly work-oriented. Apollo 17 astronaut Harrison Schmitt (2009) stated that, "We were busy as explorers (...) you try to take advantage of the time you have". Jerry Linenger (2000 p. 90) wanted to complete all mission goals, and even go beyond if possible. He stated "(...) and if I had to become something of a robot, so be it". Reinhold Ewald (2009) replied to the questions, what he has done after working, : "Well, continued working (...)".

Working will remain the main objective in extra-terrestrial habitats and interiors need to be efficiently designed. The functional design of the space craft interior as well as its facilities is as important as on Earth. Due to the many different users of the same space and hardware it is difficult to meet all the requirements.

Orientation in Microgravity

"One does not trust Newton's laws" (Ewald, 2009). Although most astronauts reported that after a few days they can move quite easily in microgravity, living and working in microgravity is different.

Lebedev describes his sensations outside the Salyut space station as follows: I felt more secure, near the station. The farther I floated away from the station, the stronger my sensations became. It was like being on the edge of a balcony; the more you bend over it, the stronger the sensations of height and fear of falling you have. The fear of falling is different up here – the fear is that of being separated from the station and getting lost in space." (Lebedev, 1990 p. 161)

Natural cues, like a defined up and down or sunlight coming from 'above' are missing in a microgravity environment. On Soviet and Russian space stations a colour system was introduced to help the astronauts orient themselves spatially. However, the ability and preferences of astronaut's to orient themselves varies. Lebedev used the "details of the compartment or the interior of the station to reorient yourself" (Lebedev, 1990 p. 55). He also suggested that "in the future the handrails should be marked with different colours according to the station planes; then it would be easier to figure out your position (...) [outside the station] I spent so much time just looking around the station" (Lebedev, 1990 p. 150). Skylab astronauts reported no problems in the areas with a clear "one-g orientation" (OWS Crew Deck and Experiment areas), but had more difficulties in areas with a "random orientation" (MDA) and translating from the EVA Airlock into the Dome area of the OWS (NASA [Bull.7], 1974 p. 20, 35). In those areas they created their own coordinate system, wherein the location of the feet defined "down" (NASA [Bull.18], 1975 p. 1)

Jean- Pierre Haigneré (2009) reported, that he used a sort of "spatial matrix of transfer between the old and the new system" but even after staying 6 months in the space station Mir "the way we were orientating ourselves was the same way we do since we are born". To illustrate the fact that his brain was still working in 'Earth-mode,' he told a story. He once looked out of the space station window to take pictures of Earth for about 20 minutes, while the space station rotated about 90°, without him noticing it. When he stopped taking pictures, he 'suddenly' found himself in a vertical orientated space station, and had the immediate feeling of falling or being inside a "well". Even if this sensation lasted only a few seconds it shows "how resistant we are by our education or our culture or our experience of life on Earth". Clervoy (2009) reported that the use of colours for orientation on the Mir space station worked well, but he personally would "enjoy being in a module, that is really 3D, where there is no ceiling or floor."

Dedicated Work Areas

On the Mir space station most of the activities were done in the core module, exceptions being scientific work in the adjacent scientific modules. Astronauts conducted experiments, slept, prepared and ate food and did exercise in the same module. Haigneré (2009, Interv.) reported about "confrontations between the activities" and urged to have a "dedicated area for dedicated activities". Similar complaints were reported from Shuttle astronauts (Linenger, 2000; Burrough, 1998).

Skylab astronauts requested a dedicated table for work, as it was not included in the initial design. The Wardroom table on Skylab was not very useful to conduct the "frequent and extensive paperwork activities" (NASA [Skylab LL], 1974 p. 9).

Lighting

Adequate lighting is very important for working and living in space. Historically spacecraft lights were very dim, due to limited available electrical power. For Apollo astronaut Harrison Schmitt (2009), that was not a problem, "if there had been a warning light or caution light, then that [light] would have been bright." In addition the station orbits Earth within 90 minutes, thus changing from high light levels to darkness very quickly.

Mobility Aids and Restraints

To support astronauts' work and life in microgravity restraints are used to secure equipment or people and for moving around.

Apollo restraints included special armrests, handholds, Velcro on the floor and a restraint while operating the Lunar module (NASA [Bull.10], 1974,p. 2). Skylab astronauts used various types of restraint. The grid floor served as a major personnel and equipment restraints. Different translations modes were recognized during the Skylab missions, depending on the available volume. In small areas, such as the crew quarters and the experiment area, astronauts orientated their body vertically and used their arms to push them from one place to another. In the large area of the dome astronauts 'flew' with their head or feet first towards the desired destination (NASA [Bull.1], 1974 pp.4-6]. In doing that "nearly anything was usable as a kickoff point or as a grabbing point" (NASA [Bull.1], 1974 p. 8).

Chair-type restraints were used on Salyut space stations and on Skylab. They proved to be unnecessary within a microgravity environment (NASA [Bull.10], 1974). Handrails are not needed as mobility aids, but are needed for stability (NASA [Bull.1], 1974 p. 19).

Today multiple restraints are provided. Equipment restraints include standardized payload restraints for the rack system, such as the seat track and mobile restraints, such as tether straps, bungees, equipment bags, equipment anchors, cable ties, panel covers, a portable computer system desk and Velcro. Personnel restraints include the foot restraints, torso restraints and other crew restraints.

Stowage and Object Management

Apollo astronaut Harrison Schmitt stated: "we had a manifest where everything was. We could look it up in an index and find it (...) The people on the ground knew better than we did, where things were and if we couldn't find something we just asked them, where it is, and they would know where it was" (Schmitt, 2009).

Since then, living and working in a space station became more complex, and storage and managing of objects have a long history of being a big challenge in extra-terrestrial habitats. In the interview, Clervoy pointed out that the number

7.7 SUMMARY OBSERVATION

one challenge by far – in terms of living in space - is just managing objects: "Where are things and where to put things - Continuously" (Clervoy, 2009).

That orientation and managing objects is different in a microgravity environment was first experienced by the Salyut cosmonauts. Lebedev (1990 p. 51) wrote in his diary: "If you take a frame of the equipment and try to put it in what seems to be the right place, but it won't fit. So you start rotating things in your head, trying to remember how the interior was orientated when you removed the equipment".

Astronauts and cosmonauts have used various ways of storing and managing objects. Apollo 11 astronaut Armstrong emphasized the importance of individual and interim stowage options, as they only had one large storage space for all crewmembers (NASA [Debriefing A11], 1969 p. 149). Salyut cosmonauts stored unused equipment behind the panels; Things they needed for daily life were stored visible on the wall. For Lebedev (1990 p. 228) "the most convenient way to store something in weightlessness" was in a regular linen bag. "You open it and everything inside floats like fish in a tank." Skylab astronauts considered the available temporary stowage insufficient (NASA [Bull.2, 1974 p. 12; NASA [Bull.3, 1974 p. 4). French-Mir astronaut Haigneré also reported difficulties finding objects during his time on the space station Mir. "There was a lot of mess", he said, and the strategy they applied was "to give any object a specific place [in order to] find it very easily and rapidly. According to Portree (1995 p. 106), the Mir storage containers took up a lot of the wall space. Haigneré (2009) reported that all modules of Mir were used to store equipment that was not used anymore and only a small corridor was left free.

For the International Space Station a new stowage system has been developed. It consists of stowage racks, lockers, trays, Aisle Stowage Containers, Resupply Stowage Platforms, and soft cargo bags (NASA [ISS], 1998 p. 292). Colour codes are used for identification of items. Also in use is the 'Inventory Management System' (IMS). This electronic stowage documentation is used to record stowage location assignment and changes including equipment transfer using a bar code system. Another not very sophisticated but reliable technique is the use of Velcro and Grey tape. They seem to be the main tools for restraining things. "Everything has Velcro on it, even the pen, the fork (...)" (Clervoy, 2009).

Changing Users and Resupply

The changing users of the spacecraft and the regular re-supplying of the space station add another challenge. "Astronauts have to manipulate hundreds of objects every day and quite often they don't go back exactly where they were taken" (Clervoy, 2009). When the Space Shuttle arrives with new equipment, only a short time is available (about 1 week) to load the re-supplies and organise them in the ISS, at the same time loading the Space Shuttle with waste from the ISS. Problems with re-supply and storing equipment were also reported on the Mir space station. The crowded space station made it very difficult to move out old equipment out and load in new equipment (Dudley-Rowley, 2006). The development of automated transfer vehicles to re-supply the station brought advancement. The European Automated Transfer Vehicle (ATV) docks automatically and stays in orbit for about 5 months. "The crew goes in, only when they need and the resupply process is spread over its stay" (Clervoy, 2009).

Finding and Losing Objects

In a micro-gravity environment, objects are not easy to find because they don't fall down like they would on Earth. In his diary, Lebedev describes a technique of how he found lost things. The cosmonauts would release a balloon, to see where it would float and then they would search at the place where the balloon went (Lebedev, 1990, p. 135).

Things get lost on Earth too, but in an extra-terrestrial habitat, objects are limited and resupply difficult. Apollo 17 astronauts reported loss of equipment. "He [Eugene Cernan] had dropped our only remaining pair of scissor in the melee [on the lunar surface]. We had left one pair behind with Ron [Evans], who had still not found his, and now ours had vanished into the lunar soil." At this time the scissors were the only way to cut open the plastic food packages (Cernan, et al., 1999 p. 323).

Skylab astronaut Jack Lousma stated, that "things seem to just disappear very easy (...) usually they show up in a day or two but sometimes that isn't soon enough" (NASA [Bull.1] p. 55). Many objects have been lost and are still being lost - some temporarily some forever. In 2008 astronaut Heidemarie Stefanyshan-Piper lost her tool-bag during an ISS extra-vehicular activity, due to an accident (MSNBC, 2008). According to Cohen (2011, 2009 pp. 16-18) the most common and frequent cause of losing tools and parts during orbital EVAs is crossing the day-night terminator. Either the EVA crew experience a huge, blinding flash of sunlight as they pass to the day side, or they are plunged suddenly into darkness on the night side. In both situations, it takes several minutes for their eyes to adjust.

Standardized Interfaces

The changing user of this type of habitat makes the management of objects and storage even more complex. When for example a 'Visiting Crew' arrives at the station, they are expected to adapt their way of handling things to that of the main crew. "You don't like to be disturbed" (Clervoy, 2009). The Skylab astronauts requested standardization of the work environment as follows: "There should be standardizations imposed on the design of all displays and controls used by the flight-crew, i.e. switches, indicator lights, control knobs" (NASA [Skylab LL], 1974 pp. SLL_2-44).

According to Bluth (1987 p. 27), "90% of all equipment in the Mir space station was designed for ease of replacement, repair and maintenance". On the US side of ISS, the mechanical interfaces are standardized insofar as any device to be attached to the front of a rack in U.S. module can also be attached to a rack in the Columbus or Kibo module. The light bulbs used are also the same, but can have different acronyms in the U.S. module or the Columbus and Kibo module (Clervoy, 2009). Another difficulty is the use of two systems of units. The European module, the Russian module and the Japanese module are set up in the SI system, but the US modules are set up in Imperial units. Therefore metric tools and inches tools are available. Also, the US conversions of SI metric units to decimal inches and vice versa to install racks and equipment on SpaceLab metric conversions were often incorrect, which resulted in holes being drilled in the wrong places in the racks and other mistakes (Cohen, 2011).

Adjustments of the Work Area

Although the Salyut cosmonauts made progress with the adjustment of their working environment, they had a rather brute approach. "When a crew arrives, the cosmonauts may rearrange things to make themselves feel at home. This kind of work usually requires moving a lot of material and equipment. Sometimes this requires sawing of metal, which not only litters the station but also takes lots of time and effort" (Lebedev, 1990 p. 137). In 1995 on Mir, cosmonauts used a machete

7.7 SUMMARY OBSERVATION

to remove the shower for new equipment. (Burrough, 1999 p.86)

Today, the internal layout of the U.S., Japanese and European modules of the ISS is modular and reconfigurable.

At a smaller scale, cosmonauts and astronauts have always adapted their environment, just like they do at home, but with a minimum of available possibilities. Lebedev describes in his diary (1990 p. 259), that they "had decorated the station with balloons and hung up Lenin's portrait" on the 65th anniversary of the October revolution. Photos illustrate that astronauts and cosmonauts still use private images and presents to "decorate" their (temporary) home – the International Space Station. Thus the appearance of the working area changes over the years depending on the crew on-duty.

Autonomy in Work Schedules

"The missions were planned, but there was a lot of open time. We were doing a lot of different things that were not planned for (...), because, we didn't know what to plan for. We had a lot of time for spontaneous observations and discovery. That's what exploration is all about. You have to be careful not to program too much" (Schmitt, 2009).

In contrary, the evidence show, that throughout human space exploration history, crewmembers have pushed for more autonomy. Compared to today, early astronauts and cosmonauts are often reported to have had a military like organizational regime, thus following strictly the schedule depicted. Apollo astronauts, Salyut and Mir cosmonauts had to follow a strict schedule. Life on-board was constantly commanded by mission control. But there are some references that relate to the fact, that these humans did not always follow the "line". Lebedev wrote in his diary (1990 p. 166) about an incident when he asked FCC to postpone their exercises for ten (!) minutes in order to finish a geological survey. They had an argument and although he finally got the permission to continue the experiment, he felt very upset long after. "All in all today I felt rather sad, because so many things had built up inside me, and I remembered so many things."

According to Jones, the third crew of Skylab astronauts turned off the radio, as a protest to heavy workload. They refused to talk with Houston Mission Control and declared that day for an unscheduled day off (Jones, et al., 2002 p. 238). However Skylab astronaut Gerald Carr (2000, p. 48) tells the story differently. He said that on their day off they forgot to configure the radio in the correct way, and mission control couldn't get through to them.

Also the Salyut cosmonauts had their way to fight against the authority from the ground. They did not tell everything to ground (cf. Linenger, 2000), which is also illustrated in Valentin Lebedev's Diary. The Salyut cosmonauts had to do plant experiments with the Oasis Greenhouse. One day they were sent some onion bulbs for a biological experiment. Instead of planting them, they ate the onions with some bread "right away". "They were delicious," Lebedev wrote in his diary (1990 p. 113), telling ground control, that they were growing well. The story blew, when Lebedev exaggerated telling the biologists, that the onions even had shoots. Onions had never bloomed before in space, so they finally had to tell the excited biologist the truth.

One reason for the cosmonaut's autonomy despite scheduled work might be that Salyut and Mir cosmonauts had no constant communication with ground at the time. They could only communicate via line-of-sight circuits. As the Soviets could use only the stations on their home territory and on ships at sea, their spacecraft were out of radio contact for at least one-third of each orbit, and often more. During the fire on Mir in 1997, the crew was out of communication range with mission control in Moscow for about half an hour. Only the ham radio would have worked, but that meant "broadcasting to the blind" (Portree, 1995 p. 106). At this time "only 20 minutes were available for two-way communication per 90-minute orbit"(Foale, 1999 p. 96). The cosmonauts had to take a decision without mission control in this situation. They decided not to leave the statio and fought the fire successfully.

Today, at the ISS, astronauts have the freedom to call any telephone on Earth at any time. But a constant communication line also means constant monitoring from ground, as Ewald refers to: "The speed of the ventilators tells mission control if somebody is in a module or not" (Ewald, 2009). This is because the ECLSS console in Mission Control monitors the oxygen uptake rate in each volume which varies with the amount of crewmembers.

Experiments are scheduled to be technologically and scientifically supported by ground personnel. Astronauts have to exercise 2h per day. As there is only limited exercise equipment on-board, also these times are scheduled. "Well, we have a scheduling program on board that has in it all of the details that we need to know in order to do the day's work. It tells us when we should go to sleep, when we should get up, when we should exercise, when to eat our meals, when and what information we need to do our tasks." (Magnus, 2009)

On-going Training

Salyut and Mir cosmonauts stayed a long term in orbit. As astronauts could not be trained for every eventuality in advance, they used an 'on-the-job-training' in orbit. This was conducted either via visiting crews or with manuscripts and even models. Lebedev wrote in his diary: "Our flight is so long that a lot of our training is forgotten, our skills get lost, and a man comfortable in his surroundings, begins to relax" (Lebedev, 1990 p. 228).

"How to make astronauts not need to memorize too many things" (Clervoy, 2009), remains an issue today. Astronaut instructors from ESA-EAC confirm that "as the mission duration will increase dramatically, it will be very difficult for any ground management or organization, to perfectly prepare all operations and procedures that will be performed by the crew, and furthermore, it will be very hard to foresee all the operations or activities that will be performed" (Aguzzi, et al., 2009).

Jean- François Clervoy believes that the selection criteria and the training methods for future astronauts travelling to Mars will be different from today. "We need people able to be self-sustainable as an individual and as a group (...) both skills need to be there." (Clervoy, 2009)

7.8 COMPARISON USABILITY MATRIX

	APOLLO	SALYUT	SKYLAB
Availability and Equipment	Command Module, Lunar Module, Lunar Roving Vehicle, ALSEPs	Work Compartment and scientific instruments	Experiment area, Telescope Mount, Multiple Docking Adapter
Spatial Arrangement	Main work on the lunar surface	Work next to other functions in the Work Compartment	Experiment work area in the Orbital work shop; Large doorways between compartments
Object Management	Integrated; one large storage space for all crewmembers	Storage behind the panels and on the walls	Mix of integrated (habitability) and "tuck anywhere" approach (work)
Ergonomic Safety	Restraints in the modules; Floor covered with Velcro	Floor covered with Velcro to aid moving around; On-board training for skill maintenance; Interior had soft surfaces	Triangular grid on wall and ceilings/floors to ease moving around

SHUTTLE	MIR	ISS
Crew Cabin, payload bay, scientific hardware	Dedicated science modules	A variety of international science modules and facilities; Regular update of equipment
Internal layout can be adjusted according mission objectives	Main activities in the main module	Different science modules and equipment
Rack system, rigid and flexible containers	Similar to Salyut; Built-in stowage compartments; Storage added up	Stowage Rack system, Inventory Management System
Multiple restraints	Multiple restraints; After a few years the outside was "covered with little holes" from micro-meteorites	Multiple restraints

7.9 COMPARISON LIVABILITY MATRIX

	APOLLO	SALYUT	SKYLAB
Territoriality and Privacy	No privateness in the module	Little privateness in the module	Large space station offered multiple possibilities
Sensory perception	Rich sensory perception on the lunar surface, viewing Earth; Music was used for motivation, music wake-up call	Cosmonauts decorated the station with photographs and drawings; Music all day; Technical smell	Acceptable and comfortable temperature, pleasant acoustic environment, no odours; Lighting was dim
External relations	Windows to the outside	Salyut 7 had about 20 windows	One big window in wardroom, small 360° windows in Orbital Workshop
Internal relations	All in one module	Small and large diameter compartment; Good overview of the station	Large open interior space; Two levels separated with grid floor; Good overview of work area

SHUTTLE	MIR	ISS
Little privateness, but two levels	Different modules offered multiple possibilities	Different modules offer multiple possibilities
Acceptable for short missions	After a few years the outside panels degraded and more heat came in (humidity); Smelled like a 'cellar' or 'cave'	Seems comfortable; Varies
6 optical windows and 2 aft-windows to observe payload bay, airlock	A lot of windows in all modules	Many windows (Zvezda: 14 windows), optical window for Earth observation (Nadir, Destiny), 360° degree window (6 windows, Cupola)
Two levels (Flight Deck and Middeck)	Different modules with different diameters; Modules arranged around Transfer Compartment	Different modules; Standardized and clear interior concept (Racks); Most of the modules arranged along one axis

7.10 COMPARISON FLEXIBILITY MATRIX

	APOLLO	SALYUT	SKYLAB
Spatial Flexibility	Mobile with the Lunar Rover	Inflatable Airlock; Removable partitions between the instrument and the living area	Large work area; Predesigned and flexible storage locations
Object Flexibility	Deployable roving vehicle; Scissors were used for food opening and as tools on EVA	Permanent evaluation and development of hardware, much onboard equipment could be repaired by cosmonauts ; Foldable table	Some objects were multi-usable, deployable containers
Individual Flexibility	“That’s what exploration is all about” (Schmitt)	Activities were scheduled, but they did not tell everything mission control; Cosmonauts were inventive on equipment, e.g. made a focus ring for the camera and adjusted the penguin suit	Unscheduled maintenance activities

SHUTTLE	MIR	ISS
Additional module can be added to increase space	Cosmonauts worked in different modules	Astronauts work in different modules; A variety of modules, Russian and Western Style
Airlock system can be reconfigured depending on the mission; Seats have different configurations during launch and on orbit	Equipment was updated some by "sawing"	Racks can be relocated; Different tools for different modules (metric and inch)
Restricted and strict time schedule	On-the-job training in orbit	Astronauts can use different modules to work, specific experiments in dedicated modules; Strict time schedule

7.11 DESIGN DIRECTIONS

The most important findings related to the human activity 'WORK' are summarized and structured in the form of design directions. These directions summarize the main points in a short sentence. The paragraph following refers to the empirical background of experiences made in extra-terrestrial habitats.

USABILITY

Provide dedicated areas and equipment

Special activities need special equipment and adequate lighting. Windows need shutters. Work zones for operation and science work shall be separated from other activity zones, especially from where people rest or sleep.

Integrate standardized interfaces

With regard to future long-term missions special attention should be paid to the standardization of tools, interfaces and surfaces of instruments in terms of maintenance and redundancy.

Design for easily managable stowage

The design of a good storage management and design concept is of high priority, especially for longer missions, where mission objectives can change. This includes associated facilities as well as the design of appropriate equipment. Individual habits of the users need to be taken into account. Also things to be stored add up over time, as on Earth.

Provide multiple restraints

Skylab astronauts reporting that "any available hardware became a natural restraint" (NASA [Bull.1], 1974 p. 10) is important information for the planning of future habitats. Adequate means for plentifula and sufficient personal and object restraints as well as translation in microgravity are required.

Design ergonomically

In order to design efficient and functional workstations, they need to be designed ergonomically. The posture in microgravity is different, but also individual preferences need to be taken into account. Orientation of workspace can change, but each specific workstation needs to be aligned with its associated equipment.

LIVABILITY

Integrate personal work areas

Personal work areas in addition to common operation and science areas need to be integrated.

A:

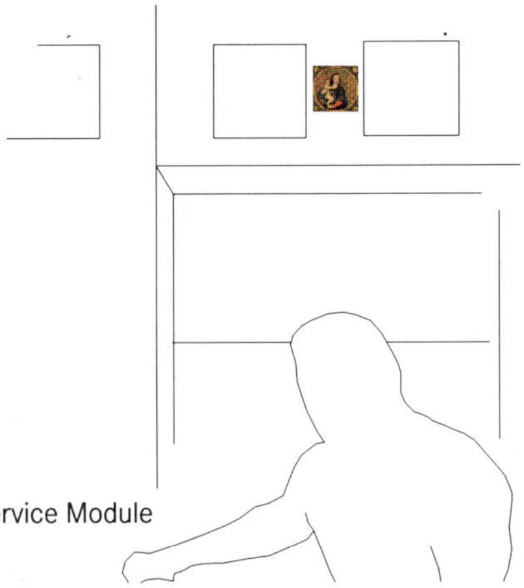

Zvezda Service Module
2000

A: Personalization of Space (2000 - 2008) *(credit A: Author)*

FLEXIBILITY

Integrate extension possibilities

Inflatable or deployable volumes can (temporarily) extend operation activities and habitable volume.

Allow reconfiguration of the work area

The ability to adjust the work environment to individual needs and to the activity to be carried out enhances productivity. In addition to small adaptations by changing crewmembers such as marking the crew's territory, larger adjustments are likely to be made on long-term missions. The number of people to be accommodated may change, or new and more advanced equipment may have to replace old equipment, etc., as was seen during the lifetime of the Mir space station. A design that allows adjustment to not yet known objectives is a benefit for the long-term use of any habitat, but especially for a space station or planetary habitat, where late layout changes are not an easy task.

Integrate on-going-training

New training concepts such as on-board training via astronaut colleagues or with the help of new media technologies will need to be integrated.

Allow individual adjustment

A variety of flexible work surfaces would support individual activities. Flexibility to mission-related adjustments will become even more important with time. In this context, adjustable lighting that distributes adequate light according to tasks is of high importance, because different work activities require different lighting.

Provide autonomy for the users

For future missions autonomy in decision making and some 'blanks' in the work schedule will play an important role.

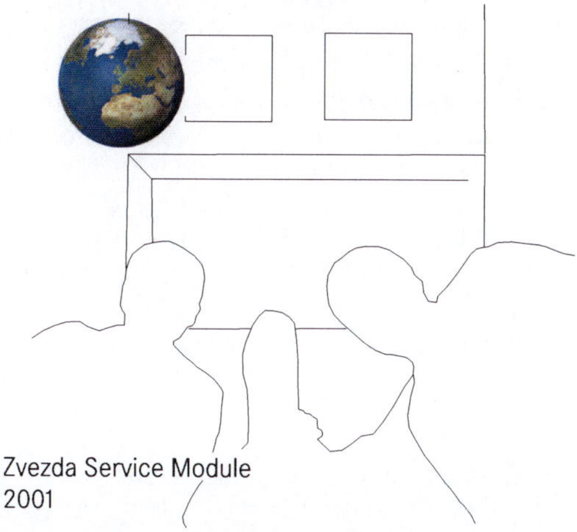

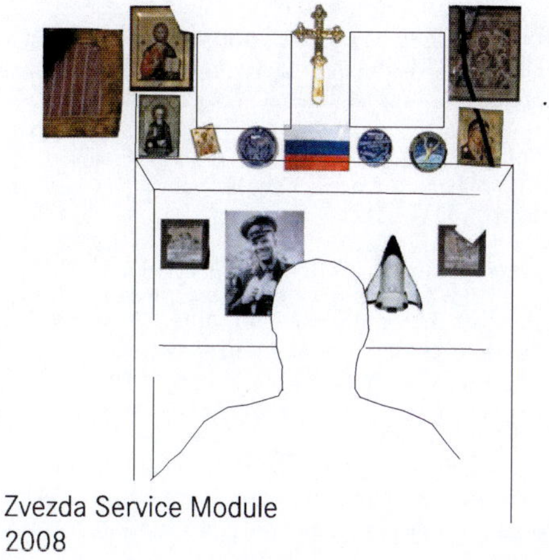

7.12 REFERENCES

Aguzzi, Manuela, Bosca, Riccardo and Müllerschkowksi, Uwe. 2009. Astronaut training in view of the future: A Columbus payload instructor perspective. [ed.] Elsevier. Acta Astronautica. 66, 2009, pp. 401-407.

B.J.Bluth, M. Helppie. 1986. Soviet Space Stations as Analogs. Washington D.C., US : NASA Headquarters, 1986. NASA Grant NAGW-659.

Beatie, Donald A. 2001. Taking Science to the Moon - Lunar Experiments and the Apollo Program. Baltimore, Maryland, U.S. : The Johns Hopkins University Press, 2001. 0-8018-7440-8.

Burrough, Bryan. 1999. Dragonfly - NASA and the crisis aboard Mir. London, GB : Fourth Estate Limited, 1999. ISBN 1 84115 087 8.

Bursch, Daniel W. 2002. Expedition Four - 120 Day Report. NASA. [Online] April 2, 2002. [Cited: 8 2, 2009.] http://spaceflight.nasa.gov/station/crew/exp4/120days.html.

Carr, Gerald P. 2000. Oral History Transcript - Gerald P. Carr. [interv.] Kevin M. Rusnak. Huntsville, Arkansas, 25 October 2000.

Cernan, Eugene and Don, Davis. 1999. The last man on the Moon. USA : St. Martin's Griffin, 1999. ISBN 0-312-19906-6.

Chamitoff, Greg. 2008. Greg Chamitoff's Journal. [Online] 8 February 2008. [Cited: 20 April 2008.] http://www.nasa.gov/mission_pages/station/expeditions/expedition17/journal_greg_chamitoff_3.html.

Clervoy, Jean- François. 2009. Transcript not published [interv.] Sandra Häuplik-Meusburger. Paris, France, June 2009.

Cohen, Marc. M. 2009. From Apollo LM to Altai: Design, Environments, Infrastructure, Missions, and Operations. AIAA-2009-6404

Cohen, Marc M. 2011. not published. Email, 2011.

Connors, Mary M., Harrison, Albert A. and Akins, Faren R. 1999. Living Aloft, Human Requirements for Extended Spaceflight. Washington, D.C., USA : NASA, Ames Research Center, 1999. SP-483 / ISBN: 1-4102-1983-6.

Dudley-Rowley, Marilyn. 2006. The MIR Crew Safety Record: Implications for Future Space Colonization. San Jose, California : Space 2006, 2006. AIAA 2006-7489.

Ewald, Reinhold. 2009. Transcript not published. [interv.] Sandra Häuplik-Meusburger. Cologne, Germany, July 2009.

Foale, Colin. 1999. Waystation to the Stars - The Story of Mir, Michael and Me. London, GB : Headline Book Publishing, 1999. ISBN 0 7472 6331 0.

Gibson, Edward R. 2000. Oral History Transcript - Edward G. Gibson. [interv.] Carol Butler. 1 December 2000.

Haigneré, Jean-Pierre. 2009. Transcript not published. [interv.] Sandra Häuplik-Meusburger. Paris, France, June 2009.

Jones, Thomas D. and Benson, Michael. 2002. The complete idiot's guide to NASA. U.S. : Marie Butler-Knight, 2002. ISBN: 0028642821.

Lebedev, Valentin. 1990. Diary of a Cosmonaut - 211 Days in Space. United States and Canada : Bantam Air & Space Series, 1990. Vol. Vol.4. ISBN: 0-553-28778-8.

Linenger, Jerry M. 2000. Off The Planet - Surviving five periolous months aboard the Space Station MIR. U.S.A. : McGraw-Hill, 2000. ISBN 0-07-136112-X.

Lousma, Jack R. 2001. Oral History Transcript - Jack Lousma. [interv.] Carol L. Butler. Houston, Texas, USA, 7 March 2001.

Magnus, Sandra. 2009. Sandra Magnus' Journal . [Online] NASA, 6 January 2009. [Cited: 10 Februar 2010.] http://www.nasa.gov/mission_pages/station/expeditions/expedition18/journal_sandra_magnus_6.html.

MSNBC. 2008. Spacewalker loses tools in orbit. [Online] 18 November 2008. [Cited: 20 January 2009.] http://www.msnbc.msn.com/id/27784400/.

NASA [Bull.1]. 1974. Skylab Experience Bulletin No.1: Translation Modes and Bump Protection. Houston, Texas, US : NASA, Lyndon B. Johnson Space Center, 1974. JSC-09535.

NASA [Bull.3]. 1974. Skylab Experience Bulletin No. 3: Architectural Evaluation for Sleeping Quarters. Houston, Texas, US : NASA, Lyndon B. Johnson Space Center, 1974. JSC-09537.

NASA [Bull.5]. 1974. Skylab Experience Bulletin No. 5: Inflight Maintenance as a Viable Program Element. Houston, Texas, US : NASA, Lyndon B. Johnson Space Center, 1974. JSC-09539.

NASA [Bull.7]. 1974. Skylab Experience Bulletin No. 7: An Overview of IVA Personal Restraint Systems. Houston,

Texas, US : NASA, Lyndon B. Johnson Space Center, 1974. JSC-09541.

NASA [Bull.10]. 1974. Skylab Experience Bulletin No. 10: Body Restraint Systems. Houston, Texas : NASA, Lyndon B. Johnson Space Center, 1974. JSC-09544.

NASA [Bull.18]. 1975. Skylab Experience Bulletin No. 18: Evaluation of Skylab IVA Architecture. Texas, Houston, US : NASA, Lyndon B. Johnson Space Center, 1975. JSC-09552.

NASA [Debriefing A11]. 1969. Apollo 11 - Technical Crew Debriefing. Houston, Texas, USA : NASA, Manned Spacecraft Center, 1969. 276.

NASA [ISS]. 1998. International Space Station Familarization. Houston Texas, USA : Lyndon B. Johnson Space Center, 1998. TD9702A.

NASA [Shuttle Reference]. 1981. Space Shuttle News Reference. Houston, Texas. : U.S. Government Printing Office, 1981. 723-000/562.

NASA [Skylab LL]. 1974. Lessons Learned on the Skylab Program. Houston, Texas : NASA, Lyndon B. Johnson Space Center, 1974. JSC-09096.

NASA [Skylab]. 1977. Skylab, our first space station. Washington D.C., US : NASA, Scientific and Technical Information Office, 1977. Prepared by George C. Marshall Space Flight Center, Edited by Leland Belew. NASA-SP-400.

NASA [SS PressKit]. Space Shuttle Press Kit: Stowage Description rigid Containers. [Online] [Cited: 10 10 2009.] www.shuttlepresskit.com/scom/224.pdf.

Oefelein, Bill. 2006. Bill Oefelein's STS-116 Mission Blog. NASA. [Online] 19 December 2006. [Cited: 5 December 2008.] http://www.nasa.gov/mission_pages/shuttle/shuttlemissions/sts116/oefelein_blog.html.

Portree, David S.F. 1995. MIR Hardware Heritage. Houston, Texas : NASA, Lyndon B. Johnson Space Center, Information Services Division, 1995. NASA RP 1357.

Prunariu, Dumitru-Dorin. 2011. Transcript not published. [interv.] Sandra Häuplik-Meusburger. Vienna, Austria, 2011.

Schlegel, Hans Wilhelm. 2009. Transcript not published. [interv.] Sandra Häuplik-Meusburger. Cologne, Germany, 2009.

Schmitt, Harrison. 2009. Transcript not published. [interv.] Sandra Häuplik-Meusburger. Vienna, Austria, 2009.

Tognini, Michel-Ange Charles. 2009. Transcript not published. [interv.] Sandra Häuplik-Meusburger. Cologne, Germany, July 2009.

Viehböck, Franz. 2009. Transcript not published. [interv.] Sandra Häuplik-Meusburger. Berndorf, Austria, 2009.

Walz, Carl. 2002. Expedition Four - Spacewalking, ISS Style. NASA. [Online] 09 04 2002. [Cited: 22 09 2009.] http://spaceflight.nasa.gov/station/crew/exp4/eva/spacewalking.html.

Wise, James. 1988. The Quantitative Modelling of Human spatial Habitability. Ames Reseach Center : NASA Contractor Report, 1988. 177501.

Woods, David and O'Brien, Frank. 2008. Apollo 15 Flight Journal - Day 11: Worden's EVA Day. Corrected Transcript and Commentary Copyright © 2000 by W. David Woods and Frank O'Brien. [Online] NASA History Divison, 18 12 2008. [Cited: 20 01 2009.]

Worden, Alfred M. 2000. Oral History Transcript - Alfred M. Worden. [interv.] Rebecca Wright. 26 May 2000.

PART 8: HUMAN ACTIVITY

LEISURE

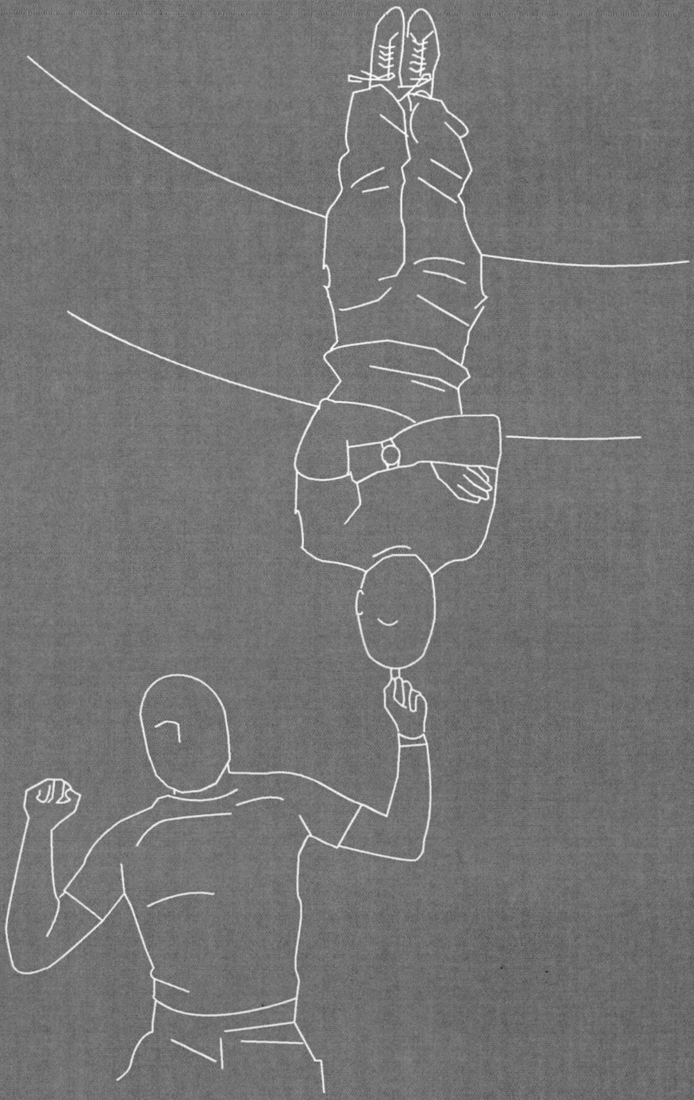

The category 'Leisure' includes the sub-activities associated with exercise and free-time activities.

8.1
LEISURE APOLLO

CONCEPT Music, Exer-Genie
LOCATION CM/LM, Lunar surface
SPECIFIC Unique activities on the lunar surface

RÉSUMÉ LEISURE

Exercise

The Apollo astronauts did some exercise during their flights for about 10 to 30 minutes. They did some gymnastics with a device called 'Exer-genie' (NASA [Debriefing A11], 1969 p. 147). Gene Cernan preferred to 'run in place', sitting in the commander's couch holding on tightly. Thus he would train his arms and legs and obtain an efficient heart rate. (NASA [A17], 1973 p. 27-16, 27-17, 27-39)

Free-Time Activities

Although the priority for other than mission-related tasks was low, Apollo astronauts had some "free time", especially during the lunar-Earth transit. Apollo astronauts had a battery operated cassette recorder and listened to music for entertainment. Dick Gordon reported that they used it "during dull, boring hours during trans-lunar coast and trans-earth coast" (NASA [Debriefing A12], 1969). They suggested using the power of the spacecraft instead of batteries.

During the Apollo 14 mission, astronaut Alan Shepard brought a golf club and hit two golf balls on the lunar surface to see how far it would go. It was his own personal humorous experiment. It was very popular back home.

Some astronauts were hopping and singing between their experiments on the lunar surface. A video shows Harrison Schmitt singing 'I was strolling on the Moon (...)' (NASA, 1972).

A:

B:

A: Apollo 16 commander John W. Young leaps from the lunar surface as he salutes the U.S. flag
B: "Earthrise" as seen from Apollo 11 mission crew in 1969
C: Astronaut looking at Earth *(credit A, B: NASA; C: Author)*

ASTRONAUTS' EXPERIENCES

USABILITY

Gene Cernan (Apollo 17): *"The whole time, we stayed busy, but joking around was a wonderful stress reliever."* *(Cernan, et al., 1999 p. 326)*

Mitchell Edgar (Apollo 14): *"I noticed it [tiring in the legs and back] throughout the flight, diminishing toward the end. (...) It felt good to pull on the Exer-Genie and straighten those muscles out." (NASA [Debriefing A 14], 1971)*

LIVABILITY

Alan Shepard (Apollo 14): *"What a neat place to whack a golf ball!" (Shepard, 1998)*

Al Worden (Apollo 15): *"We played games inside the spacecraft on the way out, where we'd take something in the middle of the spacecraft and start spinning. It's amazing; it just kept spinning and spinning and spinning." (Worden, 2000)*

Gene Cernan (Apollo 17): *"Stripped down to our liquid-cooled underwear, we had a quick dinner, debriefed with the guys on Earth via a private radio loop, and played with some of the rocks we had stowed in the cabin boxes. How amazing." (Cernan, et al., 1999 p. 328)*

FLEXIBILITY

Harrison Schmitt (Apollo 17): *"Just being there. Every day had an exciting aspect to it." (Schmitt, 2009)*

Gene Cernan (Apollo 17): *"I could really run [in place] at different speeds and for long durations, and that's the way I did all my exercise." (NASA [A 17], 1973 p. 27-17)*

C:

8.2 LEISURE SALYUT

CONCEPT Bicycle ergometer, treadmill, pressure suits, vacuum trousers, muscle-exercise equipment

LOCATION Large Diameter Compartment, on the ceiling

ORIENTATION Vertical and horizontal orientation

RÉSUMÉ LEISURE

Exercise

The cosmonauts were allowed some free choice in the type of exercises they did. Cosmonauts exercised twice a day for about 1.5 hours. Exercise equipment on Salyut stations included a bicycle ergometer, a treadmill or running track, pressure suits ('Penguin suit') or vacuum trousers ('Chibis') and muscle-exercise equipment (B.J.Bluth, 1987 pp. I-85,86; OTA, 1983 p. 52). The bicycle ergonometer was first flown on Salyut 4 and was also used for generating electricity (Portree, 1995 p. 71). It was located on the ceiling in the Working Module *[Fig. B]*.

On Salyut 7, cosmonauts changed to a 2-hour per day exercise-period. Exercise equipment on Salyut 7 included the bicycle ergometer, treadmill, running track, pressure suits, vacuum trousers and muscle exercise equipment. The bicycle ergometer could be used by turning the pedals with feet or hands and was individually adjustable (B.J.Bluth, 1987). A treadmill or running track was included, where cosmonauts spent about one hour.

A:

Muscle equipment, like a rubber absorber with strap loops and a sting stretcher was also used regularly, especially prior to EVA for conditioning purposes (B.J.Bluth, 1987 p. III-33). The station also offered improved medical facilities.

When the Indian cosmonaut Sharma visited the station, the crew gained some experience with yoga. In addition to its minimal space consumption, according to Bluth, the experiments showed that "yoga might prove beneficial for overcoming problems that occur because of weightlessness" (B.J.Bluth, 1987 p. I-87).

Free-Time Activities

According to Bluth, the Russians recognized that leisure time had a stimulating effect on the cosmonaut's mental state and fitness, and would therefore contribute to the work efficiency of the crew (cf. B.J.Bluth, 1987). Leisure activities were scheduled according to the flight stage, the cosmonaut's personal interest and the psychological crew climate. Leisure activities included the following (B.J.Bluth, 1987 p. III-28; Portree, 1995 p.67, 84):

- Two-way television for private communication
- Video cassettes of films, concerts and cartoons
- playing games, such as chess
- reading books and listening to music
- Self-education
- Sketch book
- taking pictures and looking out *[Fig. A]*

In addition surprise leisure activities were organized for each crew, and private letters, newspapers and gifts were sent with the Progress freighters. One cosmonaut received a guitar during a 140 day mission (B.J.Bluth, 1987 p. I-161).

A: Salyut 6 cosmonaut taking a picture of Earth
B: D. Prunariu (with the ergometer on the ceiling) and V. Savinykh on board Salyut 6, Mai 1981 *(credit A: Spacefacts, J. Becker; B; Dumitru-Dorin Prunariu)*

ASTRONAUTS' EXPERIENCES

USABILITY

Valery Ryumin (Salyut 6): 175 and 185 day mission: "I hate our exercises. Loved it on Earth. But here, each time I have to force myself. ***Boring and monotonous, and heavy work****. But you realize you need it to keep in shape so you grin and bear it."*
(B.J.Bluth, 1987 p. III-34)

Valentin Lebedev (Salyut 7): "The sweat doesn`t trip down, (...) and wiggles like a jellyfish as I move."
(Lebedev, 1990 p. 133)

"We are supposed to have a rest day tomorrow, but the radiogram indicates that an experiment had been planned. We absolutely refused to do it. I told them, It's about time that we start to rest and exercise during the rest days."
(Lebedev, 1990 p. 237)

LIVABILITY

Dumitru-Dorin Prunariu (Salyut 6): "we had a vacuum cleaner with back exhaust air, sold in Eastern European stores under the name 'Raketa', that looked like a rocket (...), fed by a long cable from the station's electrical network, and sometimes (...) you just took the vacuum cleaner between the legs, turned it on and then you flew like a rocket inside the station." (Prunariu, 2011)

Valentin Lebedev (Salyut 7): ***"Swimming in weightlessness is lots of fun****. We swim in this huge aerial aquarium of a station like space amphibians."*
(Lebedev, 1990 p. 216)

FLEXIBILITY

Valentin Lebedev (Salyut 7): "I like to play around with the biological experiments in the silence after working fuss is over. They calm me down. It is a time now for thinking and dreaming."
(Lebedev, 1990 p. 123)

"Sometimes it is very hard to force yourself to do. We like the treadmill the most, because we can do such a variety of exercises on it."
(Lebedev, 1990 p. 168)

"In fact, we've even made up some new exercises of our own."
(Lebedev, 1990 p. 188)

"A rest day. These days are the most difficult ones. *Each of us does whatever he wants. I took a video of the Earth's horizon."*
(Lebedev, 1990 p. 238)

B:

8.3 LEISURE SKYLAB

CONCEPT Ergometer, treadmill, bodymass, recreational equipment
LOCATION OWS, Experiment area
DIMENSIONS Experimental exercises in microgravity

RÉSUMÉ LEISURE

Exercise

The bicycle ergometer *[Fig. B]* used by the Skylab astronauts was located in the Experiment area and had triangular restraints on the pedals for the specially designed shoes, a belt for the shoulder harness and bungies tied to the floor to prevent "flying off" (Weitz, 2000). Later the restraint for the ergometer was removed (NASA [Bull.10], 1974). Its noise also bothered communication between crewmembers (NASA [Skylab CS], 1974 p. 117). The second Skylab crew installed a treadmill.

The Lower Body Negative Pressure Device (LBNP) was installed in the Experiment area. The device embraced the astronauts' lower body *[Fig. A]* and was used for training and monitoring the cardiovascular system during space flight. (NASA [LBNPD], 2002; NASA [Bull.7], 1974 p. 10)

Skylab astronauts were inventive with exercises and experimented with body movements that were only possible in microgravity. They did somersaults in the open Dome area, 'flew' from one side to the other or ran centrifugally in a circle on the "track" formed by the stowage lockers. *[Fig. C, D]*

Free-time Activities

Skylab astronauts had a tape player in the wardroom and one in the Apollo Telescope Mount. Other recreational equipment, Skylab astronauts were provided included dart sets, playing cards, balls, books, binoculars, exercise equipment, and a tape player (NASA [Skylab], 1977). Astronauts reported that the darts set did not work, but they played with the balls "for fun about once a week" (NASA [Bull.3], 1974 p. 8].

Much of their free time was used for "looking to Earth".

A:

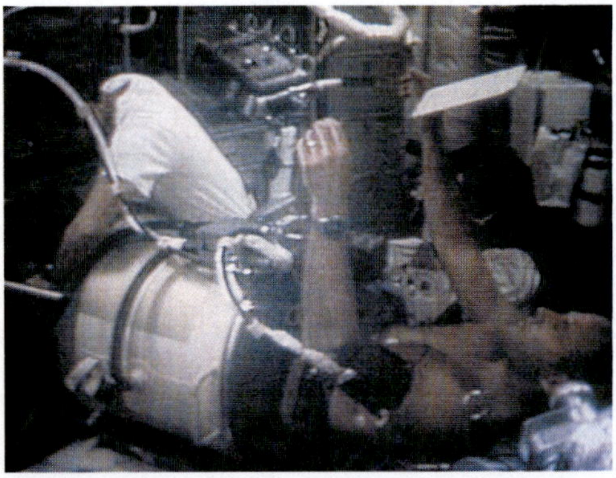

B:

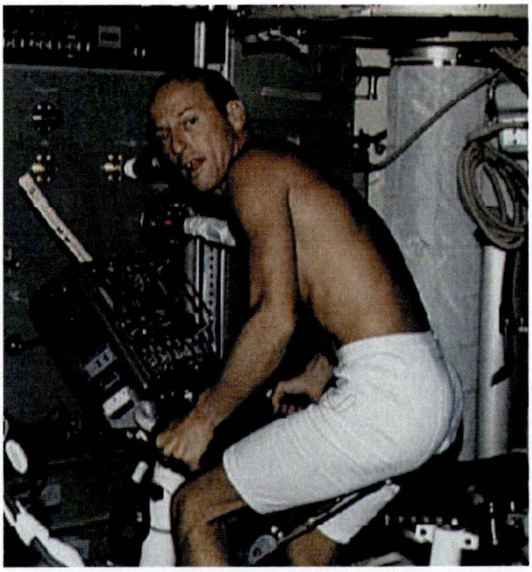

A: Jack Lousma in the Lower Body Negative Pressure Device (LBNPD)
B: Charles Conrad on the bicycle ergometer in the OWS
C: Jack R. Lousma doing acrobatics in the dome area of the OWS
D: Alan Bean doing acrobatics in OWS dome area

(credit A, B, C, D: NASA)

ASTRONAUTS' EXPERIENCES

USABILITY

Gerald Carr (Skylab 4): *"We weren't getting the right kind of leisure time that would allow us to do the right kind of job." (Carr, 2000 p. 47)*

Edward G. Gibson (Skylab 4): *"Unfortunately, even though we were up there for 84 days, we were kept pretty busy, so we didn't have a great deal of time to look out the window as much as we liked." (Gibson, 2000)*

LIVABILITY

Edward G. Gibson (Skylab 4): *"Oh yes.* ***Looking out the window, yes. Yes, we all did.****"*
(Gibson, 2000 p. 70)

Joe Kerwin (Skylab 2): *"Even just running around the ring lockers or throwing the ball around in the evening for 15 minutes makes you more relaxed."*
(NASA [Bull.4 p. 16)

Charles Conrad (Skylab 2): *"I liked the illusion of hanging (...) it made me think I was sitting up and reading."*
(NASA [Bull.4], 1974 p.16)

FLEXIBILITY

Paul Weitz (Skylab 2): *"That [bicycle ergometer] was the most ungodly, cumbersome, unwieldy, unworkable device, so we soon threw that out.* ***We took the seat off****, because you could maintain position by hanging onto the handlebars and just having your feet fixed to the pedals."*
(Weitz, 2000 p. 37)

C:

D:

8.4 LEISURE SHUTTLE

CONCEPT Ergometer, treadmill, dynabands
LOCATION Middeck
SPECIFIC Short missions

RESUMÉ LEISURE

Exercise

The exercise area is located in the Middeck or Flight deck. Available exercise equipment includes an ergometer, a treadmill, and dynabands. Astronauts on short term missions have to exercise 30 minutes per day. The treadmill *[Fig. B]* is stowed during launch and landing and installed via "quick-disconnects" on the floor of the Middeck. On-orbit astronauts use a waist belt to restrain themselves to the device, so they can 'run' in orbit. (NASA [Exercise], 2002) It produces relatively high vibration levels on the orbiter. (NASA [LSDA], 2010)

The cycle ergometer *[Fig. A]* is restrained to the middeck floor during launch and landing. On- orbit it can be installed on the flight deck or middeck. Cycling shoes are provided. (NASA [CE], 2011)

Free-Time Activities

In addition to the preferred leisure activity of 'looking out the window,' astronauts listen to music and watch movies.

The astronaut scientist Story Musgrave did the first shuttle spacewalk on the Shuttle mission STS-6 in 1983. According to Lenehan (2004) he described spacewalking as feeling like a ballerina on opening night". Although an EVA is not a leisure activity, Lenehan (2004 p. 191) found some time and "did a joyous handstand on the rim of Challenger's cargo doors".

A:

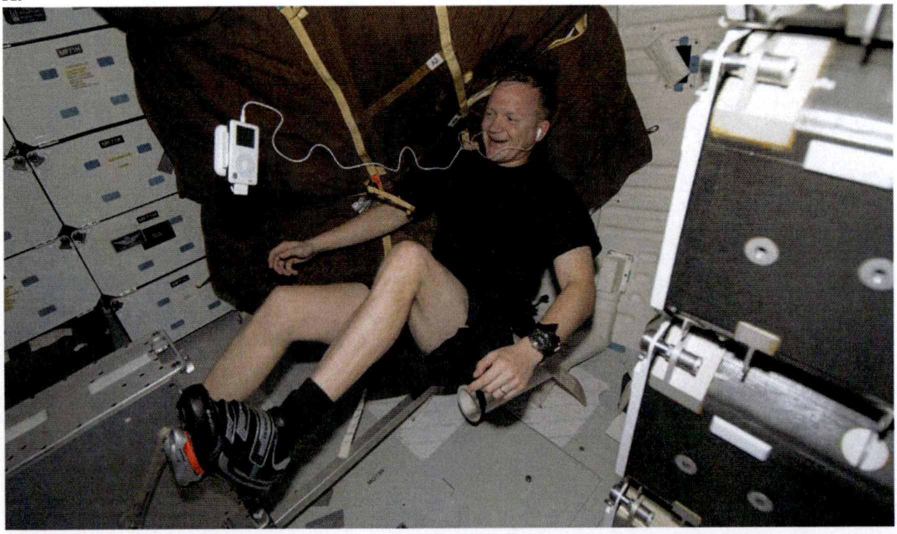

A: Astronaut Eric Boe, STS-126 pilot, exercises on a bicycle ergometer on the middeck of Space Shuttle Endeavour
B: Jean-François Clervoy on the ergometer looking out of the window
C: Astronaut Clervoy onboard the Shuttle with a rubik cube and toy rocket

(credit A: NASA; B, C: NASA, Jean-François Clervoy)

ASTRONAUTS' EXPERIENCES

USABILITY

Jim Voss (NASA): *"(...) He [Story Musgrave] was just floating very slowly over towards the wall. He had one of these pins in each hand and, as I got closer, I saw that he was floating towards the holes that these pins went in and he "docked" with the wall! (...) it was really humorous."*
(Lenehan, 2004 p. 211)

LIVABILITY

Gerhard Thiele (ESA): *"I know that I wanted to take a book onboard, and they asked what that should be for – I didn't take the book in the end."*
(Thiele, 2010)

FLEXIBILITY

Jean-François Clervoy (ESA): *" (...) the human body and spirit are uniquely free to float and reflect high above Earth, in weightlessness. We should learn to adapt to this new physical and psychological environment because there resides our far future."*
(Clervoy, 2010)

B:

C:

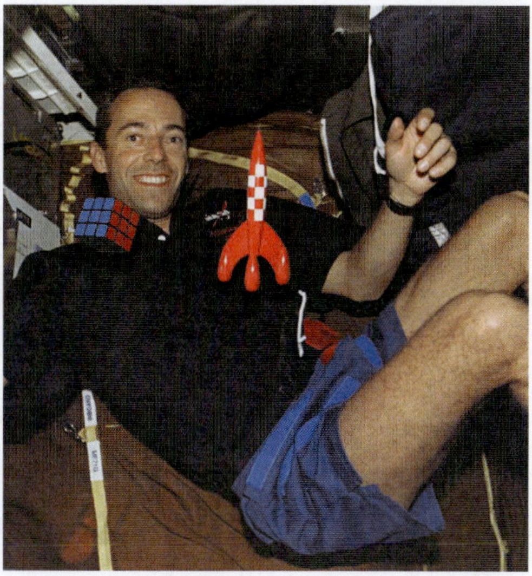

8.5 LEISURE MIR

CONCEPT	Treadmill, ergometer, running track, vacuum trousers, muscle-exercise equipment
LOCATION	Base Block, Kristall
ORIENTATION	Horizontal and vertical location

RÉSUMÉ LEISURE

Exercise

The space station Mir was equipped with a bicycle ergometer and a treadmill *[Fig. A]* in the Base Block. The ergometer could be retracted into the floor. An additional treadmill was installed in the Kristall module. While exercising cosmonauts and astronauts could watch videos or listen to music. Before return to Earth, cosmonauts wore a lower body negative pressure device, called 'Chibis' vacuum trousers *[Fig. C]* to force blood down to the lower limbs and to make the heart work harder to pump blood and to increase the heart rate (Salmon, 1997).

Free-time Activities

In theory, Mir cosmonauts had their weekends off, but in practice they often worked in their "free time." The varied reasons for these "sabbotniks" were to finish an experiment, to maintain the space station, do some housekeeping or because of "self-motivation" (Portree, 1995). The leisure activities were structured around the cosmonaut's individual needs and desires. Sometimes "surprise" leisure activities were organized (B.J.Bluth, 1987). A combination of colour and music was used to provide relaxation and also to maintain a sense of the Earth's seasons. (B.J.Bluth, 1987, Foale, 1999)

Art in Space

On 22 May 1993 a 3-dimensional sculpture was launched to the Mir space station. The sculpture 'Cosmic Dancer' *[Fig. B]* was created by artist Arthur Woods. It was made of welded aluminium and weighed exactly 1 kg with measurements of about 35 x 35 x 40 cm. It was painted in bright yellow-greenish colours to contrast with the Mir interior environment and to "offer an aesthetic contribution to the cosmonauts living quarters" (Woods, 1993-2009). This space art project stayed on board until Mir was deorbited.

Cosmonaut Alexander Polischuk enjoyed interacting with 'his' sculpture and when not playing with it, he fixed it next to his kayutka so it didn't get lost and to have it within reach.

A:

B:

A: Astronaut Shannon Lucid exercises on a treadmill in the Mir space station Base Block module
B: Cosmonaut Gennadi Mannakov releasing the Cosmic Dancer in the Progress resupply vehicle docked to the Mir space station, May 1993

C: Reinhold Ewald wearing the ‚Chibis' vacuum trousers in the Spektr Module
(credit A: NASA JSC; B: Copyright 1993-2011, Arthur Woods; C: DLR/Reinhold Ewald)

ASTRONAUTS' EXPERIENCES

USABILITY

Alexander Kaleri (Mir NASA 2): [to John Blaha] *"John, you must understand, here we have to just not work constantly,* ***we have to find time to rest*** *in such a way we can restore our energy."*
(Burrough, 1999 p. 108)

Valeri Korzun (Mir NASA 2): [about John Blaha] *"He didn't talk to us, he just worked."*
(Burrough, 1999 p. 110)

Jerry Linenger (Mir NASA 3):
"(...) strapped to the treadmill, I felt like I was running with someone sitting on my shoulders."
(Linenger, 2000 p. 178)

"Exercise was unwelcome to my space-adjusted body. It took all the willpower and self-discipline that I could muster (...)."
(Linenger, 2000 p. 179)

LIVABILITY

Alexander Polischuk (Russian):
"we circled around it [Cosmic Dancer] and it also moved freely as it wanted and it looked like it circled around us for some reason. ***That we can really call dancing!****"*
(Polischuk, 1993)

Mike Foale (NASA):
"Every time I get mail from you [his wife], it is like getting a present, or chocolate, for which I have a major craving."
(Foale, 1999 p. 80)

FLEXIBILITY

Franz Viehböck (Austromir):
"(...) Every module had windows and there were plenty of possibilities to look out. So it worked, according to the position of the station, one could take an adequate window."
(Viehböck, 2009)

C:

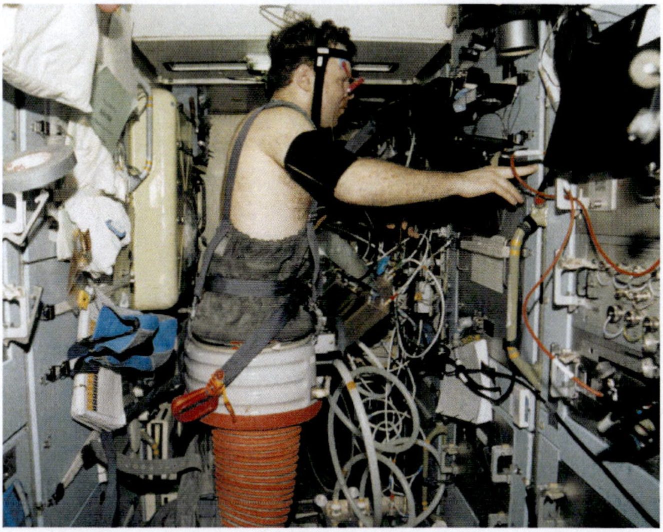

8.6 LEISURE ISS

CONCEPT Treadmill, ergometer, muscle resistive devices
LOCATION Service Module, Node 1, Node 3, U.S. Lab
SPECIFIC Horizontal

RÉSUMÉ LEISURE

Exercise

Astronauts exercise two hours per day using cardio and muscle resistive devices. The station is provided with six different exercise facilities. A treadmill and ergometer *[Fig. D]* for cardiovascular exercise is located in the Service module Zvezda. The Resistive Exercise Device (ARED) for muscle strengthening is located in the Node 1 module (Unity). A second treadmill (COLBERT) is located in Node 3 (Tranquillity). (NASA [ISS], 2010) The Flywheel exercise device to counteract muscle atrophy and bone loss is located in the Columbus module and is stowed when not in use (ESA, 2009). A cycle ergometer is located in the U.S. Lab (NASA [STS-128], 2009). Additional exercise equipment includes muscle exercisers, gymnastic balls and grip masters. The exercise facilities have an integrated vibration isolation system to prevent disturbances to scientific experiments.
Facilities may change, because exercise devices are constantly improved.

Free-time Activities

Popular leisure activities performed on orbit include the use of passive media like audio, newspapers, letters, books, magazines, television and movies. Other activities include diary keeping, filming and the favourite window gazing of course *[Fig. A]*. The cupola *[Fig. B, C]* is a dome-shaped module with seven windows. It was installed in 2010 and is used for observing external activities. (ESA [Cupola], 2010)

Communication with ground personnel, family or friends and also live conferences are important.

B:

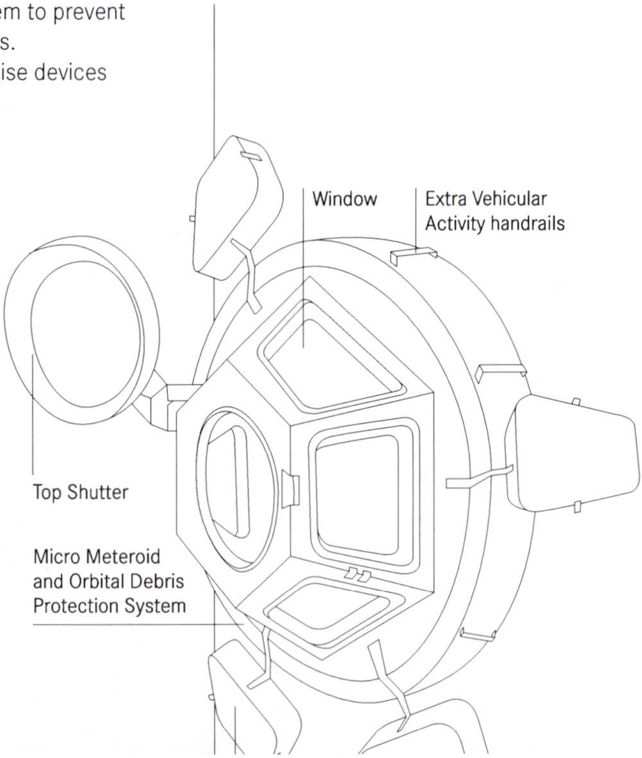

A:

A: Astronauts J. Williams uses a still camera at window in the Zvezda module
B: The Cupola
C: Fyodor Yurchikhin took this picture of the cupola from the window of the Russian Docking Compartment (Airlock)
D: James S. Voss reads a book, while exercising on the cycle ergometer in the Zvezda Service Module
(credit A, C, D: NASA; B: Author, based on NASA documents)

ASTRONAUTS' EXPERIENCES

USABILITY
Michel Tognini (ESA): *"He [astronaut at the ISS] called me on my cell phone (...) and we talked about activities onboard ISS. I was hearing him like you. And he was in space flying all over the world." (Tognini, 2009)*

Peggy Whitson (NASA): *"I always feel more relaxed after working out. While I have never been a big believer in that whole endorphin thing, I do get a sense of satisfaction from working out that positively lifts my attitude. So for me, exercise is not only a critical physical component to life up here, **it has an important psychological component too**." (Whitson, 2008)*

LIVABILITY
Jeff Williams (NASA): *"You can never tire of looking at the part of God's creation we call Earth. Travelling around the globe every 90 minutes provides lots of opportunity to view the geography, oceans, cloud formations, sunrises and sunsets, thunderstorms, city lights and many other things in vivid detail."*
(Williams, Jeff, 2007)

FLEXIBILITY
Greg Chamitoff (NASA): *"(...) In the meantime we have a few new large gaps of open space, which means there's suddenly an opportunity to do some "**advanced acrobatics**." (Chamitoff, 2008)*

C:

D:

8.7 SUMMARY OBSERVATION

	APOLLO	SALYUT	SKYLAB
Concept	Muscle exercise equipment; Music, jumping on the lunar surface, looking at Earth	Ergometer, treadmill, pressure suits, muscle exercise equipment, yoga; Video, games, books, 'surprise' activities	Ergometer, treadmill, body mass, recreational equipment
Review	Joking was a wonderful stress reliever. Astronauts played with collected rocks and listened to a radio loop	Cosmonauts preferred free choice of types of exercises; Cosmonauts made up their own exercises; Bicycle was used to generate electricity (Salyut 4)	Evaluation and inventing of 'zero gravity acrobatics'
Potentials	Efficient leisure devices were requested	Physical exercise shall not interfere with other crewmembers activities; Cosmonauts invented their own exercise	Ample space for movement seems important; Restraints were removed

Diagram

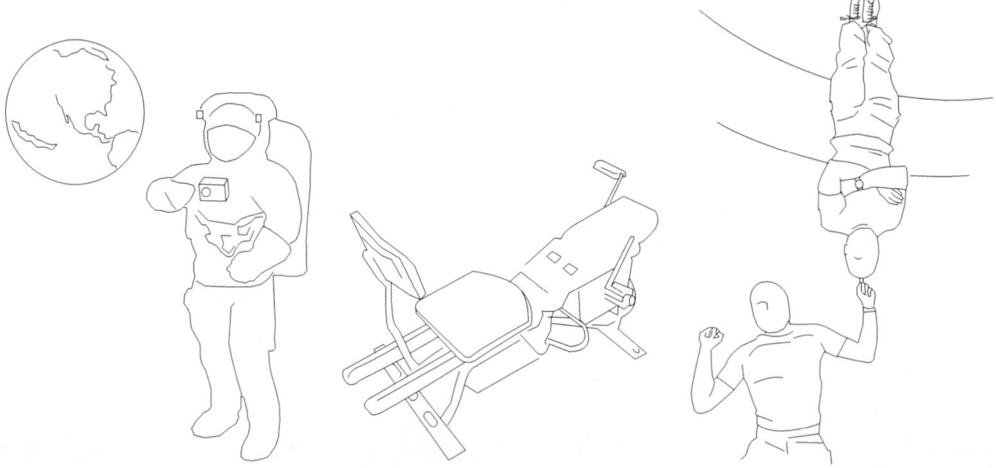

SHUTTLE	MIR	ISS
Ergometer, treadmill, dynabands	Treadmill, ergometer, vacuum trousers, muscle-exercise equipment; Music, instruments, 'surprise' leisure activities	Treadmills, ergometer, muscle resistive devices; Wide variety of exercise equipment on board
Less exercise and leisure on short term missions	Exercise and leisure activities are important countermeasures to work load and stress	No large space, except empty modules (arrival of Kibo)
Astronauts are inventive with games in weightlessness	Space can be enhanced with artistic interventions	Need for tools against boredom (future long-term missions); Use of empty 'not occupied' space

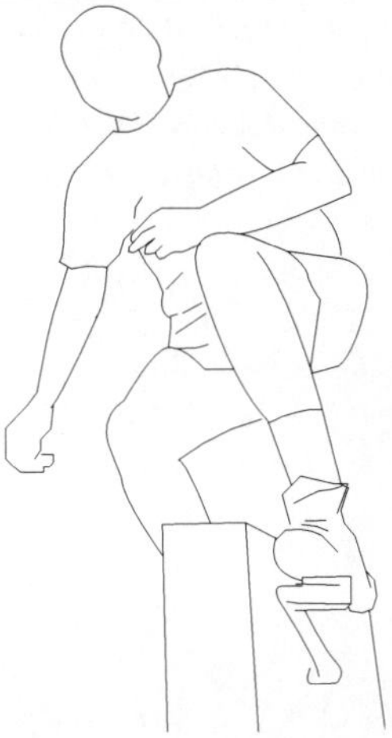

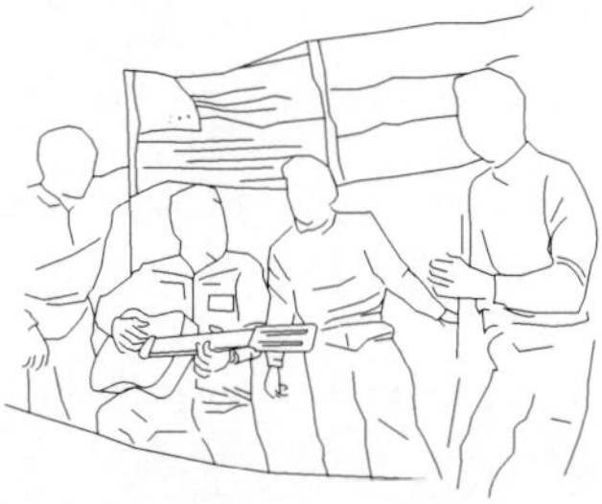

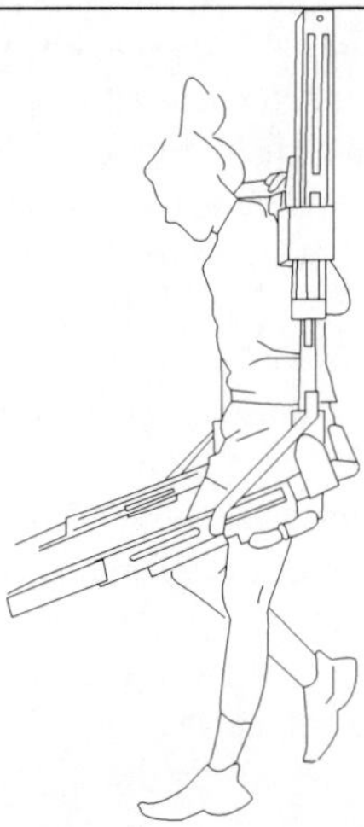

8.7 SUMMARY OBSERVATION

Exercise in Space

Living and working in microgravity puts strain on the human entity–physically, psychologically and socially. On Earth exercise is a well-known countermeasure and stress relief from work. In space astronauts have to exercise everyday to mitigate the deconditioning effects of living and working in a microgravity environment (NASA [ISS], 2010).

Exercise has become "a flight requirement for maintaining crew health [for missions] lasting more than five days." (NASA [LSDA], 2010) In his book 'Waystation to the Stars' Colin Foale (1999 p. 82) writes how glad he was, that his son [Shuttle-Mir astronaut Michael Foale] got time for exercise because of his mental and physical well being. He also reports that his Russian colleagues had too much work and had almost no spare time for exercise (p. 90, 100). Since then, exercise became a scheduled activity and various equipment was tested. Most common is the treadmill and ergometer.

Early during the Salyut mission, Russians learned that physical exercise can disturb other crew members and experiments through vibrations (B.J.Bluth, 1987). Experiences on Skylab and the Shuttle confirmed that potential effect. In addition astronauts have complained about conflicting placement of the devices close to other activity zones. (Haigneré, 2010; Burrough, 1998)

Recreation Activities

Historically astronauts do not have a lot of leisure time. Most of their free time is used for sleeping and exercising. The remaining small amount of free time on board is used for distraction, relaxation and interactive countermeasures to workload and other stressors such as social and sensory monotony.

Subjected to high workloads under a tight schedule within a confined environment, astronauts have drawn on leisure activities imported mostly from Earth. Popular leisure activities documented to-date have concentrated on passive perusal of media like records, audio cassettes, newspapers, letters, books, magazines, television, and movies (Kelly Alan D., 1994). Other activities include diary keeping, socialization with ground via communication technology, writing articles, communication with friends and relatives on Earth and Earth and space observation.

Leisure activities, like reading a book are considered a personal activity. They usually take place where astronauts sleep. Skylab astronauts complained that they did not have an adequate place to store their music tapes or books (NASA [Bull.3], 1974 p. 39). Salyut astronauts had their music tapes stored with elastic bands along the walls in the main module. Only recently all members of the permanent crew have been provided with a personal storage area for individual recreation items next to the place where they sleep (private crew quarters).

Watching a movie and "after-work-talking" during dinner is a welcomed social recreation activity. The Mir crew often watched a film together. Michael Foale invited his Russian colleagues for a film, because the Russian one was broken. During the film, they had private discussions about their lives back home and personal attitudes, which was good for 'team building'.

Leisure activity number one has always been to look out the window.

Windows

Astronauts have always looked for the nicest places with windows. According to Reinhold Ewald, the window was his preferred place in the space station, although he slept in the better protected Kristall module (Ewald, 2009). Mike Collins preferred the treadmill in Kristall, because it was cooler but also because of its windows it was "a marvellous place for Earth observation" (Foale, 1999 p. 77).

Nobody would question having a window in a house on Earth. In space it's different; the integration of windows has been a delicate topic.

Although the Mercury, Gemini and Apollo missions proved, that "adequate viewing windows" were of high value for the mission, there was "appreciable opposition when the wardroom window was proposed for Skylab" (NASA [Skylab LL], 1974 p. SLL1_7). The opponents argued that it was too expensive, that the development of it would take too long, it would weaken the structure and at last it seemed not to be essential to mission success. The window in the wardroom was finally integrated and appreciated by the astronauts.

Another difference compared to windows on Earth is, that one doesn't always see the desired view. This is because the space station "rolls" around the Earth (barbecue roll), in order to prevent thermal problems (Butler, 2002). Sometimes astronauts could not see much out of this window in Skylab but in addition the MDA had four windows in a 90-degree angle, from where the astronauts could always see some of the Earth. For the astronaut Weitz "the best windows" were in the MDA (Weitz, 2000).

Also the Salyut space stations and Mir had multiple windows installed for scientific purposes. Additionally on Mir, the private crew quarters had windows to the outside, which was favoured by the cosmonauts, but spaces with windows are also less protected from incoming radiation. The design of the ISS has not implemented windows in the private crew quarters. The seven-window-observatory-module 'Cupola' offers views of robotic and docking operations as well as provides a 360° window to observe Earth.

The positive effects of having windows are acknowledged (Kaplan, July 2001). Concerning future Moon or Mars habitats, Haigneré stated: "Nevertheless it's better to see the Moon and the Sun than being in a closed room without seeing anything. Sometimes you need reference. You need something which is natural, not artificial things around you" (Haigneré, 2009).

A Three-Dimensional Space

Due to the absence of gravity, three-dimensional space becomes more important than two dimensional floor area. Available space is measured in m^3 rather than in m^2. "You don't live on a surface, but in a volume" (Haigneré, 2009). Lebedev provides anecdotal evidence of the richness of living in a 3d-environment; he mentions "in the same space you can see many different interiors, depending on the position of your body. In other words, it's as though you could imagine several different rooms from one furnished room" (Lebedev, 1990 p. 26).

"Weightlessness is your friend," Clervoy (2009) stated in an interview, "You can use it to make your life easier. I personally feel frustrated, when I come back to Earth – to not be able to put furniture or my stuff on all surfaces. Because (...) in space you can use all surfaces". In 1982 cosmonaut Lebedev (1990 p. 55) was convinced that "this special feature of weightlessness will definitely be used in the architecture of future

8.7 SUMMARY OBSERVATION

orbital and interplanetary piloting systems (...)" He added, "by arranging living and working compartments in one [three-dimensional] space, various functional and aesthetic compositions will be created in a round interior." In reality, this unique feature has not been used to its full extent. All space stations so far have been designed in respect to how people live on Earth. Most space stations have been designed based on a fixed vertical with a few exceptions in the interiors such as a bicycle on the ceiling.

One reason for this "vertical" are Earth-bound training facilities. Astronauts need to be prepared for their missions using a spaceflight simulator in which a 0g environment cannot be simulated. Ewald (2009) is of the opinion, that "as long as we HAVE to train on Earth for space travel, it will be a compromise based on what we [astronauts] have to be prepared for and what makes sense to have as optimal living and working environment."

Unique and Experimental Activities

Astronauts have always preferred to experiment with leisure activities in microgravity.
Besides communication and conversation, the more active leisure activities are those that, due to the unique physical conditions, are only feasible in reduced gravity. These include the playful mastering of various states of body position and motion, and improvising games with common objects, such as duct-tape and chocolate balls.

Salyut 6 cosmonauts even played using the vacuum cleaner 'Raketa' not only for cleaning but also as a 'rocket' to fly through the station (Prunariu, 2011). Quite recently with the arrival of the new and empty KIBO module to the International Space Station, the same euphoria of moving in weightlessness could be observed. (NASA [Youtube], 2008)

Intimate Behaviour

The first woman entering space was Valentina Tereshkova as a single pilot in Vostok 6 in 1963. Much later in 1982 Svetlana Savitskaya became the second woman in space. She was assigned a one-week visiting mission to the 211-day Salyut 7 mission (Shayler, et al., 2005). This was the first mission that included both men and women. According to Shayler, her five fellow male cosmonauts were shaving every day and tried to help her with her experiments. In return Savitskaya was cooking for them and the cosmonauts ate "with renewed pleasure" (Shayler, et al., 2005 p. 309).

In 1991 the astronauts Jan Davis and Mark C. Lee were the first married couple to serve together on a mission (STS-47). No details about their experiences together on orbit have been released by NASA or by the couple themselves.

Sex is also possible without a partner. In his book 'Lift off' Michael Collins writes about the advice of a Skylab doctor to masturbate in order to prevent the possibility of getting infected prostate glands. No other reliable reference on this topic has been found. So far the topic of intimate behaviour has been avoided by national space agencies.

A: Development of games for microgravity *(credit A: Häuplik-Meusburger, Aguzzi, Peldszus)*

A:

8.8 COMPARISON USABILITY MATRIX

	APOLLO	SALYUT	SKYLAB
Availability and Equipment			
	Mission=work; Exercise equipment in the CM	Ergometer, treadmill, pressure suits, vacuum trousers, muscle-exercise equipment; Video, games, books	Ergometer, treadmill, body mass, recreational equipment
Spatial Arrangement			
	All in the same module; On the lunar surface	Exercise equipment in the large-diameter Work Compartment on the ceiling: vertical and horizontal orientation	Experiment work area in the lower level of the Orbital work shop; Free exercises in the 'dome'
Object Management			
	Minimal devices	Fixed location, some exercise equipment could be stowed when not in use	Fixed location for exercise equipment
Ergonomic Safety			
	'Balance' in spacesuit	Floor covered with Velcro to aid moving around; Restraints for exercise equipment; Interior had soft cladding	Harnesses and restraints for exercise equipment

SHUTTLE	MIR	ISS
Ergometer, treadmill, dynabands; Music, toys	Treadmill, ergometer, vacuum trousers, muscle-exercise equipment; Music, instruments, books	Treadmills, ergometer, muscle resistive devices, wide variety of exercise equipment on board
In the Middeck	Treadmill in the Kristall module and in the Base Block; ergometer in the Base Block	In different modules; Service module Zvezda, Node 1, Node 3 (some equipment can be relocated)
Stored during launch and landing	Some exercise equipment could be retracted into the floor when not in use	Some equipment is stowed when not in use (Rack system)
Ergonomic design and multiple restraints	Restraints for exercise equipment	Ergonomic design and multiple restraints

8.9

COMPARISON LIVABILITY MATRIX

	APOLLO	SALYUT	SKYLAB
Territoriality and Privacy			
	Limited space inside the CM and LM; Jumping and playing golf on the Lunar surface	Exercise area in the Work compartment, next to other activities	Large space station offered multiple possibilities (Dome); Exercise in main work area
Sensory perception			
	Rich sensual perception on the lunar surface; Viewing Earth was used for relaxation; Music	Cosmonauts decorated the station with photographs and drawings; Planting, Music	'Space' acrobatics in the biggest space ever; Astronauts wished for more colour variations
External relations			
	Multiple windows	Multiple windows	Wardroom window
Internal relations			
	All in one module	Not separated from other activities	Next to work activities and shower

SHUTTLE	MIR	ISS
Middeck and Flight Deck, next to other activities	In different modules, but next to other activities	In different modules, next to work activities
Inventing games in microgravity	Art in space	Wide variety of exercise and leisure activities; Viewing Earth
Shuttle windows	Multiple windows in various locations to view Earth	Varies, 360° view from the cupola
Not separated from other activities	Not separated from other activities	Large station and many modules

8.10 COMPARISON FLEXIBILITY MATRIX

	APOLLO	SALYUT	SKYLAB
Spatial Flexibility	High on the lunar surface, limited in the module	All functions in the same module	Dedicated exercise area; Free movements in the 'Dome' of the OWS
Object Flexibility	Minimal; Astronauts used materials found in-situ (rocks)	Cosmonauts adjusted equipment (e.g. penguin suit); Surprise gifts from ground	Ergometer was adjusted
Individual Flexibility	On the lunar surface, singing, hopping, playing golf between work tasks; Playful experiments with microgravity	Activities were scheduled, but they did not tell everything to mission control; Some free choice in the type of exercises; Playful experiments with microgravity	Unscheduled microgravity acrobatics in the Dome

SHUTTLE	MIR	ISS
Exercise equipment in the Middeck	Exercise equipment is in the Base Block and Kristall	A variety of exercise devices in different modules variety of modules
Exercise equipment can be stowed; Microgravity Games	Bicycle ergometer could retract into the floor	Most of the exercise equipment can be relocated
Short mission (work = priority); Playful experiments with microgravity	Playful experiments with microgravity	Multiple exercise equipment, but scheduled; Large new modules are preferred for 0g acrobatics (Kibo)

8.11 DESIGN DIRECTIONS

The most important findings related to the human activity 'LEISURE' are summarized and structured in the form of design directions. These directions summarize the main points in a short sentence. The paragraph following refers to the empirical background of experiences made in extraterrestrial habitats.

USABILITY

Integrate exercise and leisure activities

Exercise and recreation is important during space missions, especially for long durations. Space needs to be provided for individual leisure activities as well as group activities. Adequate storage should be available.

Separate exercise from other activities

Exercise areas need to be separated from other activity zones, mainly the area where food is prepared and eaten and where astronauts sleep or rest.

Integrate restraints

Within a microgravity environment, restraints that stabilize the astronaut during exercise activities and to look out should be considered. Additional restraints for equipment (e.g. Camera next to the window) are needed.

Use the three-dimensional space

At least one part of the habitat should provide enough space to allow astronauts to experience the unique microgravity environment.

LIVABILITY

Integrate windows

Windows not only support the crew in observing diverse mission objectives but also offer the crewmembers of the confined habitat various sights, both in and out. Anecdotal references from astronauts show their importance for habitability. The design should allow a good view to the outside, not only for scientific reasons but also for personal pleasure. Multiple options for observing the outside and an appropriate viewing angle certainly support the habitability inside.
Windows should have enough space around it to allow others pass by and to allow multiple body positions.

Use Media Technology for Habitability

New design solutions are required for mission scenarios where the inclusion of windows or the view outside is limited, for instance during deep space transfers, or where lingering in front of windows is not advised, such as in lunar surface habitation due to higher radiation impact (surrogate views).

Design for social recreation

An area that accommodates all crewmembers and offers adequate space for group leisure activities (watching a movie, games, etc.) needs to be integrated.

Consider intimate behaviour

So far, this topic has not been considered for integration into the design. As sexual intercourse and related intimacy is more than likely to happen in a long-term mission, this issue should be considered in future design of sleep areas and facilities, and restraints.

FLEXIBILITY

Integrate playful leisure activities

With the advent of more complex long duration missions, leisure activities will also have to address psychological and didactic issues (social interaction and stimulation) and training (alertness, concentration, dexterity). In space there is still a lack of applications specifically designed for the integrated situation of microgravity and long duration flight (cf. A Game for Space, Häuplik-Meusburger et al., 2010)

Integrate autonomy of the users

In view of long-term missions astronauts require some autonomy for mission tasks. This could be trained for today, if one part of schedule foresees individual work activites. That probably requires more exercise equipment.

Design for intimate friendship

Private crew quarters alone are not enough. An adequate volume for intimacy/privacy between two or three crewmembers is needed in addition to a space that accommodates all crew members.

8.12 REFERENCES

B.J.Bluth, M. Helppie. 1986. Soviet Space Stations as Analogs. Washington D.C., USA : NASA Headquarters, 1986. NASA Grant NAGW-659.

Butler, Aaron M. 2002. NASA Mission Control Anwers Your Questions. [Online] 01 2002. http://spaceflight.nasa.gov/feedback/expert/answer/mcc/sts-113/11_23_20_01_179.html.

Burrough, Bryan. 1999. Dragonfly - NASA and the crisis aboard Mir. London, GB : Fourth Estate Limited, 1999. ISBN 1 84115 087 8.

Carr, Gerald P. 2000. Oral History Transcript - Gerald P. Carr. [interv.] Kevin M. Rusnak. Huntsville, Arkansas, 25 October 2000.

Cernan, Eugene A. 2007. Oral History Transcript - Eugene A. Cernan. [interv.] Rebecca Wright. Houston, Texas, USA, 11 December 2007.

Chamitoff, Greg. 2008. Greg Chamitoff's Journal. [Online] 8 February 2008. [Cited: 20 April 2008.] http://www.nasa.gov/mission_pages/station/expeditions/expedition17/journal_greg_chamitoff_3.html.

Collins, Michael. 1989. Liftoff: The Story of America's Adventure in Space. USA : Grove Pr, 1989. ISBN-10: 0802131883.

Clervoy, Jean-François. 2009. [interv.] Sandra Häuplik-Meusburger. Transcript not published. Paris, France, June 2009.

ESA . 2009. ESA ISS Science & System - Operations Status Report - Increment 19. [Online] 8 May 2009. [Cited: 5 July 2009.] http://www.esa.int/esaMI/Columbus/SEMZLAZVNUF_0.html.

ESA [Cupola]. 2010. Cupola, ISS Observation Module. [Online] 9 February 2010. http://www.esa.int/esaMI/node3/SEM1RFSJR4G_0.html.

Ewald, Reinhold. 2009. Transcript not published. [interv.] Sandra Häuplik-Meusburger. Cologne, Germany, July 2009.

Foale, Colin. 1999. Waystation to the Stars - The Story of Mir, Michael and Me. London, GB : Headline Book Publishing, 1999. ISBN 0 7472 6331 0.

Gibson, Edward G. 2000. Oral History Transcript - Edward G. Gibson. [interv.] Carol Butler. 1 December 2000.

Haigneré, Jean-Pierre. 2009. Transcript not published. [interv.] Sandra Häuplik-Meusburger. Paris, France, June 2009.

Häuplik-Meusburger, Sandra, Aguzzi, Manuela and Peldszus, Regina. 2010. A Game for Space. Acta Astronautica. Volume 66, 2010, Issues 3-4, pp. 605-609.

Kaplan, Rachel. July 2001. The Nature of the View From Home: Psychological Benefits. Environment and Behaviour. July 2001, Vol. Vol. 33, No. 4, pp. 507-542.

Kelly Alan D., Kanas Nick. 1994. Leisure Activities in Space - A Survey of Astronauts and Cosmonauts. Acta Astronautica. Vol. 32, 1994, No. 6., pp. 451-457.

Lebedev, Valentin. 1990. Diary of a Cosmonaut - 211 Days in Space. United States and Canada : Bantam Air & Space Series, 1990. Vol. Vol.4. ISBN: 0-553-28778-8.

Lenehan, Anne E. 2004. Story, The Way of Water. USA : The Communications Agency, 2004. ISBN 0 9752286 0 9.

Linenger, Jerry M. 2000. Off The Planet - Surviving five periolous months aboard the Space Station Mir. U.S.A. : McGraw-Hill, 2000. ISBN 0-07-136112-X.

NASA. 1972. Singing on the Moon: Jack Schmitt and Gene Cernan sing „I was strolling ...". http://www.youtube.com/watch?v=wo3-fuYKWB4, 1972.

NASA [A17]. 1973. Apollo 17 - Technical Crew Debriefing. Houston, Texas, US : National Aeronautics and Space Administration, Manned Spacecraft Center; Prepared by Training Office, Crew Training and Simulation Division, 1973. MSC-07631.

NASA [Bull.3]. 1974. Skylab Experience Bulletin No. 3: Architectural Evaluation for Sleeping Quarters. Houston, Texas, US : NASA, Lyndon B. Johnson Space Center, 1974. JSC-09537.

NASA [Bull.4]. 1974. Skylab Experience Bulletin No. 4: Design Characteristics of the Sleep Restraint. Houston, Texas, US : NASA, Lyndon B. Johnson Space Center, 1974. JSC-09538.

NASA [Bull.7]. 1974. Skylab Experience Bulletin No. 7: An Overview of IVA Personal Restraint Systems. Houston, Texas, US : NASA, Lyndon B. Johnson Space Center, 1974. JSC-09541.

NASA [Bull.10]. 1974. Skylab Experience Bulletin No. 10: Body Restraint Systems. Houston, Texas : NASA, Lyndon B. Johnson Space Center, 1974. JSC-09544.

NASA [Bull.18]. 1975. Skylab Experience Bulletin No. 18: Evaluation of Skylab IVA Architecture. Texas, Houston, US : NASA, Lyndon B. Johnson Space Center, 1975. JSC-09552.

NASA [CE]. 2011. Life Sciences Data Archive: Cycle Ergometer Hardware Description. [Online] 26 01 2011. [Cited: 30 01 2011.] http://lsda.jsc.nasa.gov/scripts/hardware/hardw.cfm?hardware_id=172.

NASA [Debriefing A11]. 1969. Apollo 11 - Technical Crew Debriefing. Houston, Texas : NASA, Manned Spacecraft Center, 1969. 276.

8.12 REFERENCES

NASA [Debriefing A12]. 1969. Apollo 12 - Technical Crew Debriefing. Mission Operations Branch - Flight Crew Support Division. Houston, Texas, USA : Manned Spacecraft Center, 1969. p. 308.

NASA [Debriefing A14]. 1971. Apollo 14 Technical Crew Debriefing (U). prepared by: Mission Operations Branch Flight Crew Support Division. Houston, Texas : Manned Spacecraft Center, 1971. February 17, 1971.

NASA [Exercise]. 2002. Human Space Flight: Exercise Equipment. [Online] 04 07 2002. [Cited: 05 10 2010.] http://spaceflight.nasa.gov/shuttle/reference/shutref/crew/exercise.html.

NASA [ISS]. 2010. Reference Guide to the International Space Station. Assembly Complete Edition. Washington, DC, US : Library of Congress Cataloging-in-Publication Data, 2010. NP-2010-09-682-HQ

NASA [LBNPD]. 2002. Biomedical Results from Skylab: Lower Body Negative Pressure Device . [Online] 28 02 2002. [Cited: 15 01 2011.] http://lsda.jsc.nasa.gov/books/skylab/appAla8.html.

NASA [LSDA]. 2010. Life Sciences Data Archive: Evaluation of the Ergometer Vibration Isolation System (DTO 658). [Online] 10 2010. [Cited: 01 12 2010.] http://lsda.jsc.nasa.gov/scripts/experiment/exper.cfm?exp_index=851.

NASA [Skylab LL]. 1974. Lessons Learned on the Skylab Program. Houston, Texas : NASA, Lyndon B. Johnson Space Center, USA, 1974. JSC-09096.

NASA [Skylab]. 1977. Skylab, our first space station. Washington D.C., US : NASA, Scientific and Technical Information Office, 1977. Prepared by George C. Marshall Space Flight Center, Edited by Leland Belew. NASA-SP-400.

NASA [STS-128]. 2009. STS-128 ‚Racking Up New Science' - Space Shuttle Press Kit. [Online] 2009. http://www.nasa.gov/pdf/379392main_STS-128_Press_Kit.pdf.

NASA [Skylab CS]. 1974. MSFC Skylab Crew Systems Mission Evaluation - Technical Memorandum. Alabama : George C. Marshall Space Flight Center, 1974. Technical Memorandum. NASA TM X-64825.

NASA [Youtube]. 2008. Youtube: Astronauts enter the new Japanes module Kibo. [Online] 2008. http://www.youtube.com/watch?v=mCH0y-KwhbU.

Polischuk, Alexander Cosmonaut. 1993. Cosmic Dancer. [interv.] translated from the audio recording made on MIR space station. http://www.cosmicdancer.com/cosmonaut.php. 1993.

Portree, David S.F. 1995. MIR Hardware Heritage. Houston, Texas : NASA, Lyndon B. Johnson Space Center, Information Services Division, 1995. NASA RP 1357.

Prunariu, Dumitru-Dorin. 2011. Transcript not published. [interv.] Sandra Häuplik-Meusburger. Vienna, Austria, 2011.

OTA. 1983. SALYUT - Soviet Steps Toward Permanent Human Presence in Space. Washington D.C. : U.S. Congress, Office of Technology Assessment, 1983. Andelin, John; Naismith, Nancy; Mills, William; Rogers, Thomas F.; Chandler, Philip P.. OTA-TM-STI-14.

Salmon, Andy. 1997. Science on-board the Mir space station 1986-94. Journal of the British Interplanetary Society. Vol. 50, 1997, Vol. No. 8, pp. 283-295.

Schmitt, Harrison. 2009. Transcript not published. [interv.] Sandra Häuplik-Meusburger. Vienna, Austria, 2009.

Shayler, David J. and Moule, Ian. 2005. Women in space - Following Valentina. Berlin, Heidelberg, New York : Springer-Praxis Books in Space Exploration, 2005. ISBN 1-85233-744-3.

Shepard, Alan B. 1998. Oral History Transcript - Alan. B. Shepard. [interv.] Roy Neal. 20 February 1998.

Thiele, Gerhard. 2010. Transcript not published. [interv.] Sandra Häuplik-Meusburger. Vienna, Austria, December 2010.

Tognini, Michel-Ange Charles. 2009. Transcript not published. [interv.] Sandra Häuplik-Meusburger. Cologne, Germany, July 2009.

Viehböck, Franz. 2009. Transcript not published. [interv.] Sandra Häuplik-Meusburger. Berndorf, Austria, 2009.

Weitz, Paul J. 2000. Oral History Transcript - Interview with Paul J. Weitz. [interv.] Rebecca Wright. 26 March 2000.

Whitson, Peggy. 2008. Peggy Whitson's Journal - Exercise in Space. NASA. [Online] 18 January 2008. [Cited: 5 February 2009.]

Williams, Jeff. 2007. Jeff William's Journals: Getting Settled in. [Online] 23 November 2007. [Cited: 15 August 2008.] http://www.nasa.gov/mission_pages/station/expeditions/expedition13/journals_williams_2.html.

Woods, Arthur. 1993-2009. Cosmic Dancer. [Online] 1993-2009. http://www.cosmicdancer.com/.

Worden, Alfred M. 2000. Oral History Transcript - Alfred M. Worden. [interv.] Rebecca Wright. 26 May 2000.

PART 9:

APPENDIX

APPENDIX

A1 ACRONYMS AND ABBREVIATIONS

AL	Airlock
ALSEP	Apollo Lunar Surface Experiment Package
ARED	Advanced Resistive Exercise Device
ASI	Italian Space Agency
ATM	Apollo Telescope Mount (Skylab)
Canadarm	Manipulator arm on Space Shuttle provided by CSA
Colbert	Combined Operational Load Bearing External Resistance Treadmill
Columbus	ESA Laboratory Module (ISS)
CSA	Canadian Space Agency
CSM	Command and Service Module (Apollo)
CM	Command Module (Apollo)
D	Diameter
ECLSS	Environmental Control and Life Support System
EF	Exposed Facility (JEM, ISS)
ELM-ES	Experiment Logistics Module (JEM, ISS)
EMU	Extra-Vehicular Mobility Unit (Apollo)
ESA	European Space Agency
EVA	Extra-Vehicular Activity
FGB	Functional Cargo Block or Zarya (ISS)
HV	Habitable Volume, usable space
IMS	Inventory Management System
ISS	International Space Station
JAXA	Japan Exploration Agency, the successor to NASDA
JEM	Japan Experiment Module, Kibo (ISS)
Kibo	means 'Hope', thematic name for the JEM (ISS)
Kosmos	Soviet lab module used with Salyut 7
L	Length
LCVG	Liquid Cooling and Ventilation Garment
LBNPD	Lower Body Negative Pressure Device
LEO	Low Earth Orbit
LER	NASA Lunar Electric Rover, successor name to the SPR
LM	Lunar Module, the Apollo Lunar Lander
LOI	Lunar Orbit Insertion

LPR	Lunar Pressurized Rover (Alenia)
LRV	Lunar Roving Vehicle (Apollo)
LSS	Life Support System
M	Mass
MAR	Middeck Accomodation Rack (Shuttle)
MDA	Multiple Docking Adapter (Skylab)
Mir	Mir Space Station, thematic name meaning World or Peace
MLI	Multi Layer Insulation
Module	Internally pressurized element intended for habitation
MPLM	Multi-Purpose Logistic Module
NASA	National Aeronautics and Space Administration
NASDA	National Space Development Agency of Japan, the predecessor to JAXA
NSTS	National Space Transportation System, formal name for the Space Shuttle
OWS	Orbital Workshop (Skylab Space Station), also called the Saturn Workshop
PLSS	Portable Life Support System
RSA	Russian Space Agency
Roscosmos	Russian Federal Space Agency, formerly Rosaviakosmos
Salyut	Soviet Space Station program meaning Salute or . . . .
Soyuz	Soviet/Russian Freighter
SPR	NASA Small Pressurized Rover original name of the LER
SM	Service Module (Apollo)
SS	Space Station
SSRMS	Space Station Remote Manipulator System (ISS)
STS	Space Transportation System/Space Shuttle
TeSS	Temporary Sleep Station (ISS)
TEI	Trans-Earth injection
TLI	Trans-lunar injection
Progress	Soviet/Russian Freighter
V	Volume
Zarya	or Functional Cargo Block (ISS)
Zvezda	Service Module (ISS)

APPENDIX

A2 FACTS APOLLO

APOLLO MISSIONS

Apollo 7
Walter Schirra, Donn Eisele, Walter Cunningham
1968 October 11 - 22
Duration: 10d 20h 9m 3s
First manned test of spacecraft

Apollo 8
Frank Borman, James Lovell, William Anders
1968 December 21 - 27
Duration: 6d 03h 00m 42s
First manned flight in deep space

Apollo 9
James McDivitt, David Scott, Russell Schweickhart
1969 March 3 - 13
Duration: 10d 1h 0m 54s
First lunar vehicle configuration flight

Apollo 10
Thomas Stafford, John Young, Eugene Cernan
1969 May 18 - 26
Duration: 8d 0h 4m 23s
Separation and docking of LM and CSM in lunar orbit

Apollo 11
Neil A. Armstrong, Michael Collins, Edwin Aldrin
1969 July 16 - 24
Duration: 8d 3h 18m 35s
First manned lunar landing

Apollo 12
Charles Conrad, Michael Collins, Edwin Alan Bean
1969 November 14 - 21
Duration: 10d 4h 36m 25s

Apollo 13
James Lovell, John Swigert, Fred Haise
1970 April 11 - 17
Aborted, no lunar landing
LM was used as 'lifeboat' during return to Earth

Apollo 14
Alan Shepard, Stuart Roosa, Edgar Mitchell
1971 January 31 - February 9
Duration: 9d 0h 1m 58s

Apollo 15
David R. Scott, James B. Irwin, Alfred M. Worden
1971 July 26 - August 7
Duration: 12d 17h 12m
Lunar surface stay time: 66.9h
LRV (used for the first time) traversed: 27.9km

Apollo 16
John W. Young, Thomas K. Mattingly II, Charles M. Duke Jr.
1972 April 16 - 27
Duration: 11d 1h 51m
Lunar surface stay time: 71h
EVAs: 3; total 20h 14m
LRV traversed: 26.7km

Apollo 17
Eugene A. Cernan, Ronald E. Evans, Harrison H. Schmitt
1972 December 7 - 19
Duration: 12d 13h 52m
Last mission to the Moon

Main Sources:
(Young, 2007 pp. 283- 290; NASA KSC [Apollo], 2000; Eckart, 1999)

APOLLO ARCHITECTURE

Launch Vehicle: Saturn IB, Saturn V
Crew Transportation: Saturn V, Apollo CSM

Command Module (CM)
Crew, spacecraft operation systems, re-entry equipment
Geometry: conical pressure vessel
Pressurized Volume: 10.34m^3
"Unoccupied" volume: 5.94m^3
Diameter: 3.9m (max)
Height: 3.65m
Mass: 5.6t

Service Module (SM)
Consumables (Oxygen, Water, Helium, Fuel cells, fuel), main propulsion system
Geometry: cylindrical pressure vessel
Diameter: 3.9m
Length: 7.6m
Mass: 23.3t

Lunar Module (LM)
Two-stage vehicle (Ascent stage, Descent stage)
Pressurized Volume: 6.65m^3
Habitable Volume: 4.5m^3
Height: 7.04m
Width at tanks: 4.22m
Width at footpads: 9.45m
Mass: 15.1 - 16.4t

Geometry Descent Stage: octagonal prism
Descent stage Width: 4.22m
Descent stage Height: 1.7m
Geometry Accent Stage: irregular
Ascent Stage Height: 2.8m
Ascent Stage Width: 4.0 - 4,3m
Ascent stage Mass: 4.8t
Descent stage Mass: 11.6t

Lunar Roving Vehicle (LRV)
Deployable electric lunar roving vehicle
Length deployed: 3.1m
Length collapsed: about 1.5m
Width: 1.83m
Height: 1.14m (max)
Mass: 210kg

Main Sources:
(Cohen, 2009 p. 33; NASA [Constellation Program], 2008; Young, 2007 pp. 29-56; NASA [LRV], 2005; Eckhart, 1996; NASA [Apollo 17 LM], 1972; NASA [Apollo 11 CSM], 1969; NASA [Apollo 11 LM], 1969)

APPENDIX

A3
FACTS SALYUT

SALYUT MISSIONS

Launch Vehicle: Proton
Crew Transportation: Soyuz

Salyut 1
Civil, 3 crewmembers
In orbit: 1971 April 19 – October 11 (175d)
Days occupied: 24d

Expedition Soyuz 10 failed docking
Expedition Soyuz 11
Georgi Dobrovolski, Viktor Patsayev, Vladislav Volkov
1971 June 6 – 29
Crew died on re-entry

Salyut 2
In orbit: 1973 April 3 – May 28 (54d)
Days occupied: 0
Failed

Salyut 3
Thought to be military (Almaz) space station
In orbit: 1974 June 24 – 1975 January 24 (213d)
Days occupied: 15d

Expedition Soyuz 14
Yuri Artyukhin, Pavel Popovich
1974 July 3 – 19
Expedition Soyuz 15 failed docking

Salyut 4
Civil
In orbit: 1974 December 26 – 1977 February 2 (770d)
Days occupied: 28d, 62d

Expedition Soyuz 17
Georgi Grechko, Aleksei Gubarev
1975 January 11 – February 10
Expedition Soyuz 18
Pyotr Klimuk, Vitali Sevastyanov
1975 May 24 – July 26

Salyut 5
Thought to be military (Almaz) space station
In orbit: 1976 June 22 – 1977 August 8 (412d)
Days occupied: 49d, 16d

Expedition Soyuz 21
Boris Volynov, Vitali Zholobov
1976 June 6 – August 24
Expedition Soyuz 24
Viktor Gorbatko, Yuri Glazkov
1977 February 7 – 25

Salyut 6
Civil
In orbit: 1977 September 29 – 1982 July 29 (1764d)
Days occupied: 683d

Expedition EO-1 (96d)
Yuri Romanenko, Georgi Grechko
1977 December 10 – 1978 March 16
Visiting Expedition EP-1 (6d)
Visiting Expedition EP-2 (8d)

Expedition EO-2 (140d)
Vladimir Kovalyonok, Aleksandr Ivanchenkov
1978 June 15 – November 2
Visiting Expedition EP-3 (8d)
Visiting Expedition EP-4 (8d)

Expedition EO-3 (175d)
Vladimir Lyakhov, Valery Ryumin
1979 February 25 – September 3

Expedition EO-4 (185d)
Leonid Popov, Valery Ryumin
1980 April 9 – October 11
Visiting Expedition EP-5 (8d)
Visiting Expedition EP-6 (4d)
Visiting Expedition EP-7 (8d)
Visiting Expedition EP-8 (8d)
Expedition EO-5 (13d)

Expedition EO-6 (74d)
Leonid Kizim, Oleg Makarov, Gennady Strekalov
1981 March 12 - May 26
Visiting Expedition EP-9 (8d)
Visiting Expedition EP-10 (8d)

Salyut 7
Civil
In orbit: 1982 April 19 - 1991 February 7 (3216d)
Days occupied: 816d

Expedition EO-1 (211d)
Anatoli Berezovoy, Valentin Lebedev
1982 May 13 - December 10
Visiting Expedition EP-1 (8d)
Visiting Expedition EP-2 (8d)

ExpeditionEO-2 (150d)
Vladimir Lyakhov, Aleksandr Aleksandrov
1983 June 27 - November 23

Expedition EO-3 (237d)
Leonid Kizim, Vladimir Solovyov, Oleg Atkov
1984 February 8 - October 2
Visiting Expedition EP-3 (8d)
Visiting Expedition EP-4 (12d)

Expedition EO-4-1a (169d)
Visiting Expedition EP-5 (9d)
Visiting Expedition EO-4.2 (65d)

Expedition EO-5 (125d)
Leonid Kizim, Vladimir Solovyov
1986 March 13 - July 16

Main Sources:
(B.J.Bluth, 1987; OTA, 1983 p. 25; NewScientist, 1976; Wikipedia [Salyut])
Habitable Volume: 90-100m
Max. Diameter: 4.15m
Length: 13 - 15mv
Mass: 19t

SALYUT ARCHITECTURE

Salyut 1-5 had one docking port.
Salyut 6 and 7 had two docking ports.

Salyut 7
Pressurized Volume: ~ 90m^3
Diameter: 2.0 - 4.15m
Length: ~13.1m

Transfer Compartment
Docking port, 7 portholes, space suit storage, scientific equipment
Diameter: 2.0 - 2.9m
Length: 3m

Small Diameter Work Compartment
Diameter: 2.9m
Length: 3.5m

Large Diameter Work Compartment
Diameter: 4.15m
Length: 2.7m

Intermediate Compartment
2 portholes for observation, second docking unit
Diameter: 4.15m (2.0m)
Length: 2.2m

Main Sources:
(Eckhart, 1996; Portree, 1995; B.J.Bluth, 1987 p. 18; OTA, 1983; Vasilyev, et al., 1974)

APPENDIX

A4
FACTS SKYLAB

SKYLAB MISSIONS

Launch Vehicle: Saturn V
Crew Transportation: Saturn V, Apollo CSM

Skylab
In orbit: 1973 May 14 – 1979 July 11 (6y 58d)
Days occupied: 28d, 59d, 84d

Skylab 1
Launch

Skylab 2
1973 May 25 - June 22
Charles Conrad, Joseph P. Kerwin, Paul J. Weitz
Duration: 28d 0h 49m 49s

Skylab 3
1973 July 28 - September 25
Alan Bean, Owen K. Garriott, Jack R. Lousma
Duration: 59d 1h 9m 4s

Skylab 4
1973 November 16 - 1974 February 8
Gerald P. Carr, Edward G. Gibson, William R. Pogue
Duration: 83d 4h 38m 12s

Main Sources:
(NASA KSC, 2000; Compton, et al., 1983 pp. 374-375; NASA [Skylab], 1977; NASA [OWS], 1974)

SKYLAB ARCHITECTURE

Habitable Volume: 320m^3 without Apollo CMS (available data varies, according to Compton)
Max. Diameter: 6.6m
Length: ~26m

Multiple Docking Adapter (MDA)
Docking, controls, telescope
Working Volume: 32.3m^3
Diameter: 3.1m
Length: 5.3m

Orbital Workshop (OWS)
Primary living and working area, experiment, laboratory, storage
Working Volume: 270.2m^3
Diameter: 6.6m
Length: 14.7m

Airlock Module
Working Volume: 17.3m^3
Diameter: 3.1m (1.65 lock compartment)
Length: 5.4m (2.15 lock compartment)

Apollo Telescope Mount (ATM), Instrument Unit, Apollo CMS

Main Sources:
(Loewy, 2002; NASA KSC, 2000; NASA [History], 1997; Compton, et al., 1983 pp. 374-375; NASA [Skylab], 1977; NASA [OWS], 1974; NASA [Bull.2], 1974)

APPENDIX

A5
FACTS SPACE SHUTTLE

SPACE-SHUTTLE MISSIONS

Partially re-usable launch and re-entry system
Elements: External tank, Solid Rocket Boosters, STS Orbiter Vehicle
Shuttles: Challenger, and Columbia; operational: Discovery, Atlantis, and Endeavour

First Flight:
1981 April 12 (Columbia)
First Shuttle Mission:
1981 April 12 -14 (STS-1)
First Shuttle operational Mission:
1982 November 11 – 16 (STS-5)

Shuttle Missions
(ordered according launch date)

1981: STS-2, STS-1
1982: STS-5, STS-4, STS-3
1983: STS-9, STS-8, STS-7, STS-6
1984: STS-51A, STS-41G, STS-41D, STS-41C, STS-41B
1985: STS-61B, STS-61A, STS-51J, STS-51I, STS-51F, STS-51G, STS-51B, STS-51D, STS-51C
1986: STS-51L, STS-61C
1988: STS-27, STS-26
1989: STS-33, STS-34, STS-28, STS-30, STS-29
1990: STS-35, STS-38, STS-41, STS-31, STS-36, STS-32
1991: STS-44, STS-48, STS-43, STS-40, STS-39, STS-37
1992: STS-53, STS-52, STS-47, STS-46, STS-50, STS-49, STS-45, STS-42
1993: STS-61, STS-58, STS-51, STS-57, STS-55, STS-56, STS-54
1994: STS-66, STS-68, STS-64, STS-65, STS-59, STS-62, STS-60
1995: STS-74, STS-73, STS-69, STS-70, STS-71, STS-67, STS-63
1996: STS-80, STS-79, STS-78, STS-77, STS-76, STS-75, STS-72
1997: STS-87, STS-86, STS-85, STS-94, STS-84, STS-83, STS-82, STS-81
1998: STS-88, STS-95, STS-91, STS-90, STS-89
1999: STS-103, STS-93, STS-96
2000: STS-97, STS-92, STS-106, STS-101, STS-99
2001: STS-108, STS-105, STS-104, STS-100, STS-102, STS-98
2002: STS-113, STS-112, STS-111, STS-110, STS-109
2003: STS-107
2004: -
2005: STS-114
2006: STS-116, STS-115, STS-121
2007: STS-120, STS-118, STS-117
2008: STS-126, STS-124, STS-123, STS-122
2009: STS-129, STS-128, STS-127, STS-125, STS-119
2010: STS-132, STS-131, STS-130

2011: STS-133, STS-132, STS-134

STS-80
Longest flight with a Shuttle: 17d 15h 53m 18s

STS-134 scheduled
Last shuttle mission

Main Sources:
(NASA [Space Shuttle], 2010)

SPACE-SHUTTLE ARCHITECTURE

Length: 37.2m
Height: 17.2m
Wingspan: 23.8m
Cargo max.: 28t

Mission Length: 7-17 days, typical

Forward Fuselage
Nose, pressurized crew compartment (Flight Deck, Middeck, Airlock), Lower Deck
Volume: 65.8m^3 (with airlock in the middeck)
Middeck Floorplan Length: 4m
Middeck Floorplan Width: 2.7 (front) - 3.7m (rear)

Mid Fuselage
Payload Bay, Payload Bay Doors with radiator panels
Length: 18.3m
Width: 5.4m
Height: 3.96m plus Payload Bay Doors
Diameter: 4.6m

Aft Fuselage
Not pressurized (Orbital manoeuvring systems, main engines, external tank)
Length: 5.5m
Width: 6.7m
Height: 6.1m

Main Sources:
(NASA [ISS], 2006 p. 42; NASA [Shuttle], 2006; NASA [Orbiter], 2003; NASA [Teacher], 1986; NASA [Shuttle Reference], 1981)

APPENDIX

A6 FACTS MIR

MIR MISSIONS

Launch Vehicle: Proton
Crew Transportation: Soyuz, Space Shuttle
Automatic docking of modules

In orbit: 1986 February 20 – 2001 March 23 (5511d)
Days occupied: 4594d

First Expedition: Mir EO-1
Leonid Kizim, Vladimir Solovyov
1986 March 13 – July 16 (125d)
50 days on Salyut and 75 days on Mir

Expeditions EO-2, EO-3, EO-4
Constantly occupied: 1987 February 5 – 1989 April 27

EO-3/EO-4
Valeri Polyakov spent 437d 18h on Mir and set the world record

Expedition:
EO-5, EO-6, EO-7, EO-8, EO-9, EO-10, EO-11, EO-12, EO-13, EO-14, EO-15, EO-16, EO-17, EO-18, EO-19, EO-20, EO-21, EO-22, EO-23, EO-24, EO-25, EO-26, EO-27
Constantly occupied: 1989 September 5 – 1999 August 28

EO-9/EO-10
1991 May 18 – 1992 March 25
Sergei Krikalyov spent 311d 20h 1m on Mir while the Soviet Union collapsed

EO-15/EO-16/E0-17
1994 January 8 – 1995 March 22
Valeri Polyakov spent 438 days in once (longest manned spaceflight)

EO-18
1995 March 14 – July 7
Norman Thagard is the first American on Mir, returned with the STS-71 first Shuttle flight to dock to Mir

EO-21/EO-22
1996 March 22 – 1996 September 26
Shannon Lucid stayed 188 days in orbit (longest woman flight duration)

EO-28
2000 April 4 – June 16
Last mission

Main Sources:
(Harland, 2005; Culbertson, 2004; Larson, et al., 1999 pp. 356-357; Portree, 1995; B.J.Bluth, 1987)

MIR ARCHITECTURE

Pressurized Volume: ~380m^3
33m x 31m x 27.5m
Max. Diameter: 4.35m
Mass: 130 - 140 t
Number of Solar arrays: 11
Area of solar arrays: 224m
Docking ports: 4
Berthing ports: 4

Core Module (Base Block)
Living quarters, 2 docking ports, 4 berthing ports
Launch: 1986 February 20 (unmanned)
Pressurized Volume: 90m^3
Diameter (max.): 4.15m
Length: 13.13m
Mass: 20.9t

Kvant 1
Astrophysics Module
Launch: 1987 March 31
Pressurized Volume: 40m^3
Diameter: 4.15m
Length: 5.8m
Mass: 11t

Kvant 2
Extension Module
Launch: 1989 November 26
Pressurized Volume: 61.3m^3
Diameter: 4.35m
Length: 13.73m
Mass: 18.5t

Kristall
Technology Module
Launch: 1990 May 31
Pressurized Volume: 60.8m^3
Diameter: 4.35
Length: 13.73m
Mass: 19.6t

Spektr
Geophysical Module
Launch: 1995 May 23
Pressurized Volume: 61.9m^3
Diameter: 4.35m
Length: 12m
Mass: 19.6t

Priroda
International Ecology Research Module
Launch: 1996 May 23
Pressurized Volume: 66m^3
Diameter: 4.35m
Length: 12m
Mass: 19.7t

Main Sources:
(ESA, 2001; Larson, et al., 1999; Portree, 1995; B.J.Bluth, 1987)

A7 FACTS ISS

ISS MISSIONS

Launch Vehicle: Proton, STS
Crew Transportation: Soyuz, Space Shuttle

Expedition 1
William Shepherd, Yuri Gidzenko, Sergei Krikalev
2000 October 31 - November 2

Expedition 2
Susan Helms, James Voss, Yury Usachev
2001 March 8 - August 20 (147d)
Denis Tito becomes the first space tourist

Expedition 3
Frank L. Culbertson, Mikhail Tyurin, Vladimir Dezhurov
2001 August 10 - December 15

Expedition 4
Yury I. Onufrienko, Daniel W. Bursch, Carl E. Walz
2001 December 5 - 2002 June 15

Expedition 5
Valery Korzun, Peggy Whitson, Sergei Treschev
2002 June 7 - December 2

Expedition 6
Kenneth Bowersox, Donald Pettit, Nikolai Budarin
2002 November 23 - 2003 May 3

Expedition 7
Yuri Malenchenko, Ed Lu
2003 April 25 - October 27

Expedition 8
Michael Foale, Alexander Kaleri, Pedro Duque
2003 October 18 -April 29

Expedition 9
Gennady Padalka, Mike Fincke, Andre Kuipers
2004 April 18 - October 23

Expedition 10
Leroy Chiao, Salizhan Sharipov, Yuri Shargin
2004 October 13 - 2005 April 24

Expedition 11
Sergei Krikalev, John Phillips, Roberto Vittori
2005 April 14 - October 19

Expedition 12
Bill McArthur, Valery Tokarev, Gregory Olsen
2005 September 30 - 2006 April 8

Expedition 13
Thomas Reiter, Pavel Vinogradov, Jeffrey Williams
2006 March 29 - 2006 September 28

Expedition 14
Sunita Williams, Michael Lopez-Alegria, Mikhail Tyurin
2006 September 18 - 2007 April 21

Expedition 15
Fyodor Yurchikhin, Oleg Kotov, Clayton Anderson
2007 April 7 - October 21

Expedition 16
Yuri Malenchenko, Peggy Whitson, Sheihk Muszahphar Shukor, Leopold Eyharts, Garrett Reisman, Dan Tani
2007 October 10 - April 19
The Columbus module - the main European ISS component - arrives

Expedition 17
Sergei Volkov, Oleg Kononenko, Greg Chamitoff
2008 April 8 - October 23
The first part of Japan's bus-sized Kibo lab arrives on shuttle Discovery

Expedition 18
Michael Fincke, Yury Lonchakov, Koichi Wakata, Sandra Magnus, Greg Chamitoff
2008 October 14 - 2009 April 8

Expedition 19
Gennady Padalka, Michael Barratt, Koichi Wakata, Charles Simonyi
2009 March 26 - October 11

Expedition 20
Gennady Padalka, Michael Barratt, Koichi Wakata, Timothy L. Kopra, Roman Romanenko, Frank De Winne, Robert Thirsk, Nicole P. Scott
2009 May 27 - October 11

Expedition 21
Frank De Winne, Robert Thirsk, Roman Romanenko, Nicole P. Scott, Maxim Suraev, Guy Laliberte
2009 October 11 - December 1

Expedition 22
Jeffrey N. Williams, Maxim Suraev, Oleg Kotov, Soichi Noguchi, Timothy Creamer
2010 December 1 - March 18

Expedition 23
Oleg Kotow , Michail Kornijenko, Tracy Caldwell-Dyson, Alexander Skworzow, T. J. Creamer, Soichi Noguchi
2010 March 18 - June 2

Expedition 24
Alexander Skworzow, Shannon Walker, Doug Wheelock, Tracy Caldwell-Dyson, Michail Kornijenko, Fjodor Jurtschichin
2010 June 2 - September 25

Expedition 25
Douglas Wheelock, Oleg Skripochka and Alexander Kaleri, Shannon Walker, Fyodor Yurchikhin
2010 September 25 - November 30
The ISS has become the longest habitable space outpost ever built.

Expedition 26
Scott Kelly, Oleg Skripochka, Alexander Kaleri, Dmitry Kondratyev, Paolo Nespoli, Catherine Coleman
2010 October 7 - March 2011

Expedition 27
Dmitry Kondratyev, Andrey Borisenko, Catherine Coleman, Alexander Samokutyaev, Paolo Nespoli, Ron Garan
March 2011

Expedition 28
Andrey Borisenko, Alexander Samokutyaev, Mike Fossum, Satoshi Furukawa, Ron Garan, Sergei Volkov
March 2011

Main Sources:
(RSA, 2010; NASA [Exp], 2009; Cohen, 2009)

APPENDIX

A7 FACTS ISS

ISS ARCHITECTURE

At assembly complete:
Pressurized Volume: 935m^3
Max. Diameter: 4.5m at the Mir Core
Length, overall: 74m
Width: 110m
Mass: 419t
Array surface area: 2,500m (110kW)

FGB (Zarya)
Functional Cargo Block
Launch: 1998 November 20
Volume: 71.5m^3
Diameter: 4.1m
Length: 12.8m

Pressurized mating Adapters (PMA 1+2)
For docking with the Space Shuttle and Russian Modules
Launch: 1998 December 4 (PMA 1,2), 2000 October 11 (PMA 3)
Width: 1.37 – 1.9m
Length: 1.86m

Zvezda
Service Module
Launch: 2000, July 11
Diameter: 4.25m
Length: 13.1m

Pressurized mating Adapter (PMA 3)
For docking with the Space Shuttle and Russian Modules
Launch: 2000 October 11

Nodes
Node 1 (Unity), 2 (Harmony), and 3 (Tranquillity)
Connecting elements
Length: 5.5m (Node 1), 6.7m (Node 2, 3)
Width: 4.3m

U.S. Lab (Destiny)
Research Module
24 equipment racks for control of ISS systems and for scientific research
Launch: 2001 February 7
Pressurized Volume: 120m^3
Length: 8.5m (9.2m)
Width: 4.3m

U.S. Airlock (Quest)
Used with Russian and U.S. spacesuits
Launch 2001 July 12
Length: 5.5m
Width: 4m

ESA Columbus Module
European Research Module
Volume: 75m^3 (total), 25m^3 (habitable)
Length: 6.87m (total), 6.1m (habitable)
Diameter: 4.48m
Total volume of payload racks: 25m

JEM/KIBO
Japanese Experimental Module/Pressurized Module (PM)
23 standard equipment racks for controlling experiment and systems
Length: 11.2m
Diameter: 4.4m (inner 4.2m)

Main Sources:
(NASA [ISS], 2010; Cohen, 2009; ESA [Columbus], 2008; JAXA, 2007; NASA [ISS], 2006; Thirkettle, et al., 2002; Kitmacher, 2002; Larson, et al., 1999; NASA [ISS], 1998)

APPENDIX

A8
REFERENCES

B.J.Bluth, M. Helppie. 1987. Soviet Space Stations as Analogs. NASA Headquarters Washington : NASA Grant, 1987. NAGW-659.

Cohen. 2009. From Apollo LM to Altair: Design, Environments, Infrastructure, Missions, and Operations. Pasadena, California : AIAA SPACE 2009 Conference & Exposition, 2009. AIAA 2009-6404.

Compton, W. David and Charles D., Benson. 1983. Living and Working in Space, A History of Skylab. Washington, US. : National Aeronautics and Space Administration, The NASA History Series, 1983. NASA SP-4208.

Culbertson, Frank L. Jr. Director, Shuttle-Mir (Phase 1) Program. 2004. Spaceflight Nasa - The Meanings of Mir. [Online] 4 April 2004. [Cited: 28 March 2008.] http://spaceflight.nasa.gov/history/shuttle-mir/references/r-documents-mirmeanings.htm.

Eckart, Peter. 1999. The Lunar Base Handbook - An Introduction to Lunar Base Design, Development and Operations. U.S.A. : The McGraw-Hill companies, Inc., 1999. Vol. Space Technology Series. ISBN 0-07-240171-0.

Eckhart, Peter. 1996. Spaceflight Life Support And Biospherics. U.S.A. : Space Technology Library, 1996. ISBN 1-881883-04-3.

ESA [Columbus]. 2008. European Columbus Laboratory. ESA - ISS Human Spaceflight and Exploration. [Online] 8 August 2008. [Cited: 5 10 2009.] http://www.esa.int/esaHS/ESAFRG0VMOC_iss_0.html.

ESA. 2001. Mir FAQs - Facts and history. [Online] European Space Agency, 21 February 2001. [Cited: 2 March 2009.] http://www.esa.int/esaCP/ESA28WTM5JC_Life_0.html.

Harland, David Michael. 2005. The story of the spacestation MIR. Berlin : Springer Praxis Books in Space Exploration, 2005. ISBN: 0387230114 .

JAXA. 2007. Kibo Handbook. [Online] Japan Aerospace Exploration Agency (JAXA) Human Space Systems And Utilization Program Group, 2007. [Cited: 12 November 2009.] http://kibo.jaxa.jp/en/mission/1j/.

Kitmacher, Gary H. 2002. Design of the Space Station Habitable Modules. Houston, Texas, USA : The World Space Congress, 2002. p. 18. IAC-02-IAA.8.2.04.

Larson, J. Wiley and Pranke, Linda K. 1999. Human Spaceflight - Mission Analysis and Design. New York : Mc Graw Hill, 1999. ISBN 0-07-236811-X.

Loewy, Raymond. 2002. Never Leave Well Enough Alone. : The Johns Hopkins University Press, 2002. ISBN 0801872111.

NASA [Apollo 11 CSM]. 1969. Apollo 11 Command and Service Module (CSM). National Space Science Data Center (NSSDC). [Online] 1969. http://nssdc.gsfc.nasa.gov/nmc/spacecraftDisplay.do?id=1969-059A. NSSDC ID:1969-059A.

NASA [Apollo 11 LM]. 1969. Apollo 11 Lunar Module/EASEP. [Online] 1969. http://nssdc.gsfc.nasa.gov/nmc/masterCatalog.do?sc=1969-059C. NSSDC ID: 1969-059C.

NASA [Apollo 17 LM]. 1972. Apollo 17 Lunar Module/ALSEP. [Online] 1972. [Cited: 5 January 2010.] http://nssdc.gsfc.nasa.gov/nmc/spacecraftDisplay.do?id=1972-096C. NSSDC ID: 1972-096C.

NASA [Bull.2]. 1974. Skylab Experience Bulletin No. 2: Architectural Evaluation for Airlock. Houston, Texas, US : NASA, Lyndon B. Johnson Space Center, 1974. JSC-09536.

NASA [Constellation Program]. 2008. NASA Facts - Constellation Program. Houston, Texas : Lyndon B. Johnson Space Center, 2008. FS-2008-09-007-JSC.

NASA [ECLSS]. 2009. Environmental Control and Life Support System (ECLSS), System Engineering Workshop, Presenter: L. Peterson : Life Sciences Department. ISU SSP 2009, Ames Research Center, USA

NASA [Factsheet]. 2006. Living and Working in Space - Factsheet. Houston, Texas : NASA Lyndon B. Johnson Space Center, 2006. FS-2006-05-019-JSC.

NASA [History]. 1997. NASA Facts: A History of U.S. Space Stations. Houston, Texas : Lyndon B. Johnson Space Center, 1997. IS-1997-06-ISS009JSC.

NASA [ISS]. 1998. International Space Station Familarization. Houston, Texas, US : Lyndon B. Johnson Space Center, 1998. TD9702A.

NASA [ISS]. 2006. Reference Guide to the International Space Station. [ed.] Gary, editor Kitmacher. Washington, DC, US : Library of Congress Cataloging-in-Publication Data, 2006. ISBN 0-9710327-2-6.

NASA [ISS]. 2010. Reference Guide to the International Space Station. Assembly Complete Edition. Washington, DC, US : Library of Congress Cataloging-in-Publication Data, 2010. NP-2010-09-682-HQ

NASA [Exp]. 2009. International Space Station - Space Station Crew. [Online] 1 December 2009. [Cited: 5 December 2009.] http://www.nasa.gov/mission_pages/station/expeditions/index.html.

NASA [LRV]. 2005. The Apollo Lunar Roving Vehicle. Goddard Space Flight Center. [Online] 15 November 2005. [Cited: 24 January 2009.] http://nssdc.gsfc.nasa.gov/planetary/lunar/apollo_lrv.html.

NASA [Orbiter]. 2003. Space Shuttle Basics - Orbiter Statistics. [Online] NASA, 25 06 2003. [Cited: 02 05 2009.] http://spaceflight.nasa.gov/shuttle/reference/basics/orbiter/index.html.

NASA [Shuttle Basic]. 2005. Space Shuttle Basics. [Online] 2005. http://www.spaceflight.nasa.gov/shuttle/reference/basics/index.html.

NASA [Shuttle Reference]. 1981. Space Shuttle News Reference. s.l. : U.S. Government Printing Office, 1981. 723-000/562.

NASA [Shuttle]. 2006. Space Shuttle - The Orbiter. [Online] 5 March 2006. [Cited: 10 December 2009.] http://www.nasa.gov/returntoflight/system/system_Orbiter.html.

NASA [OWS]. 1974. MSFC Skylab Orbital Workshop Vol. III. Alabama : Skylab Program Office, George C. Marshall Space Flight Center, 1974. Technical Memorandum. NASA TM X-64813.

NASA [Skylab]. 1977. Skylab, our first space station. Washington D.C., US : NASA, Scientific and Technical Information Office, 1977. Prepared by George C. Marshall Space Flight Center, Edited by Leland Belew. NASA-SP-400.

NASA [Space Shuttle]. 2010. Space Shuttle Mission Archives. [Online] NASA, 27 May 2010. [Cited: 30 November 2009.] http://www.nasa.gov/mission_pages/shuttle/shuttlemissions/list_main.html.

NASA [Teacher]. 1986. Teacher in Space Project. Pasadena, California, US : NASA STI Facility, 1986. N86-70963.

NASA KSC [Apollo]. 2000. Project Apollo. [Online] 25 August 2000. http://science.ksc.nasa.gov/history/apollo/apollo.html.

NASA KSC. 2000. Project Skylab. [Online] 25 November 2000. [Cited: 15 October 2008.] http://science.ksc.nasa.gov/history/skylab/skylab.html.

NewScientist. 1976. Cosmonauts meet up with their space station. New Scientist. Vol 72, 1976, Vol. No 1009, 15 July.

OTA. 1983. SALYUT - Soviet Steps Toward Permanent Human Presence in Space. Washington D.C. : U.S. Congress, Office of Technology Assessment, 1983. Andelin, John; Naismith, Nancy; Mills, William; Rogers, Thomas F.; Chandler, Philip P.. OTA-TM-STI-14.

Portree, David S.F. 1995. MIR Hardware Heritage. Houston, Texas : NASA, Lyndon B. Johnson Space Center, Information Services Division, 1995. NASA RP 1357.

RSA. 2010. Russian Federal Space Agency; ISS: 10 Years of Human Space Mission. [Online] 1 November 2010. [Cited: 15 December 2010.] http://www.federalspace.ru/main.php?id=2&nid=10717.

Thirkettle, A., et al. Februar 2002. ISS: Columbus. ESA Bulletin. Februar 2002, 109, pp. 27-34.

Vasilyev, M. P., et al. 1974. Salyut Space Station in Orbit, Translation of ‚Salyut na Orbite`, Mashinostroyeniye Press, Moscow, 1973, 160 pages. Washington, D.C. : NASA, National Aeronautics and Space Administration, 1974. NASA TT F-15,450.

Wikipedia [Salyut]. Salyut program. [Online] [Cited: 2009 February 2009.] http://en.wikipedia.org/wiki/Salyut_program.

Young, Anthony. 2007. Lunar and Planetary Rovers, The Wheels of Apollo and the Quest for Mars. Berlin, Heidelberg, New York : Springer-Praxis Books in Space Exploration, 2007. ISBN 0-387-30774-5.

INDEX

INDEX

MIX
Papier aus verantwortungsvollen Quellen
Paper from responsible sources
FSC® C105338

Printed by Books on Demand, Germany